S0-BIP-397

WITHDRAWN

The Social and Cultural Construction of Risk

Essays on Risk Selection and Perception

In addition to its French and Italian origins, the English word risk *has two other possible derivations which are interesting to contemplate: the Vulgar Latin word* resecum, *meaning danger, rock, risk at sea, and the Greek word* rhiza, *cliff, through the meaning "to sail round a cliff." Though unattested, these classical sources are appealing to the editors for the visual image they evoke: Ulysses and his men making their perilous voyage between Scylla and Charybdis in the Strait of Messina.*

Technology, Risk, and Society

An International Series in Risk Analysis

The Social and Cultural Construction of Risk

Essays on Risk Selection and Perception

Edited by

Branden B. Johnson
Michigan Technological University

and

Vincent T. Covello
National Science Foundation

D. Reidel Publishing Company

A MEMBER OF THE KLUWER ACADEMIC PUBLISHERS GROUP 

Dordrecht / Boston / Lancaster / Tokyo

Library of Congress Cataloging-in-Publication Data

CIP

The Social and cultural construction of risk: essays on risk selection and perception / edited by Branden B. Johnson and Vincent T. Covello.

p. cm. — (Technology, risk, and society; v. 3)

Bibliography: p.

Includes index.

ISBN 1-55608-033-6

1. Risk management. 2. Risk. 3. Risk—Social aspects. 4. Risk assessment. 5. Environmental impact analysis. 6. Technology—Risk assessment. I. Johnson, Branden B. II. Covello, Vincent T. III. Series.

HD61.S59 1987

306 — dc 19

87-25393

CIP

Published by D. Reidel Publishing Company,
P.O. Box 17, 3300 AA Dordrecht, Holland.

Sold and distributed in the U.S.A. and Canada
by Kluwer Academic Publishers,
101 Philip Drive, Norwell, MA 02061, U.S.A.

In all other countries, sold and distributed
by Kluwer Academic Publishers Group,
P.O.Box 322, 3300 AH Dordrecht, Holland.

Printed in The Netherlands

Contents

Introduction

The Social and Cultural Construction of Risk: Issues, Methods, and Case Studies

Vincent T. Covello and Branden B. Johnson

Risks to health, safety, and the environment abound in the world and people cope as best they can. But before action can be taken to control, reduce, or eliminate these risks, decisions must be made about which risks are important and which risks can safely be ignored. The challenge for decision makers is that consensus on these matters is often lacking. Risks believed by some individuals and groups to be tolerable or acceptable — such as the risks of nuclear power or industrial pollutants — are intolerable and unacceptable to others. This book addresses this issue by exploring how particular technological risks come to be selected for societal attention and action. Each section of the volume examines, from a different perspective, how individuals, groups, communities, and societies decide what is risky, how risky it is, and what should be done.

The writing of this book was inspired by another book: *Risk and Culture: An Essay on the Selection of Technoloqical and Environmental Dangers*. Published in 1982 and written by two distinguished scholars — Mary Douglas, a British social anthropologist, and Aaron Wildavsky, an American political scientist — the book received wide critical attention and offered several provocative ideas on the nature of risk selection, perception, and acceptance.

According to Douglas and Wildavsky, modern Americans perceive the world to be an inhospitable and frightening place. Americans are afraid of "Nothing much . . . except the food they eat, the water they drink, the air they breathe, the land they live on, and the energy they use" (1982: 10). These fears, rather than becoming diminished with time, are becoming greater. Despite increases in average life expectancy, reductions in the frequency of catastrophic events, and assurances by authorities that "the health of the American people has never been better" (U. S. Surgeon General 1979), survey data indicate that most Americans believe that life is getting riskier. A poll by Louis Harris and Associates (Harris 1980), for example, indicates that approximately four-fifths of all Americans agree that ". . . people are subject to more risk today than they were 20 years ago." Only 6 percent believe that there is less risk.

B. B. Johnson and V. T. Covello (eds.), The Social and Cultural Construction of Risk, vii—xiii.

These and other data on risk perceptions prompted Douglas and Wildavsky to ask two fundamental questions: Why do people emphasize some risks while ignoring others? And more specifically, why have Americans singled out industrial pollution as a principal source of concern?

In answering these questions, Douglas and Wildavsky made the basic assumption that societies selectively choose risks for attention. More specifically, Americans — as well as people in all other societies, including primitive tribes and highly industrialized nations — select just a few risks for attention and ignore a vast array of others. Each society has its own distinctive portfolio of risks that are believed to be worthy of concern; each society highlights some risks and downplays others; and each society institutionalizes means for controlling some risks and not others.

Douglas and Wildavsky also make the challenging and provocative claim that people do not focus on particular risks simply in order to protect health, safety, or the environment. The choice also reflects their beliefs about values, social institutions, nature, and moral behavior. Risks are exaggerated or minimized according to the social, cultural, and moral acceptability of the underlying activities.

According to this view, risk is not an objective reality; instead, "the perception of risk is a social process" (1982: 6). Increasing concerns about industrial risks cannot be explained by individual psychology or by objective reality (i.e., by what available scientific evidence can justify): instead, they can only be understood through social and cultural analysis and interpretation.

Douglas and Wildavsky argue that selection is necessary because it is impossible for people to be aware of all risks. Just thinking about the hundreds of thousands of risks that a person might conceivably be exposed to is alarming. Since no one can attend to everything, and since individuals cannot look in all directions at once, people must decide which risks to fear most, which risks are worthy of attention and concern, which risks are worth taking, and which risks can be ignored.

Douglas and Wildavsky point out that the risks that are finally selected for attention are not necessarily chosen because the scientific evidence is solid. Moreover, in some cases the risks that are selected have little relation to real danger and may be among the least likely to affect people. What societies choose to call risky is largely determined by social and cultural factors, not nature.

To illustrate some of these points, the authors note that the Lele tribe of Zaire are susceptible to many diseases and illnesses — e.g., gastroenteritis, tuberculosis, leprosy, ulcers, barrenness, and pneumonia. However, the Lele focus on only three risks: being struck by lightning, barrenness, and bronchitis. When these events do occur, they are usually arttributed to some moral transgression or defect rather than to a physical cause.

In other societies, other risks are selected for concern and attention. In the United States, for example, the dominant risks to health are those associated with cardiovascular disease, lung cancer due to cigarette smoking, and automobile accidents (Covello and Mumpower 1985). However, in recent years Americans have focused much of their attention and resources on the risks of cancer due to industrial pollution. This focus has persisted despite a fragile consensus among scientists that there has been a slight decrease in current cancer rates — if one discounts lung cancer, which is closely associated with smoking, and one corrects for the greater number of older people in the current U.S. population. Furthermore, as Philip Handler (1979), a former President of the U.S. National Academy of Sciences, notes: ". . . 'pollutants' of all sorts may contribute to — rather than cause — perhaps five percent of all cancers."

Douglas and Wildavsky go beyond the assertion that risks are selectively chosen to a specific argument about why some risks are chosen and others are not. Much of this argument is based on earlier anthropological work by Douglas (1966) on beliefs about impurity, danger, pollution, and taboo. Douglas argued in this work, for example, that it would be a mistake to view all food prohibitions — such as the injunction in the Old Testament against eating pork — as simply forerunners of modern food regulations. Such food prohibitions often serve a variety of purposes, including the affirmation of ethical norms, distinguishing one group from another, and symbolically bringing order into a chaotic world by classification.

Building on this earlier work by Douglas, Douglas and Wildavsky point out that people's concerns and fears about different types or risks can often be more accurately seen as ways of maintaining social solidarity than as reflecting health or environmental concerns. As such, health or environmental concerns should never be taken at face value. Instead, the analysis must look further to discover what social norms, policies, or institutions are being defemded or attacked. To illustrate this point, Douglas and Wildavsky observe that the Hima, a cattle herding tribe in Uganda, believe that cattle will die if the animal comes into contact with women or if a person eats farm produce while drinking cow milk. The authors explain these beliefs by noting the need and desire of the Hima to maintain the existing sexual division of labor and to prevent encroachments by a neighboring tribe of farmers.

Turning their attention from the Hima to the United States and other industrialized societies, Douglas and Wildavsky see the increasing concern about cancer risks from industrial pollution as partly the result of the rise in prominence of environmental groups. The authors claim that these groups function as contemporary sects, although with a secular rather than a religious orientation. According to Douglas and Wildavsky, environmental groups share with sects the need to expand and to make themselves attractive and distinctive. They also share with sects the need for opposition and for enemies — "a threatening evil on a cosmic scale" — in order to build group solidarity. In the United States and many other industrial nations, this enemy is industry and big business. The risks selected as a rallying point for group expansion are health and environmental risks associated with industrial pollution. These risks are used by environmental groups as a means for holding their membership together and for attacking the establishment groups that they oppose.

Most critics see this argument as the weakest part of the book (Zald 1983; Kaprow 1985). More specifically, critics have questioned Douglas and Wildavksy's arguments for why the environmental movement gained prominence in the last two decades (Zald 1983). Equally unconvincing to many critics is the claim that present-day concerns about industrial pollution, radiation, and the side effects of new drugs and chemicals are empirically comparable to concerns about pollution and impurity in primitive societies and exotic religious cults (Winner 1982; Kaprow 1985); or that environment groups have been excessive in their demands for government regulation of industrial chemicals (Nelkin 1982; Winner 1982). This is a continuing debate, however, and many of the outstanding and unresolved issues in the debate are discussed by authors in this volume (Rayner, Chapter 1; Gerlach, Chapter 5; Johnson, Chapter 6).

Despite these reservations, few critics have challenged the basic thesis of the book that societies selectively choose risks for attention. Douglas and Wildavsky's work represents a major contribution to the literature on risk by pointing to the important role of social and cultural factors in setting risk agendas and in determining which risks will be emphasized or de-emphasized.

Less noticed, but of equal scholarly significance, Douglas and Wildavsky's work also provides a complementary perspective to the reigning psychological approach to risk selection and perception. This approach, developed principally by cognitive psychologists over the last ten years, has led to many significant advances in knowledge about the dynamics of risk perception and about the cognitive limitations of individuals in processing risk information (see Slovic *et al.* 1980; Kahneman and Tversky 1984; Covello 1984). However, what has been missing from this literature has been the attempt to integrate psychological factors with social and cultural factors. Many of the chapters in this volume suggest directions for such integration.

Psychological research has suggested, for example, that people often overestimate the risks of dramatic causes of death — such as airplane accidents — and that such overestimates are partly due to the greater memorability and imaginability of such events (Lichtenstein *et al.* 1978). Any factor that makes a risk unusually memorable or imaginable — such as a recent disaster or intense media coverage — may distort risk perceptions. Clearly pertinent to this issue is the chapter in this volume on the controversy surrounding use of the pesticide EDB and media coverage of the case (Sharlin, Chapter 7). Also relevant are the chapters on social networks and the ability of such networks to reinforce or counteract the influence of psychological factors on risk perceptions (Fitchen *et al.*, Chapter 2; Fowlkes and Miller, Chapter 3). Similarly, the chapters on participation in environmental protest movements and membership in occupational communities illustrate the reinforcing or counteracting effects of such groups on risk perceptions (Walsh, Chapter 4; Ferguson, Chapter 12; Tarr and Jabobson, Chapter 13).

A second important finding from the psychological literature on risk perception is that lay people often have difficulty understanding and interpreting probabilistic information, especially when the probabilities are small and the risks are unfamiliar (Slovic *et al.* 1980; Covello 1984). Experts experience similar difficulties interpreting probabilistic information, although expert knowledge can mitigate the effects of various judgmental biases (Fischhoff *et al.* 1981; Kahneman and Tversky 1979). What this research does not address, however, and what several chapters in this volume clearly demonstrate, is the strong influence of social and cultural factors on the risk judgments and interpretations of experts (Carlson and Millard, Chapter 11; Lynn, Chapter 14; Jasanoff, Chapter 15).

A third important finding from the psychological literature on risk perception is that the way in which information about risks is presented — e.g., whether the information is framed as benefits or costs — can exert a powerful influence on risk selection, perceptions, and concerns (Kahneman and Tversky 1984). The definition of what constitutes a benefit or cost, however, is also strongly influenced by social and cultural factors. For example, economic factors and vested interests can, in some circumstances, override or subdue concerns about risks (e.g., Fitchen *et al.*, Chapter 2; Brown, Chapter 10; Carlson and Millard, Chapter 11). Similarly, factors related to the credibility and trustworthiness of institutions (e.g., Fitchen *et al.*, Chapter 2; Walsh, Chapter 4; Bronstein, Chapter 8; Gale, Chapter 9; Carlson and Millard, Chapter 11) and to ideology (Gerlach, Chapter 5; Johnson, Chapter 6; Lynn, Chapter 14) can exacerbate or mute concerns about particular risks.

Finally, one of the most important findings to emerge from the psychological literature on risk perception is that people take into consideration a large number of factors in evaluating the seriousness of a risk (e.g., Slovic *et al.* 1980; Covello, 1983, 1984). These factors include, among others, catastrophic potential, familiarity, voluntariness, and dread. This finding — that concerns about risks are based largely on these qualitative factors and only partly on expected mortality rates — is also a central thesis in this volume. But the chapters in this volume discuss several qualitative factors that are rarely if ever mentioned in the psychological literature. These include organizational affiliations, community dynamics, institutional context, ideology, and social interactions with family, friends, fellow workers, and neighbors.

The integration of these and other social and cultural factors with psychological factors poses a substantial challenge for future research. At a minimum, this integration requires that researchers recognize the value of different disciplinary perspectives on risk selection and perception. Propositions about the importance of such factors as institutional trustworthiness, for example, need to be tested intensively by researchers from psychology, social psychology, sociology, anthropology, economics, history, philosophy, law, policy analysis, management science, and political science.

In summary, what is missing from the current literature on risk selection and perception is systematic and rigorous analysis of how perceptions and judgments arise from a host of complex factors, including historical trends, underlying values, ideological currents, and the nature of social, cultural, economic, scientific, and political institutions at specific points in time and in specific places. The chapters in this volume represent an important first step in this direction. They draw on a wide range of materials, including media reports of technological controversides; anthropological discussions of purity an danger; and social and political theories of voluntary organizations, community dynamics, collective action, organizational commitment, and social change. Through the analysis of such materials, it is hoped that current debates about health and environmental risks, and the complex set of social and cultural factors underlying these debates, will be illuminated.

Note: The views expressed in this work are those of the authors and do not necessarily represent the views of the National Science Foundation.

References

Covello, V. 'The Perception of Technological Risks: A Literature Review,' *Technological Forecasting and Social Change*, 1983, 23, 285—297.

Covello, V. 'Uses of Social and Behavioral Research on Risk,' *Environment International*, June 1984, 4, 17—26.

Covello, V. and J. Mumpower, 'Risk Assessment and Risk Management: An Historical Perspective,' *Risk Analysis: An International Journal*, June 1985, 5(2), 103—120.

Douglas, M. *Purity and Danger: An Analysis of Concepts of Pollution and Taboo*. London: Routledge & Kegan Paul, 1966.

Douglas, M. and A. Wildavsky, *Risk and Culture: An Essay on Selection of Technological and Environmental Dangers*, Berkeley: University of California Press, 1982.

Fischhoff, B., P. Slovic, and S. Lichtenstein. 'Lay Foibles and Expert Fables in Judgments about Risk,' pp. 161—202 in T. O'Riordan and R. K. Turner (eds.), *Progress in Resource Management and Environmental Planning*. New York: Wiley, 1981.

Handler, P., 'Some Comments on Risk Assessment,' in *National Research Council Current Issues and Studies. Annual Report*. Washington, D.C.: National Academy of Sciences, 1979.

Harris, L. and Associates. *Risk in a Complex Society*, New York: Marsh and McLennan Companies, 1980.

Kahneman, D. and A. Tversky. 'Prospect Theory: An Analysis of Decision Under Risk,' *Econometrics*, March 1979, 47, 263—291.

Kahneman, D. and A. Tversky. 'Choices, Values, and Frames,' *American Psychologist*, April 1984, 39(4), 341—350.

Kaprow, M. 'Manufacturing Danger: Fear and Pollution in Industrial Society,' *American Anthropologist*, June 1985, 87(2), 342—356.

Lichtenstein, S., P. Slovic, B. Fischhoff, M. Layman, and B. Combs. 'Judged Frequency of Lethal Events,' *Journal of Experimental Psychology: Human Learning and Memory*, 1978, 4(6), 551—578.

Nelkin, D. 'Blunders in the Business of Risk' (review of *Risk and Culture* by M. Douglas and A. Wildavsky), *Nature*, August 19, 1982, 298, 775—776.

Slovic, P., B. Fischhoff, and S. Lichtenstein. 'Facts and Fears: Understanding Perceived Risk,' pp. 181—213 in R. Schwing and W. Albers (eds.), *Societal Risk Assessment: How Safe is Safe Enough*? New York: Plenum, 1980.

U.S. Surgeon General, *Healthy People: The Surgeon General's Report on Health Promotion and Disease Prevention*. Washington, D.C.: U.S. Government Printing Office, 1979.

Winner, L. 'Pollution as Delusion,' pp. 8, 18 in *New York Times Book Review*, August 8, 1982.

Zald, M. 'Explaining Our Fears,' *Science*, March 11, 1983, 219, 1211—1212.

List of Contributors

Janet Bronstein (Ph.D., Anthropology) is Health Care Analyst at the University of Alabama, Birmingham.

Michael S. Brown (Ph.D., City and Regional Planning) is Director of the Office of Safe Waste Management, Massachusetts Department of Environmental Management, Boston.

W. Bernard Carlson (Ph.D., History and Sociology of Science) is Assistant Professor of Humanities in the School of Engineering and Applied Science, University of Virginia, Charlottesville.

Vincent T. Covello (Ph.D., Sociology) is the Director of the Risk Assessment Program at the U.S. National Science Foundation, Washington, D.C. He is also President-Elect of the International Society for Risk Analysis.

Eugene S. Ferguson (Ph.D., History) is Professor Emeritus of History of Technology at the University of Delaware, Wilmington.

June Fessenden-Raden (Ph.D., Biochemistry) is Associate Professor of Biochemistry and of Biology and Society, Cornell University, Ithaca, New York.

Janet M. Fitchen (Ph.D., Anthropology) is Assistant Professor of Anthropology, Ithaca College, and Adjunct Assistant Professor of Science, Technology and Society, Cornell University, Ithaca, New York.

Martha R. Folkes (Ph.D., Sociology) is Associate Professor and Associate Dean of the School of Family Studies at the University of Connecticut, Storrs.

Richard P. Gale (Ph.D., Sociology) is Professor of Sociology at the University of Oregon, Eugene.

Luther P. Gerlach (Ph.D., Anthropology) is Professor of Anthropology at the University of Minnesota, Minneapolis.

Jenifer S. Heath (M.S., Toxicology and Nutrition) is a Jesse Smith Noyes predoctoral fellow in environmental toxicology at Cornell University, Ithaca, New York.

B. B. Johnson and V. T. Covello (eds.), The Social and Cultural Construction of Risk, xv—xvi.

Charles Jacobson (M.S., Applied History) is a doctoral candidate in the Applied History and Social Science program, Carnegie-Mellon University Pittsburgh, Pennsylvania.

Sheila Jasanoff (Ph.D., Linguistics; J.D.) is Professor of Science, Technology and Society at Cornell University, Ithaca, New York.

Branden B. Johnson (Ph.D., Geography) is Assistant Professor of Science, Technology, and Society at Michigan Technological University, Houghton.

Frances M. Lynn (D.P.H., Health Policy and Administration) is Assistant Professor, Department of Environmental Sciences and Engineering, and Research Associate, Institute for Environmental Studies, University of North Carolina, Chapel Hill.

Andre J. Millard (Ph.D., History) is Assistant Professor of History at Bentley College, Waltham, Massachusetts.

Patricia Y. Miller (Ph.C., Sociology) is Associate Professor of Sociology at Smith College, Northampton, Massachusetts.

Steve Rayner (Ph.D., Anthropology) is Research Associate in the Energy Division, Oak Ridge National Laboratory, Oak Ridge, Tennessee.

Harold I. Sharlin (Ph.D., Economic History) is President of HIS Associates, Washington, D.C.

Joel A. Tarr (Ph.D., History) is Professor of History and Public Policy at Carnegie-Mellon University, Pittsburgh, Pennsylvania.

Edward J. Walsh (Ph.D., Sociology) is Associate Professor of Sociology at Pennsylvania State University, University Park.

PART I

Part I

Reality, Perception, and the Social Construction of Risk

The social construction of risk viewpoint — the claim that risk is a social construct stemming primarily or wholly from social and cultural factors — raises an interesting and difficult philosophical issue. From one perspective, this view can be interpreted to imply that people are incapable of perceiving what is really dangerous since there are no actual or objective risks in the world. Risks do not exist "out there." Instead, there are only subjective perceptions of risks and the perceptions of one person are no more valid than the perceptions of any other. Risk is only what people choose to say it is. Since everything is situational, relative, and contextual, nothing has objective reality. Taken to such an extreme, the social construction of risk viewpoint appears to violate common sense and calls into question the logical and rational foundations of societal risk management (see Sederberg 1984).

For risk managers, this interpretation implies that the only choice available is an impossible choice between situational chaos and naive realism. In the first chapter of this book, however, Steve Rayner argues that the situation is not hopeless. Adopting a perspective shared by several other authors (Hirst 1984; Hollinger 1973), he argues that a case can be made for better and worse social constructions of risk.

In developing this claim, Rayner draws on a framework first put forward by Funtowicz and Ravetz (1985). The basic argument is that risk issues can be seen as moving through a sequence from "consensual science," through "clinical consultancy," to "total environmental assessment" depending on the degree of scientific uncertainty in the risk estimates and on the importance of the issue to the parties at interest. The further the issue is from scientific consensus and the greater the importance of the issue, the greater will be the significance of cultural variation and the more problematic will be standards of objective validity.

Rayner then provides several examples of how one variation of social construction of risk theory — technically known as grid/group analysis (see also Johnson, Chapter 6) — can be used to enhance societal risk management. His first example, focused on radiation issues, suggests that the integration of "individualists" into teams might lead to more responsible behavior by such individuals. Similarly, the reduction of group ties

B. B. Johnson and V. T. Covello (eds.), The Social and Cultural Construction of Risk, 3—4.

and routines among "hierarchists" might lead to equivalent results. His second example, focused on hazardous waste issues, suggests that different kinds of social groups will require different incentive packages in order to accept a hazardous waste facility, and that the social construction of risk perspective can help to identify the components of such a package.

References

Funtowicz, S. and J. Ravetz. pp. 217—231 in C. Whipple and V. Covello (eds.), *Risk Analysis in the Private Sector*, New York: Plenum, 1985.

Hirst, P. Q. 'Review Article: Witches, Relativism and Magic,' *The Sociological Reveiw*, August 1984, 32(3), 573—588.

Hollinger, D. A. 'T. S. Kuhn's Theory of Science and Its Implications for History,' *American Historical Review*, April 1973, 78(2), 370—393.

Sederberg, P. C. *The Politics of Meaning: Power and Explanation in the Construction of Social Reality*. Tucson: University of Arizona Press, 1984.

Chapter 1

Risk and Relativism in Science for Policy

Steve Rayner

Introduction

The publication of Douglas and Wildavsky's *Risk and Culture* in 1982 called the attention of risk analysts to the argument that risks are defined and perceived differently by people in different cultural contexts. Hence, in complex societies, such as the contemporary United States, it is to be expected that there will be considerable disagreement between the members of various constituencies, as to what constitutes a technological risk, as well as about how such risks should be managed at various institutional levels.

This viewpoint has met with a mixed reception in scientific and technical circles as well as among policy makers who depend on the guidance of scientific experts. There are those who reject the concept of cultural relativism in risk analysis for the same reason they have been fending it off in the sociology of science for the past decade. It is argued that the view of scientific truth as relative, that is to say context dependent, precludes a systematic cross-cultural concensus about the natural order that is independent of cultural viewpoint. Since, for example, the Japanese and the Americans agree on the number of protons in the nucleus of beryllium, we are told that relativism is either scientifically trivial in that it refers to kinds of knowledge that are not significant to science, or it is simply false. These critics are the defenders of the notion that scientific knowledge is independent from the social and cultural context that sustains a category of human activity called *science*, as well as from decisions made within science about how to organize the various activities that constitute the scientific enterprise (Agassi 1984).

This naive-realist view of science provides the rationale for the National Academy of Science report (1983) that recommends dividing risk analysis into separate stages in which facts are first established and later evaluated for their social implications. Ironically, this viewpoint is cherished equally by the academic champions of both capital and labor. Some big-business executives seem to fear that acceptance of relativism may strengthen the legal actions of community and labor activists who currently have to prove probable cause of injury according to quite stringent scientific and legal

B. B. Johnson and V. T. Covello (eds.), The Social and Cultural Construction of Risk, 5—23.

criteria. Relativism may be seen in such circles as opening the door to successful lawsuits for an endless list of imagined harms. Labor and community leaders, on the other hand, seem to fear that relativism will allow industry to reject liability claims on the basis that risk is a subjective notion, that has no objective existence in nature, and therefore cannot justify expensive compensatory or corrective action (Kaprow 1985). The error, in both camps, is to confuse cultural relativism with individual subjectivism and to suppose that differential social construction of risk precludes an intersubjective concensus based on empirical feedback from human interaction with an objective universe.

Even those who reject the naive-realist view of science, and are sympathetic to the notion that cultural relativism operates within the hallowed confines of science, as well as in politics, have questioned the usefulness of this insight to the science-for-policy process. The concept of cultural relativism is unlikely to break into the realm of applied policy analysis unless it enables us to do something more practical than appreciate the cultural diversity of perceptions of risk with much the same academic detachment that we appreciate divergent perceptions of beauty in museums of ethnic art. Sadly, many exponents of cultural relativism in the sociology of science seem to have satisfied themselves with *demystifying science* (Latour and Woolgar 1979) in just this way. We have had our fun exposing the smug pretentions of positivist science to privileged knowledge of the world, but unless we can do a better job the last laugh will rightly be on us.

The aim of this paper is, therefore, to indicate the proper place for cultural relativism in science for policy using the specific example of grid/group analysis in societal risk management. This is simultaneously a theoretical-rational problem of epistemology and a practical-empirical problem of usable knowledge (Ravetz 1984). The first problem is the status of relativistic knowledge in its relation to a world that exists independently of it, but can only be known through it. The second problem is one of demonstrating actual cases in which grid/group analysis improves the capacity of risk managers to resolve what might be the hardest category of policy conflict that they face. That is those conflicts that arise from differing perceptions of what constitutes a risk and of what measures are appropriate to manage it effectively.

Cultural Relativism and Usable Knowledge

The problem of the epistemological status of relativism in scientific knowledge is rooted in the process by which that knowledge is created, and the relationship between that process and the particular variety of relativism that is being advocated. It is interesting that critics often characterize the relativist position by the extreme argument that knowlege created in one set of social circumstances is entirely self-validating and incommunicable to members of another culture or subculture. On this basis, it is objected, any one person's version of the world has as valid a claim to be scientific truth as any other and any ludicrous proposition is believable (Agassi 1984).

However, this final solipsism is actually the very antithesis of cultural, as opposed to individual, relativism. Cultural relativism emphasizes that the validity of public knowledge depends on its relation to the context of its creation through social activities such as science, technology, religion, and even magic. By denying the possibility of directly comparing knowledge to nature, except through the culturally created categories of human thought, we are emphatically not denying the existence of any basis for validating public knowledge, of which scientific knowledge is one sort. On the contrary, we argue that public knowledge always must be evaluated as part of the social system, the laboratory, workshop, community, or sect, that creates and sustains it (Douglas 1978). The socially determined rules for establishing claims to knowledge, testing them, and evaluating them therefore become part of that knowledge, as do the rules that interdict certain kinds of inquiry (Ravetz 1984), for example, the boundary rules of academic disciplines (Douglas 1984). The point is that public knowledge can only be evaluated as a whole system, a process of production and use, and not as an artifact to be compared against nature's pattern (Bloor 1976).

To reduce this argument to the personal solipsism that our critics ascribe to us is a travesty of the cultural relativist position. However, such a reductionism enables the naive realists to attack cultural relativism on yet another front. Accepting solipsism would oblige cultural relativists to deny that knowledge created in different contexts may refer to the same thing. Yet it is clear that at least some culturally created concepts may refer to phenomena that exist independently of those concepts; although they may only be knowable through one or another version of them. That is to say that socially created categories and modes of reasoning are not the only constraints on human knowledge. Much of human knowledge is obtained through experience with forces that would continue to exist in the absence of human agency (Barnes 1974; Bloor 1976). This practical experience is interpreted through cultural categories but, in some useful sense, may be said to exist independently of those categories. It may, therefore, be described as natural feedback into the knowledge process.

Even the exponents of philosophical idealism have been obliged to confront the phenomenon of natural feedback in order to escape solipsism. Bishop Berkeley (1710) identified this experience as human perception of the ideas of the supreme spirit that were irresistable to lesser spirits who could not but perceive them as their own. The good bishop's dean misunderstood the accommodation with natural feedback that was achieved even in Berkeley's fanciful idealism. The dean refused to open the front door to his superior on the basis that Berkeley argued objects only exist insofar as they are conceived of by spirits. It should therefore have been possible for Berkeley to pass through by a simple act of thought. However, in attempting to expose the absurdity of Berkeley's position by invoking natural feedback, the dean neglected to note that Berkeley had already incorporated the concept into his own system. The door was conceived of by the same God who attends to sycamore trees in quadrangles when no-one else is about, and Berkeley was obliged to acknowledge the input.

Nature, of course, seldom presents such clearly defined signals as the door that God presented to Berkeley. Natural feedback into the knowledge process is, therefore, always subjected to the conceptual massaging imposed by existing categories of thought. The combination of natural feedback with cultural constraints on the organization of information combine to form a total knowledge system, parts of which may be overdetermined by either cultural or natural constraints at different times and places. However, both types of constraint are always present in the knowledge process and profoundly shape our most fundamental ordering concepts such as space and time (Rayner 1982). Traditional empiricism has tended to reify the feedback process, according the artifactual status of objective knowledge to the information that it is seen to provide, while seeking to separate the culturally determined components of knowledge and reducing them to the status of subjective values. The convenience of the fact/value dichotomy is clearly attractive to exponents of science for policy who seek clear and simple solutions to complex problems (Cohen 1985). Alas it does grievous violence to our ability to find real solutions.

It should be noted that there is nothing in this argument to preclude the transferability of knowledge from one context to another, although such arguments have been made by cultural relativists (Wittgenstein 1953), Such a transfer of knowledge, however, would itself transform the recipient culture and in some way, trivial or significant, inevitably alter the social relations within it. In these cases, the interesting question for the risk analyst is not whether different social systems can converge on common definitions of problems at some useful level. Modern world history has indisputably demonstrated that this is possible, though often difficult. Rather we should start by asking what is at stake for those involved in developing a cross-cultural concensus, and how flexible is the particular knowledge process to enable it to accept or resist the change?

Fortunately, the field of risk analysis has already produced a model of the production of scientific knowledge that enables us to address these questions. Funtowicz and Ravetz (1985) have described three kinds of science predicated by two variables, *systems uncertainty* and *decision stakes*. Whereas systems uncertainty contains the elements of inexactness, uncertainty, and ignorance encountered in technical studies; decision stakes involves the costs and benefits of the various policy options to all interested parties. This model generates three kinds of science, each with its own style of risk assessment (Figure 1).

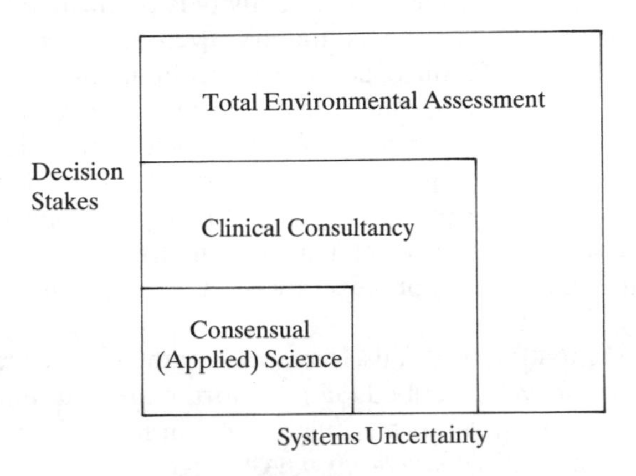

Fig. 1. Funtowicz and Ravetz's three kinds of science.

Low systems uncertainty and decision stakes describe situations in which data bases are large and reliable, and the technical community largely agrees on appropriate methods of investigation. Funtowicz and Ravetz call this *applied science.* I prefer the term *consensual science* because the adjective *applied* is commonly used to describe scientific activities that are designed to produce information for practical technical purposes, even where decision stakes and systems uncertainty are higher. The consensus referred to here is achieved, in part, by the low decision stakes. Controversies about scientific facts are unlikely to be heated where the symbolic loads that such facts carry are either well established or unimportant. Knowledge is likely to have a very strong component of natural feedback based on long-term practical interaction between the social systems represented here and the nonhuman universe. The variations in perspective on risk emphasized by cultural relativism are likely to be minor within this framework.

When both systems uncertainty and decision stakes are considerable, but professional expertise is still a useful guide to action, Funtowicz and Ravetz define a different style of activity, the *clinical* mode of technical consultancy. This kind of activity involves the use of quantitative tools, supplemented explicitly by experienced qualitative judgment. The exercise of this judgment increases the decision stakes for the consulting scientist and begins to bring to the fore differences of interpretation rooted in perhaps competing institutional, educational, and disciplinary cultures. There is some kind of unstable balance or alternation of overdetermination between natural feedback and cultural constraints on the knowledge process throughout this kind of activity.

Finally, when decision stakes and systems uncertainty are very high, Funtowicz and Ravetz present us with a scientific style they term *total environmental assessment*. This kind of activity is permeated by qualitative judgments and value commitments. Inquiry, even into technical questions takes the form largely of a dialogue, which may be in an advocacy or even an adversary mode. Although the proportion of risk assessments that fall into this mode is only a tiny proportion of the whole, they are often those of greatest political significance. As the authors of this typology note, total environmental assessment provides the most plausible opportunity for the application of a cultural-relativist perspective, for here the social constraints on the knowledge process are clearly dominant over natural feedback.

Given the compatibility of this framework with the version of cultural relativism that I have described, it is unfortunate that Funtowicz and Ravetz present their model as a critique of cultural relativism, which they disparagingly term *social reductionism*. Although not naive realists themselves, they share the realists' concern that recognizing the cultural construction of risk will lead to social irresponsibility in facing threats to life and limb arising out of technology and the environment. I have argued above that this concern is misplaced (except insofar as any theory can be abused by the unscrupulous). However, Funtowicz and Ravetz (p. 20) describe their model as a solution to "the difficulty created by the contradiction between the ideal of public-knowledge science and the characteristics of the problems encountered in risk assessment, without falling into sectarian relativism or social reductionism."

Rather than being an antidote to cultural relativism, the distinction of three kinds of science more properly defines those instances where the role of cultural variation in knowledge is, respectively, trivial, integral, and dominant as we move from consensual science, through clinical consultancy to total environmental assessment. The role of natural feedback varies inversely to that of cultural constraints through the same progression.

There are certain parallels between the three kinds of science just described and Weinberg's (1972) distinction between science and trans-science. It is instructive to note these in passing, partly to clarify the Funtowicz—Ravetz model, and partly to enable me to emphasize that although natural feedback and cultural constraint may predominate at opposite ends of the knowledge spectrum, neither element will ever be entirely absent.

Initially Weinberg distinguishes trans-scientific questions as those which attempt to deal with social problems through the procedures of science, yet cannot be answered by science. In other words, these are the culturally overdetermined (social) questions, as distinct from the questions that can be answered by manipulating information that is overdetermined by natural sources and which Weinberg therefore calls scientific. To his great credit, Weinberg recognizes the fuzziness of the boundary between these two types of activity. Science is maintained by critical judgment of peers — clearly a social activity. Only those with the proper credentials are allowed to participate — questions of propriety, of course, have little to do with whether a question can be answered by science in principle. Further into his argument, Weinberg seems to be saying that science is consensual knowledge arrived at through criticism, while trans-science is preconsensual criticism, beset by uncertainty, that may or may not at some future point converge on an agreed solution (p. 219). Or, in cultural relativist terms, trans-science is where cultural variation is dominant, science is where cross-cultural concensus about knowledge has been achieved, and confirmed by natural feedback. The fuzzy boundary corresponds to the clinical mode of Funtowicz and Ravetz.

Cultural Relativism and Risk Analysis

Having indicated what I take to be a defensible epistemological justification for cultural relativism in science for policy, we are left with the question of how these general observations about the cultural foundations of human knowledge can be operationalized for societal risk analysis and management. If each culture or subculture has a unique way of looking at the world, are there not so many cultures that the array of possible interpretations is as vast operationally as if we were dealing with the solipsistic world views of individuals?

Happily this is not the case. Anthropologists have devised a variety of classificatory schemes that reduce the range of cultural variation to a useful number of culture types. Grid/group analysis is one such scheme that has received some prominance for its application to the field of risk analysis in Douglas and Wildavsky's *Risk and Culture. Grid* and *group* are organizational variables describing the cumulative effects of hierarchies and social networks among members of a social unit. Measuring each variable as high or low on a pair of orthogonal axes, we can generate four prototype visions of social life, each with its own characteristic view of the world and approaches to risk (Gross and Rayner 1985).

Risk and Culture has received justifiable criticism, such as that of Funtowicz and Ravetz (1985), Nelkin (1982), Kaprow (1985), and others, that Douglas and Wildavsky have done violence to the ethnography, in forcing the mail-order environmentalists and Nimby (not-in-my-backyard) coalitions into the same category as the universalist opponents of nuclear energy. From my own perspective, the book is flawed throughout most of the discussion of risk behavior in American society by the reduction of the two-dimensional framework of grid/group analysis to a simple center/

border dichotomy. However, these criticisms of the execution of the analysis pale beside the significance of the attempt to structure risk behavior in the framework of cultural analysis.

As a scientific work, *Risk and Culture* was clearly a product of total environmental assessment, for that is where most radical theories begin life. As such, Douglas and Wildavsky's analysis is thought-provoking and suggestive, but does not provide practical guidelines about how the underlying concepts of grid and group might be applied to actual problems of risk management. If practical applications can be defined in combination with a quantitative technique for measuring the variables, the grid/group approach itself would shift into the mode of clinical consultancy. This brings us to the second problem that I promised to address in this paper, that of demonstrating cases in which grid/group analysis can improve the capacity of risk managers to resolve problems arising from different cultural perceptions of what constitutes a risk and what measures are appropriate to resolve it.

I shall briefly outline two areas in which my own preliminary research indicates that grid/group analysis can provide a practical framework for understanding and modeling risk behavior as the basis for policy decisions. These are the fields of occupational safety and hazardous-facility siting. Both cases are characterized by significant systems uncertainty and fairly high decision stakes. Uncertainty in the first case is concerned with the long-term effects of low-level ionizing radiation insults to medical personnel. In the second case uncertainty focuses on the environmental transport of toxins out of disposal facilities, and consequent health effects.

Risk Management and Hospital Culture

A specific hypothesis of *Risk and Culture* is that people in individualist, bureaucratic, and small-group organizations will each focus on different kinds of risk — technical, environmental, or societal — according to what they perceive as being most threatening to their preferred institutional order. Douglas and Wildavsky share Thompson's (1981) assumption that members of the fourth ideal type generated by grid/group analysis lack the resources and the will to participate in risk management. However, there is a wide range of occupational categories in radiology and allied disciplines, each involving different lengths of training, job organization, and future prospects. The organizational experience of consultants, radiobiologists, specialists in nuclear medicine, radiopharmacists, radiological technicians, nurses, housekeepers, maintenance workers, and cleaning staff is each likely to be very different. Classifying each of these according to the multiple-hierarchy variables of grid and the group variables of networking and team membership, we were able to exemplify all four grid/group categories (Table I) and correlate them with different levels and

TABLE I Four organizational contexts of medical personnel exposed to ionizing radiation.

Organizational context	Competitive individualist	Bureaucratic	Egalitarian small group	Stratified individuals
Personel	Radiotherapists Radiodiagnos-ticians Specialists in nuclear medicine	Radiotech-nicians Hospital Administrators	Staff of free clinics PSR groups Coalition for medical rights for women	Maintenance staff, junior nurses, cleaners, and porters
Transactional arena	Ego-based networks	Organic groups	Mechanical groups	Atomized niches
Transactional mode	Competition	Routine procedures	Cooperation	Controlled
Decision making	Individual	Committee	Consensus	Limited by others
Driving values	Expansion	System maintenance	Equality	Survival
Focus of attention	Professional career (cure)	Routinization of procedures (standardiza-tion)	Health maintenance (prevention)	Diverse

types of concern with occupational exposure to low-level ionizing radiation (Rayner 1986).

The first of these is the competitive individualist context. Here everyone is expected to work hard at developing his own speciality which he can market to others. Organization is ego-centered. The individualist carefully maintains his own social networks with those who need his particularly rare talents. He will also cultivate contacts with specialists in other fields on whom he can call for help with tasks that are outside of his own speciality. In this context we find the successful consultant in radiotherapy, radiodiagnosis or nuclear medicine. There are high rewards at the center of individualist networks. But in all competitive status and market systems, uncertainty is an intrinsic element in social transactions. To maintain credibility, the competitive individual relies on continued success. He must always innovate, or push back the boundaries of his special skill or knowledge a little further. Decisions about risks are adaptive expert judgments constrained by the minimum of formal procedures.

Competitive pressures restrict the use of time in this context, so the perceived consequences of delay or procrastination are likely to encourage calculated risk taking. Under these conditions, we find that occupational radiogenic disorders are perceived as just one of a variety of legitimate hazards which must be accepted in pursuit of the pecuniary, status and humanitarian rewards of specialist medical practice.

In extreme cases this outlook can lead to a cavalier attitude to radiation hazards. Surgeons, traditionally the most glamorous category of medical practitioners, have earned themselves a particular notoriety among some radiation protection officers for flouting required procedures for experiments with radiopharmaceuticals. These individualists tend to have a high level of confidence in their personal control of hazards, and relatively little time for rules and regulations.

In certain other cases the risk of malpractice proceedings is more immediate to the individualist medical practitioner than personal radiation hazards. The imperatives of documenting treatment for "cover-your-ass" medicine, combined with a low perception of occupational risks, may encourage these physicians to order more X-rays than are strictly required for treatment and thus expose patients to greater radiation hazards.

The second is the bureaucratic context. This sort of organizational framework is bound by rules and an emphasis on the procedures of decision-making. Conflicting goals and interests are combined, without being resolved, in routine procedural rules. Career success consists of rising through a hierarchy of qualifications and grades. In seeking promotions, the candidate needs to convince superiors that he has successfully internalized the necessary skills and rules for their application, including safety rules. Medical administrators and radiological technicians in large hospitals often experience this sort of organizational environment.

Like other aspects of this sort of organization, risks are routinized as far as possible. Properly constituted bodies decide what dangers there are, and what procedures are appropriate to minimize them. When accidents do occur, mechanisms for finding facts, compensating victims, and reveiwing procedures, are on hand. The second-order risk of overconfidence in dealing with a potential hazard may arise out of these organizational factors. Even where hazards are perceived, the routine apportionment of time to specific tasks may tend to encourage a diminished sense of their urgency. Thus a small hazard may increase because bureaucracies respond slowly to danger.

Within this organizational framework, perception of radiological hazards is less sharp than in the individualist context. Whereas the individualist will incorporate occupational risks into his rational calculations, the bureaucrat tends to underestimate hazards or even fail to perceive them at all. Overconfidence may encourage a relaxed attitude toward safety measures, resulting in excessive exposure to medical staff or their patients. Overconfidence which leads to not wearing a dosimeter or failing to ensure adequate pelvic protection for a patient during routine X-rays can seriously upset EPA exposure guidelines. It also seems that routine reliance on exposure standards encourages some people to believe that the standard is a natural threshold below which there is no risk entailed in working with ionizing radiation. A consequence of this belief is that those personnel are exposed more than is necessary by not bothering to minimize their dosage provided that it is thought to be within the standard.

Another type is the small-group context. For organizational reasons, voluntary groups tend to be egalitarian and small (Olson 1965). Unlike the individualist context, the emphasis here is on cooperation rather than competition. The group's smallness is inimical to the development of bureaucracy. Decisions are made by consensus wherever possible. Everyone is held to be responsible for the welfare of everyone else in the group, and solidarity is the highest value. However, in the absence of dispute-resolving mechanisms, such as seniority within a bureaucracy or the entreprenurial power of the individualist, the reality of small-group organization is often far from its ideal. It is frequently disputatious, and in order to preserve its normative values of brotherly love, the resulting disruption is frequently blamed on the infiltration of outsiders (Rayner 1979). Small groups, therefore, often have built-in susceptibility to ideas of contamination by invisible and insidious influences which disrupt the tranquility of their ideal world. This description suggests that sensitivity to radiation hazards in medicine would be stronger among members of this kind of organization than among bureaucrats or competitive individualists who tend toward high levels of confidence.

Medicine does not immediately suggest itself as a haven of small-group activity, and it is certainly true that the field is dominated by the individuals and bureaucratic forms of organization. However, there are small not-for-profit clinics that are run on cooperative lines by medical staff who have eschewed the dominant patterns of medical organization. There are also physicians' and nurses' organizations, such as some local chapters of Physicians for Social Responsibility and the Coalition for Medical Rights for Women, which fit this pattern. Based on the opposition to centralized institutions in nuclear energy, we expect opposition to capital-intensive medicine here. Just as "no growth" is not an inviting rally cry with which opponents of centralized nuclear energy would expect to attract support, so outright rebuttal of potentially life-saving technology would not be likely to appeal either to patients or medical practitioners themselves. Their opposition to centralized medicine, therefore, manifests itself as an emphasis on the dangers of radiomedicine to the patient and medical staff alike.

The last of the four types is a category consisting of stratified individuals who lack the qualifications or control over goods and services necessary for participation in the individualist framework, or else they are without access to the established institutions of decision-making within a bureaucracy. People in this category tend to be the most vulnerable in any social system. They are likely to doubt the assurances of experts that activities perceived as being hazardous are, in fact, safe. In medicine, this category of workers includes junior nursing and auxiliary staff, who are not officially recognized as nuclear workers, who are not monitored for radiation exposure, but nevertheless may be exposed to low-level radiation while assisting with patients (particularly children) being X-rayed, or caring for patients who have received radioactive pharmaceutical implants for either diagnostic or therapeutic purposes. These kinds of cases are probably the most common occasions of concern for occupational radiation hazards among medical professionals in the hospitals where this research was conducted.

However, the most concern of all is voiced by nonmedical staff such as the plumbers who are called upon to maintain fume cupboards in which radiopharmaceuticals are handled or to clear "hot" wastepipes through which radioactive materials have been disposed. This category of workers also includes cleaners in radiographic or radiopharmaceutical facilities. These workers occupy the bottom rungs of the hospital hierarchy and, unlike nurses and technicians, they do not work in stable teams. Since fewer than 25 percent of American hospitals provide any sort of radiation hazard training for these sorts of personnel (NIOSH 1976), it is hardly surprising that they express the most serious concern about radiation exposure in the hospitals.

The significance of this study for hospital risk managers is related to institutional design to promote safety. For example, we have observed that both competitive individualists and bureaucrats may be prone to overconfidence with risk, but for very different reasons (Figure 2). The

GRID	Low	High
High	*Stratified Individuals* Passive worry response High concern for own radiation risks. Low satisfaction with safety measures.	*Bureaucratic* Denial response Low concern for own radiation risks. High satisfaction with safety measures.
Low	*Competitive Individualist* Incorporation response Low concern for own radiation risks. High satisfaction with safety measures.	*Egalitarian Small Group* Protest response High concern for own radiation risks. Low satisfaction with safety measures.
		GROUP

Fig. 2. Attitudes of medical personnel to occupational radiation hazards.

individualist has confidence in potentially hazardous activities that are under his personal control, while the bureaucrat may underestimate a hazard because of his confidence in the routines that are designed to afford safety. A risk manager wishing to sensitize both to the hazards that they may be overlooking in the workplace would have to adopt diametrically opposite organizational strategies in each case. It would be desirable, for example, both to increase the integration of the individualist into a stable team and to involve him in developing routines for the whole group to follow. In contrast, it would seem that the group ties of the bureaucrat should be weakened and the job restructured to reduce emphasis on routines and encourage pragmatic consideration of problems as they arise. Attempts to rectify overconfidence through institutional design that did not have the benefit of a systematic cultural relativist perspective might well adopt only one of these strategies, for example, improving the safety performance of individualists, while exacerbating the problem with the bureaucrats.

Siting Noxious Facilities

The relevance of the grid/group typology to the problem of siting noxious facilities lies in the fact that the social organization of each type of system makes its members sensitive to different aspects of the problem and leads them to favor characteristically different decision strategies. In particular, the preferred spread of liabilities and benefits and the favored means of obtaining consent from those affected by the decision will vary between the four different kinds of organization.

Policy makers in competitive individualist organizations will favor a *revealed preference* (Thaler and Rosen 1975) approach to obtaining consent — sometimes called *implicit consent* (MacLean 1986) — which allows market forces to determine planning priorities. The rationale being that people's preferences for one solution or another will be reflected accurately in how they spend their money. The problem with this approach is neatly caught by an adage among British road builders that the best soil conditions are always to be found in working-class areas. Of course, the wealthy can better afford to oppose planning applications that affect their interests. Thus, this system will also favor a loss-spreading approach to liability (Calabresi 1970), in which market mechanisms determine who bears losses. (When pure market solutions are not available, institutions of this sort will attempt to reproduce what the market would have done if it had not been impeded by high information and/or transaction costs.) Under these conditions, noxious facilities will be viewed as a legitimate and inevitable cost of the pursuit of economic and social progress. At the same time, competitive individualists will attempt to use their strength in the market place to avoid having such facilities encroach on their own personal interests, such as property values.

Members of hierarchical or bureaucratic organizations favour what is sometimes called *hypothetical consent* (Rawls 1971). Here, the citizen is assumed to have entered into a social contract with the decision-making institution, whereby he may be deemed to assent to decisions made through the rational procedures of that institution, even though he might not like the particular outcome. The acceptability of a siting decision will, therefore, be determined by appeals to the constitutionality of the institutional structures of decision making.

Policy makers in this kind of system favor redistributive taxation as a means of apportioning costs. In this way, bureaucracies make use of redistributive mechanisms to apportion liabilities in a way that seems to them to be least disruptive, not to the whole of society perhaps, but certainly to those clienteles whose stability they see as important to the survival of the institution. While confidence in the institution itself is maintained, the routinization of dangers and the diffusion of liability will tend to diminish the perception of uncertainty and the urgency of hazards as they arise. Whereas the individual competitor will incorporate these into his calculations, the bureaucratic policy maker may underestimate the costs to the host community, or even fail to perceive them at all.

Egalitarian groups favor expressed preferences that appeal to explicit judgments of ranked values (shared by the group) rather than the revealed preference or hypothetical consent approaches for determining the acceptability of controversial facilities that are favored respectively by entrepreneurs or bureaucrats. Furthermore, members of egalitarian groups will seek a moral determination of liability for unforeseen costs that appeals to those same ranked values, rather than the market or distributive approaches to loss favored by entrepreneurs or bureaucrats. The acceptability of controversial facilities will therefore be determined according to criteria that are very different from those invoked by entrepreneurs or bureaucrats.

There is less to be said about the preferences of stratified individuals. They tend not to articulate any distinctive theory of consent or of distribution. Generally, they are not decision makers themselves since most of the options for societal choice do tend to be monopolized by the other three grid/group types. It is likely that these people will express discontent that their consent is seldom, if ever, sought by any means and that they are constantly having to bear the costs of other people's errors. When threatened directly, they may form Nimby groups to protest the imposition of someone else's decision.

From the contrast between the four institutional cultures engaged in siting controversies, it is clear that policy makers within each type of sociocultural framework frequently have great difficulty in understanding the fears and objections of others, especially when the language of probabilities and size of loss is used to represent these nonprobabilistic concerns. In particular, given the variation of the desired spread of costs and the fundamentally incompatible approaches to the question of consent, each kind of institution is unlikely to be able to offer a package of compensations for accepting a noxious facility that would satisfy a clientele whose social experience is of one of the other types of social organization and corresponding culture. Yet, the provision of special compensation to communities located close to controversial facilities is now recognized by political scientists as an increasingly important requirement in the design of control systems for modern technologies (Starr 1980; Kevin 1980).

Hitherto, proposals to provide incentives to communities to host controversial facilities, such as a high-level nuclear waste facility, have concentrated on providing monetary payments (LaPorte 1978; Starr 1980) and have stressed the relationships between the Federal Government and the States concerned, rather than the host communities (Kevin 1980; Lee 1980). There are probably two main reasons for this. One is that the legal precedents for compensation are based on redressing damage actually suffered. In the absence of demonstrable loss there is some difficulty in establishing which communities around a site, and which members of those communities should receive payment (O'Hare 1977; McMahon *et al.* 1981). Another factor is Dillon's rule, that local governments are but creatures of the states who can, therefore, speak directly for them. Unfortunately, such a constitutional nicety clearly underestimates the real political independence and power of local governments.

It has recently been proposed that it may be more appropriate to design incentive packages containing more than just monetary payments (O'Conner 1980; Carnes *et al.* 1983). Furthermore, the same authors recognize that such a package should be responsive to the diverse, and possibly competing, concerns within and between the communities actually affected. The addition of a grid/group perspective would enable us to go further than just arguing that the organizational cultures of the communities must be taken into account when decision makers design packages of compensating benefits for living with something that the community deems to be undesirable. We can actually suggest a cultural basis for the design of such packages based on cultural analysis of a population's preferences for different sorts of institutional safeguards (Rayner 1984).

Competitive individualists in market cultures will require an emphasis on short-term rewards. They will look for actions designed to bring extra benefits to communities that assume costs on behalf of the wider society. These might include direct monetary payments, tax incentives, and the location of desirable projects as well as noxious or dangerous ones in the vicinity. In general, incentives that can be disposed of independently by individual householders will be preferred to those that have to be controlled by community governments such as a local city council.

Stratified individuals will probably have to put up with whatever they are offered. But, we should question the morality of taking advantage of their vulnerability. If we do not, they may turn to protest and attract the support of the egalitarian groups. Then, the policy makers' problems multiply.

Hierarchical groups or bureaucracies will require a greater emphasis on system maintenance and investments in the community that will ensure future controls are adequate and that compensation for unforeseen contingencies will be forthcoming. A suitable package might include support for the community infrastructure in the form of investment in public works, such as new schools. Land-value guarantees might be welcome as promoting community stability, while guaranteed trust funds and insurance schemes could compensate those who actually suffer in the event of an accident or an anomaly arising.

Egalitarian groups will not willingly accept any decision made on behalf of others because it would violate their most cherished ethical principles with respect to both consent and the distribution of liabilities. However, actions to mitigate the risks might help to reassure some members of such communities. Measures designed to prevent, reduce, or eliminate adverse consequences before they occur might include the provision of buffer zones, community-controlled monitoring, and the right of the community to shut down the operation if negotiated safety thresholds are breached.

Of course, few real communities will correspond overwhelmingly to only one or another of these ideal types. Real cases are likely to present the analyst and the decision maker with a mixture of two or more. However, it is possible to conceive of a situation in which planners could assemble packages of safeguards and compensations designed to satisfy the concerns and interests of each section of the community in proportion to its size and influence within the whole. As important as knowing what to include in an offering package of incentives is knowing what not to offer that might incense a significant subsection of any community. For example, offering cash on the nail to a predominantly entrepreneurial community might not turn out to be such a good idea if a significant proportion of strong-group egalitarians is also present. Egalitarians are likely to be outraged by any offer that smacks of selling the common good for the benefit of a few, or of selling the heritage of future generations for present gains. They may, however, be impressed by measures designed to protect the community as a whole, which would be unlikely to provoke the entrepreneurs who would also benefit. In such a case, an emphasis on the interests of a significant minority would probably be more profitable than the obvious course of statisfying the majority.

There is clearly much work to be done, as considerable skill and knowledge that we do not yet have will be required in putting together culturally compatible incentive packages. With the introduction of measurable grid/group scales (Gross and Rayner 1985) developing cultural profiles of communities as the basis for designing such packages of compensations for controversial facilities becomes plausible. Whether or not it becomes practicable will depend largely on how well technical and societal risk analysts are able to cooperate in persuading policy-making and risk-managing institutions to suspend their own way of looking at the issues in order to accommodate the views of those for whom they are planning.

Measuring Culture

These two examples illustrate the potential usefulness to societal risk analysis of the consistent cultural-relativist framework that is provided by grid/group analysis. We can see how people from different organizational and institutional contexts may talk past each other because they maintain different culturally conditioned perceptual filters. These admit concerns relevant to their day-to-day experience, while blocking those ideas that are irrelevant or would place obstacles in the way of their daily lives. By making the basis of this disagreement explicit, we might alert policy makers to unforeseen problems before they become political crises as well as suggest some new approaches to their solution.

Furthermore, we may be able better to understand some of the technical disagreements within the realm of clinical consultancy. Hopefully, scientific debate can begin explicitly to accept that scientists can never free themselves entirely of perceptual biases arising from cultural experience. Such biases may not make a great deal of difference to the conduct of concensual science, but they are a major part of clinical consultancy and all of total environmental assessment. Any true ideal of scientific objectivity surely must recognize the inevitability of systematic subjectivity as an objective part of the scientific process. Therefore, the best we can do in science for policy at the level of clinical consultancy, where most of this kind of research takes place, is to incorporate a systematic awareness of cultural bias into the conduct and presentation of research, remembering that there are always, at least, four ways of looking at something.

Ultimately, this enterprise depends on establishing consistent definitions of the grid and group variables, and developing our ability to make formal measurements. In *Measuring Culture* (Gross and Rayner 1985) we have attempted to provide a general model for this kind of analysis. We have advocated rendering diverse social organizations as mathematically modeled social units. Second, we have applied polythetic scales to the measurement of the multiple-hierarchy (grid) and social-network (group) aspects of that model in order that measurements made in this way can be replicated by different researchers. This is significant both because the ability to make some consistent measurements is an important feature of clinical consultancy as contrasted with total environmental assessment and because it establishes the empirical basis for falsifying grid/group analysis.

Perhaps most importantly, we have tried to provide clear criteria to resolve disagreements between analysts about such issues as whether or not Nimbys and egalitarians belong in the same category, as Thompson (1981) and Douglas and Wildavsky (1982) claim, or whether they exhibit characteristically different behavioral syndromes toward risk as argued by Funtowicz and Ravetz (1985) and Gross and Rayner (1985). Whether these criteria can reduce the systems uncertainty in grid/group analysis enough to form the basis of sufficient consensus to shift grid/group into the realm of clinical consultancy remains to be seen. The decision stakes for societal risk analysis would seem to be high enough to justify the attempt.

References

Agassi, Joseph. 'The Cheapening of Science,' *Inquiry*, 1984, 27, 167—172.

Barnes, Barry. *Scientific Knowledge and Sociological Theory*. London: Routledge and Kegan Paul, 1974.

Berkeley, George. *A Treatise Concerning the Principles of Human Understanding*. Dublin: Rhames and Pepyat, 1710.

Bloor, David. *Knowledge and Social Imagery*. London: Routledge and Kegan Paul, 1976.

Calabresi, Guido. *The Cost of Accidents*. New Haven, Connecticut: Yale University Press, 1970.

Carnes, S., E. Copenhaver, J. Sorensen, E. J. Soderstrom, J. Reed, D. Bjornstadt, and E. Peelle. 'Incentives and Nuclear Waste Siting: Prospects and Constraints,' *Energy Systems and Policy*, 1983, 7, 323—351.

Cohen, Bernard. 'Criteria for Technology Acceptability,' *Risk Analysis*, 1985, 5, 1—3.

Douglas, Mary. *Cultural Bias*. London: Royal Anthropological Institute Occasional Paper 35, 1978.

Douglas, Mary. 'A Backdoor Approach to Thinking About the Social Order.' Address to the American Sociological Association, San Antonio, Texas, 1984.

Douglas, M. and A. Wildavsky. *Risk and Culture*. Berkeley, California: California University Press, 1982.

Funtowicz, S. O. and J. R. Ravetz. 'Three Types of Risk Assessment: A Methodological Analysis,' in C. Whipple and V. Covello (eds.), *Risk Analysis in the Private Sector*. New York: Plenum, 1985.

Gross, J. L. and S. Rayner. *Measuring Culture*. New York: Columbia University Press, 1985.

Kaprow, Miriam Lee. 'Manufacturing Danger: Fear and Pollution in Industrial Society,' *American Anthropologist*, 1985, 87, 357—364.

Kevin, D. 'Federal/State Relations in Radioactive Waste Management Oceans Program.' U.S. Congress Office of Technology Assessment, Washington, D.C., 1980.

LaPorte, Todd. 'Nuclear Waste: Increasing Scale and Sociopolitical Impact,' *Science*, 1978, 207, 22.

Latour, B. and S. Wolgar. *Laboratory Life: The Social Construction of Scientific Facts*. Beverly Hills, California: Sage, 1979.

Lee, Kai. 'A Federalist Strategy for Nuclear Waste Management,' *Science*, 1980, 208, 679.
MacLean, Douglas. *Values at Risk*, Rowman and Allenheld, Totowa, New Jersey, 1986.
McMahon, R., C. Ernst, R. Myares, and C. Haymore. 'Using Compensation and Incentives When Siting Hazardous Waste Management Facilities.' U.S. Environmental Protection Agency, Washington, D.C., 1981.
National Academy of Science. *Risk Assessment in the Federal Government: Managing the Process*. Washington, D.C.: National Academy Press, 1983.
Nelkin, Dorothy. 'Blunders in the Business of Risk,' *Nature*, 1982, 298, 775—6.
NIOSH. 'Occupational Health Services for Hospital Employees.' National Institute for Occupational Safety and Health, HEW Publications, Washington, D.C., 1976.
O'Connor, W. A. 'Incentives for the Construction of Low-Level Nuclear Waste Facilities,' in *Low-Level Waste: A Program for Action*. Final Report of the National Governors Association Task Force on Low-Level Radioactive Waste Disposal, Washington, D.C., 1980.
O'Hare, Michael. 'Not on My Block You Don't,' *Public Policy*, 1977, 25, 407.
Olson, Mancur. *The Logic of Collective Action: Public Goods and the Theory of Groups*. Cambridge, Massachusetts: Havard University Press, 1965.
Ravetz, Jerome, R. 'Usable Knowledge, Usable Ignorance: Incomplete Knowledge with Policy Implications.' Address to Task Force Meeting on The Sustainable Development of the Biosphere, International Institute for Applied Systems Analysis, Laxenberg, Austria, 1984.
Rawls, John. *A Theory of Justice*. Cambridge, Massachusetts: Belknap Press, 1971.
Rayner, Steve. 'The Perception of Time and Space in Egalitarian Sects: A Millenarian Cosmology,' in Mary Douglas (ed.), *Essays in the Sociology of Perception*. London: Routledge and Kegan Paul, 1982.
Rayner, Steve. 'The Classification and Dynamics of Sectarian Forms of Organization: Grid/Group Perspectives on the Far Left in Britain,' Ph.D. Thesis, University of London, 1979.
Rayner, Steve. 'Disagreeing About Risk: The Institutional Cultures of Risk Management and Planning for Future Generations,' in S. Hadden (ed.) *Risk Analysis, Institutions, and Public Policy*. Port Washington, New York: Associated Faculty Press, 1984.
Rayner, Steve. 'Management of Radiation Hazards in Hospitals: Plural Rationalities in a Single Institution,' *Social Studies of Science*, 1986, 16, 573—591.
Starr, Chauncey. 'Risk Criteria for Nuclear Power Plants.' paper presented to the ANS/ENS International Conference, Washington, D.C., 1980.
Thaler, R. and S. Rosen. 'The Value of Saving a Life,' in N. Terleckyj (ed.), *Household Production and Consumption*. New York: National Bureau of Economic Research, 1975.
Thompson, Michael. 'To Hell With the Turkeys.' IIASA Working Paper, Laxenburg, Austria, 1981.
Weinberg, Alvin. 'Science and Trans-science,' *Minerva*, 1972, 10, 209—212.
Wittgenstein, Ludwig. *Philosophical Investigations*. Oxford: Blackwell, 1953.

PART II

Part II

Community Dynamics and the Social Construction of Risk

The focus of this section of the book is on community dynamics and the social construction of risk. The two chapters in this section examine cases where risk selection and perceptions emerge from social interactions among members of a community.

In recent years several distinctive approaches have been adopted by researchers interested in community dynamics, risk selection, and risk perception. One set of studies has focused on one-industry towns. These studies have discovered, among other findings, that people living in such towns are more likely to deny the existence of industrial pollution problems than people living in multi-industry towns (e.g., Tichenor *et al.* 1973: 62). Another set of studies has focused on the link between levels of personal satisfaction among members of a community and perceptions of risk (Kiecolt and Nigg 1982; Preston *et al.* 1983; Fischer *et al.* 1977; Galster and Hesser 1981). A third set of studies has examined the process of decision-making in small communities, focusing on who makes decisions in local communities and how items get added to the local political agenda (e.g., Crain *et al.* 1969; Crenson 1971). A fourth set of studies has focused on the influence of internal versus external factors on perceptions of risk. Coleman's work (1957), for example, suggests that sources of risk that are perceived to be internal to the community will be less feared than sources of risk that are perceived to be external to the community. Coleman's research also suggests that risk controversies often move from the specific to the general as existing interpersonal and intergroup relationships break down and as "old quarrels which have nothing to do with the original issue" are openly raised (Coleman 1957: 25). A final set of studies — such as case studies of fluoridation (Crain *et al.* 1969) and the Salem witchcraft trials of the 1690s (Boyer and Nissenbaum 1974; Demos 1982; Weisman 1984) — has focused on the complex economic, political, and cultural forces within a community that structure both the form and nature of risk selection and perception.

The first chapter in this section, by Janet Fitchen, Jenifer S. Heath, and June Fessenden-Raden, falls largely within this last category and focuses on the role of community politics and economics in the social construction of risk. Specifically, the authors describe a case involving the possible

B. B. Johnson and V. T. Covello (eds.), The Social and Cultural Construction of Risk, 27—29.

industrial contamination of drinking water in an upstate New York community. The case is notable for the extraordinary lack of interest in the problem by local residents. The authors suggest that the community's lack of interest and concern can be traced to several sources. First, local economic problems led to fears that inclusion of the site in the federal Superfund program might impose high cleanup costs on local industry and that these costs might force these industries to close down. Second, community residents expressed high levels of trust in local government officials and their ability to handle the issue. Third, residents were concerned about the apparent arbitrariness of state and federal agencies and their inability to provide cleanup resources or reliable risk estimates. Finally, local residents, businessmen, and local officials pointed to the lack of physical evidence of contamination (e.g., smell or taste) and to the length of time that had passed since the original contamination. Fitchen *et al.* point out that these various factors tended to reinforce one another but that the situation might well have been different if trust in local government had been lower or if the physical evidence of contamination were more clearly apparent.

The second chapter in this section of the book, by Martha Fowlkes and Patricia Miller, focuses on the now infamous hazardous waste site in Love Canal, New York. Most other discussions of this case have focused on interactions among government, industry, and community activists in the social construction of risk (see, for example, the chapter by Tarr and Jacobson in this volume). Fowlkes and Miller, however, take a different tack, and emphasize the influence of social ties on the social construction of risk. The chapter represents a particularly important contribution to the literature in that it underlines the importance of life cycle and social networks in the process of identifying and selecting risks for attention and concern. Fowlkes and Miller point out, for example, that the Love Canal area was not a closely knit neighborhood prior to the discovery of toxic wastes; instead, it was an area defined principally by its geographical boundaries. As the crisis developed, however, it became possible to distinguish two types of community residents: "minimalists" — people who either denied that there was a problem at all or believed that any problem that did exist was only minor in significance — and "maximalists" — people who believed that the risks were substantial and that chemical contamination might be more widespread than officially acknowledged. The most important predictors of whether people became minimalists or maximalists were the age and composition of households. These factors were, however, only predisposing and not determinative.

Fowlkes and Miller found that differences among neighbors that were ordinarily unimportant in community affairs became "lines of fission" as the crisis developed. The analysis shows, for example, that minimalists generally (a) were older and without children at home; (b) had links with the local chemical industry, (c) were relatively isolated from neighborhood social networks, despite having lived there for many years, (d) adopted a *laissez-faire* attitude toward risks generally, and (e) were strongly attached to their homes, which they viewed as their principal economic resource in old age. Maximalists, by comparison, had close links with others in the community. Many of the maximalists were young parents who shared interests and maintained frequent contact with other young parents in the area. Concerns about the welfare of their children led many of these parents to place less weight on property issues and more weight on health issues. Compared to minimalists, maximalists were also more active in seeking out risk information and gave greater credence to what they learned from others — particularly from nonofficial sources.

These two provocative chapters raise several important questions for future research. To what extent do social networks — e.g., family and friends — shape risk information and the speed with which it is transmitted? To what extent does the perception of threat as external influence individual and community perceptions of risk? Under what conditions do internal divisions within a community exacerbate conflict and disagreements about risk selection and perception?

References

Boyer, Paul and Stephen Nissenbaum. *Salem Possessed: The Social Origins of Witchcraft.* Cambridge: Harvard University Press, 1974.

Coleman, James. *Community Conflict.* Glencoe, Illinois: Free Press, 1957.

Crain, Robert L., Elihu Katz, and Donald B. Rosenthal. *The Politics of Community Conflict: The Fluoridation Decision.* Indianapolis: Bobbs-Merrill, 1969.

Crenson, Matthew A. *The Un-Politics of Air Pollution: A Study of Non-Decisionmaking in the Cities.* Baltimore: The John Hopkins University Press, 1971.

Demos, John P. *Entertaining Satan: Witchcraft and the Culture of Early New England.* New York: Oxford University Press, 1982.

Fischer, Claude S., Robert M. Jackson, C. Ann Stueve, Kathleen Gerson, Lynne M. Jones, and Mark Baldassare. *Networks and Places: Social Relations in the Urban Setting.* New York: Free Press, 1977.

Galster, George C. and Garry W. Hesser. "Residential Satisfaction: Compositional and Contextual Correlates," *Environment and Behavior,* November 1981, 13:6, 735—758.

Kiecolt, K. Jill and Joanne M. Nigg. "Mobility and Perceptions of a Hazardous Environment," *Environment and Behavior,* 1982, 14, 131—154.

Preston, Valerie, S. Martin Taylor, and David C. Hodge. "Adjustment to Natural and Technological Hazards: A Study of an Urban Residential Community," *Environment and Behavior,* March 1983, 15, 2, 143—164.

Tichenor, Philip J., Jane M. Rodenkirchen, Clarice N. Olien, and George A. Donohue. "Community Issues, Conflict, and Public Affairs Knowledge," pp. 45—79 in Peter Clarke (ed.), *New Models for Mass Communications Research.* Beverly Hills, California: Sage, 1973.

Weisman, Richard. *Witchcraft, Magic, and Religion in 17th-Century Massachusetts.* Amherst: University of Massachusetts Press, 1984.

Chapter 2

Risk Perception in Community Context: A Case Study

Janet M. Fitchen, Jenifer S. Heath, and June Fessenden-Raden

Introduction

This chapter takes a holistic, contextual, and community-based approach to understanding social construction of risks. Our research utilizes community settings as arenas in which actual localized risks are perceived and addressed.[1] The case study approach enables us to examine the dynamics of risk perception, concentrating on overt expressions and reactions over time, rather than on individual perceptions at a single point in time This naturalistic *in situ* approach reveals factors influencing risk perception that may be obscured in studies in which individual respondents are asked to rank hypothetical risks without regard for the complicating factors and actual consequences that exist in ongoing community life (see, for contrast, the psychometric research of Slovic *et al.* 1980, 1984).

The central thesis of this chapter is that risk perception is a complex and dynamic process that is influenced by the local context in which the risk is embedded and by the manner in which the risk is addressed.

The Research Context

This case study is part of a larger research project on environmental chemicals and risk management. Our specific emphasis is on public health risks stemming from chemical contamination of groundwater used for drinking water supplies. This interdisciplinary research project combines the expertise of environmental toxicology, analytical chemistry, genetics, education, geohydrology, environmental engineering, and anthropology. We presently are studying nonmetropolitan communities in New York state that are faced with decisions about their public water supply because toxic chemicals contaminate the groundwater that is used as their drinking water source. In each community the social science component of our research addresses the perceptions, responses, decisions and actions both of the resident public and community officials and of outside intervenors including various federal and state agencies and private consulting firms.

B. B. Johnson and V. T. Covello (eds.), The Social and Cultural Construction of Risk, 31—54.

We have studied these contamination situations in some depth, reconstructing earlier events, examining decisions and attitudes, and monitoring each situation as it continues to unfold. In-depth interviews with local decision makers and other community people involved in the issue have been supplemented by interviews with personnel from government agencies involved in the local problem, by informal discussions with local residents, attendance at public hearings and meetings, monitoring of local newspapers, careful researching of public documents and reports, and collection of background information on the community context.

This research seeks to understand how people in actual communities perceive the specific human health risks resulting from chemical groundwater contamination and how they respond to these risks within the contexts of their everyday lives. Before we present the case study itself, it is useful to lay out briefly the technical aspects of groundwater, its chemical contamination, and the relevant health risks.

Groundwater and Chemical Contamination

For over half of the U.S. population, groundwater is the primary source of drinking water. Ninety-five percent of all rural households are dependent on groundwater as an economically irreplaceable source of drinking water. Virtually all groundwater is vulnerable to chemical contamination. The most prominent sources of contamination are: leaching of wastes stored in landfills, ponds, or lagoons; leaking underground storage tanks; improper application of agricultural chemicals; and leaking public sewage systems and private septic systems. Industrial chemicals, pesticides, cleaning solvents, gasoline, even highway deicing salt have all seeped down through the soil to the aquifers from which groundwater is drawn.

Synthetic organic chemicals contaminate groundwater from Maine to Hawaii. Forty states have documented cases of serious contamination; scores of public water supply wells have been closed. More than 700 synthetic organic chemicals have been identified in various drinking water supplies that come from groundwater (Goyer 1984). As groundwater testing programs are implemented and tests for organic chemicals are perfected, the extent and magnitude of groundwater contamination is becoming apparent. No one knows how long or at what level these wells have been contaminated or what possible long-term health effects, if any, might be experienced by those who have been drinking, cooking with, and bathing in water contaminated with low levels of these chemicals.

Unlike surface water which flows in defined, visible areas, diluting the contaminant and moving it quickly away from a given place, groundwater moves very slowly, from inches to feet per year, with limited dilution. Dispersion of the contaminant within the aquifer is dependent on the properties of the chemical, the subsurface soil, and precipitation conditions. The complexity of the aquifer system makes it both difficult and expensive to monitor and clean up groundwater. Treatment of water at the point of use has often become the "clean-up" method of choice for removing chemical contaminants. This, however, does not address the basic problem of current and future resource contamination. Furthermore, many chemicals contaminate drinking water supplies at concentrations that cannot be seen, smelled, or tasted, and thus the presence of these chemicals may go unnoticed and untreated.

The Health Risk

The public health risk from chemical contamination of groundwater is one of unknown dimensions. Some of these chemicals can interact in the body in subtle ways that go undetected for months or years. Some chemicals may cause immediate or acute reactions such as skin rashes or dizziness; others may pose little or no health threat. The consideration of any health risk is complicated by such variables as individual susceptibility, mixtures of chemicals, and duration and route of exposure, as well as individual lifestyle. Additionally, different chemicals have different properties and may pose different risks.

Volatile organic chemicals (VOCs) such as trichloroethylene, vinyl chloride and benzene are among the most common contaminants found in groundwater. The U.S. Environmental Protection Agency (EPA) has not, as of this writing, established safe drinking water standards for VOCs other than trihalomethanes. Several states have established their own VOC guidelines: the New York State Department of Health has set safe drinking water guidelines at fifty parts per billion (ppb) for a single organic chemical not specifically covered by a standard or other guideline, and at 100 ppb for the total concentration of organic chemicals in drinking water.

Trichloroethylene (TCE) is the chemical that has contaminated both public and private water supply wells in the community described in this chapter. Trichloroethylene is widely used as a degreaser in industry, in dry cleaning, and in the synthesis of other organic chemicals. It has also been used medicinally as an anaesthetic and agriculturally as a pesticide. A small, uncharged chemical, TCE is readily absorbed through the skin, digestive tract, and lungs. Immediate health effects of TCE in humans are primarily manifested in the central nervous system and, depending on the amount of exposure, range from headaches and incoordination to unconsciousness and death. Other acute effects in humans include liver and kidney damage, irregularities in heart rhythm, and skin irritation. There are indications that even at low levels TCE exposures of long duration may cause headaches, insomnia, tremors, and nervous disorders lasting for more than a year after exposure ceases (Browing 1965; National Academy of Sciences 1977, 1980).[2]

Trichloroethylene was identified as an animal carcinogen by the National Cancer Institute in 1976. A two-year study had found that mice, but not rats, developed liver cancer after forced feeding of high levels of TCE. This study has been criticized because industrial grade TCE (containing some mutagenic impurities) was used, resulting in the possibility that it was an impurity, rather than TCE itself, that caused the cancer. The study has recently been repeated using TCE with no known impurities and the results suggested carcinogenic activity. Cancer risk assessments, however, have been based on the earlier study. These assessments indicate that if one million people consumed two liters of water with 50 ppb of TCE (New York's guideline level) every day for 70 years, between one and ten of them would develop cancer as a result of the TCE (National Cancer Institute 1976). These estimates of risk consider ingestion of TCE in "drinking" water as the only route of exposure, but in fact the total amount of TCE absorbed could be doubled by inhalation of volatized TCE, and skin absorption may increase the total amount absorbed by up to nine times over ingestion alone (Berry *et al.* 1984; Brown *et al.* 1984). As with all estimates of risk to humans based on data from other species exposed under conditions not comparable to those in which people function in their everyday lives, these human cancer estimates involve a high degree of uncertainty.

Expression of Public Concern over the Health Risk

In several of the communities we have studied thus far, where chemicals such as TCE have contaminated groundwater used for drinking water, public awareness of and concern over the health risk has been neither widespread nor sustained. People's complaints focus less on the health risk than on the nuisance of having to boil water, the off-color of the substitute water supply, the inconvenience or expense of filters, and the effect of water problems on property values. Furthermore, residents in these communities are not angrily pointing their fingers at the firms or individuals that may have caused the contamination, whether factories, farms, or private home-owners. Almost entirely absent in these cases are angry public meetings, adversarial confrontations with local officials, and citizen lawsuits.

In comparison to Love Canal (NY), Woburn (MA), and Times Beach (MO), which have become household words with clear connotations nationwide, several of the communities we have been examining seem like noncases. These "quiet" situations are hardly the stuff of television documentaries and newspaper specials. But the coincidence of our turning up several "quiet" cases makes us think that such instances must be more common than generally believed, and should be examined. ("Quietness" was not a criterion for selection of communities to study; rather, the selection was based on the fact that chemical contaminants above guideline levels had led to the closing of public supply wells.)

The question that arises is why so little public concern is expressed in some communities about health risks from toxic chemicals while so much concern is expressed in others. The answer could not lie solely in the nature or severity of the particular health risks, for we find that public perception and reaction in different communities may be quite different even when the health risks are similar. Perhaps in any community there are some individuals who are deeply concerned and others who are less concerned. But at the community level it appears that one (or sometimes more than one) overall public perception of the risk emerges, and that this public perception subsequently filters individual perceptions and may override a diversity of individual and group risk perceptions. We can thus talk of community risk perception, and of different perceptions in different communities. These intercommunity differences of risk perception demand research attention: why are there different social constructions of risk in different communities within the same society?[3] Answers to this question require a comparative approach, which we are presently undertaking (see Fitchen, forthcoming). However, a single in-depth case study is a useful first step because it indicates relevant factors that should be examined in subsequent comparative work. The case study presented in this chapter thus addresses intermediate-level questions such as the following: what factors in the specific community context surrounding a risk influence the general public perception of that risk in that community?

CASE STUDY: THE CONTAMINATION SITUATION AND COMMUNITY PERCEPTIONS OF THE RISK

This small city is a manufacturing and retail center located in a rural county in upstate New York. The city and the adjacent built-up sections of the surrounding township together comprise a fairly tight-knit community of under 25,000 people that has a small-town atmosphere and describes itself as "a family community" with "old-fashioned rural values and friendliness." For the last several years, the community has been trying to resolve a problem of TCE contamination of groundwater used for the drinking water supply. State and federal agencies have been investigating the contamination prior to making plans for cleaning up the aquifer and developing a long-range solution to the drinking water problem. As the years have gone by, local awareness of and concern over health risks, which was never very high, has waned and become sidetracked by preoccupation with the process of investigation and remediation.

Discovery of Contamination

During the 1970s, it was widely considered a forward-looking move for communities to shift from using easily contaminated surface water to tapping "pristine" underground sources for drinking water. At that time, this small city drilled several wells that local officials anticipated would enable the city to phase out dependence on river water and an aging treatment plant, and to obtain purer water at less cost. However, the new wells turned out to have an unanticipated problem. Soon after the wells were put into operation, local officials acting on a concern about a chemical spill (unrelated to TCE) near the well area, had a well sampled and the water tested by the State Department of Health. The test results, quite unexpectedly, indicated levels of TCE at three times the level specified in the state's drinking water quality guidelines. Although local officials did not consider this a major health threat, they issued a "boil water advisory." (Such a notice, usually broadcast over local radio stations and printed in local newspapers, is a customary precaution whenever water quality problems arise in a public supply; and boiling does, in fact, remove volatile organic compounds such as TCE from the water.) Customers of the city water system were told that due to the TCE found by the tests, they should boil their water before drinking it or cooking with it. No mention was made of precautions for washing or bathing.

At the time TCE was found, there were apparently no adverse health effects experienced or reported locally that people attributed to the TCE contamination. Although some people apparently doubted the need to boil their drinking water, most went along with the advice. People accepted the statement that their local officials were acting in a "precautionary" manner to protect the community's residents, and assumed that the problem would soon be solved and the nuisance of boiling water ended. However, when further testing was done, city officials and county health personnel repeated their advice to boil water, at least for drinking and cooking. They also passed along information that had been supplied by the State Department of Health concerning the health risks from TCE. The local health department's press release stated the risk from TCE only in terms of the probability of cancer (i.e., one excess cancer per million people). The press release concluded that since the affected wells had only been used for a short period, people had been exposed to only "a small portion of a small risk." There was also a statement that there is no acute effect from the "dilute solution" of TCE in the public water supply at the time.

Early Response

City officials were not convinced that the health risk was great enough to warrant closing the affected wells, despite the urging of state health officials that they do so. When the local newspaper conducted a poll on what course of action city water customers favored, however, opinion ran two to one in favor of closing the wells. These results apparently did not reflect any new concern, or reason for concern, about the health risk; they mainly reflected the fact that residents were tired of boiling their drinking water. When pressure from the State Department of Health mounted, local officials felt that they had no alternative but to close the wells. The city then reverted to using the river with heavy chlorination. There was some doubt expressed locally that the antiquated water treatment equipment would hold up, but both local officials and the public considered this a temporary situation and hoped the community could soon return to using the wells. In general, residents on the city water supply did not believe the TCE contamination had posed a serious health threat. Although no one knew how long they had been drinking water with TCE in it, people were confident that their local officials had the situation well in hand. Residents did not form any citizen action groups. There was no groundswell of public concern or anxiety.

Soon after the closing of the city wells, however, a new set of problems arose to complicate the chemical contamination situation. In the township adjacent to the city, some private wells located near the municipal wellfield were tested and found to have TCE at high levels, as much as fourteen times the state guidelines. A few people on affected private wells reported dizziness and skin irritations from such activities as taking hot showers, but these complaints received little public or media attention, and there was no major outcry about the health risk.

As weeks went by and more testing was done, it was found that the contaminant level in the public wells had dropped, but that some nearby private wells now contained much higher concentrations of TCE than found in earlier tests, with a few having over one hundred times the state guideline. Other private wells in the same area, however, showed no trace of TCE at all, while still others showed fluctuating concentrations. The contamination problem was clearly more complex and changeable than local officials and citizens had first thought. A county legislator representing the township was concerned over the spread of contamination to private wells and called a public meeting to inform citizens in the township about the problem. At this first public meeting on the contamination problem, the legislator emphasized county willingness to help citizens in the affected area. He indicated that the county would consider arranging to have private wells tested and loaning water filters to homeowners to remove the contaminant. County health officials presented a statement about the health risk from TCE, essentially repeating their earlier statement about the cancer risk from a lifetime of drinking contaminated water at the state guideline levels. While the county health commissioner indicated that ideally no one should have to drink water containing any toxic chemicals at all, he also reminded those in attendance that smoking a pack or two of cigarettes a day carries a much greater risk of cancer. Little concrete information was given about health effects, either immediate or delayed, at the higher levels actually found in some private wells. Specific instructions were given (orally) about boiling techniques appropriate for the higher levels of TCE (longer time if boiling in shallower pans). Written instructions were subsequently mailed to a wider population.

Individual response to news of higher levels of chemical contaminants in private wells varied among residents. A few dozen people living near the affected public and private wells phoned the health department to ask about the safety of their drinking water, but at that time fewer than fifteen private wells had enough TCE for local health officials to suggest that their drinking water be boiled. Those residents whose wells had tested positive asked the local health department specifically about procedures for safeguarding their health. Some people with private wells in or near the affected area boiled their water for drinking and cooking even if their own well had not yet been found to contain TCE. Others began obtaining drinking water from a relative or friend on city water (which was now taken from the river and contained no TCE), or from someone with a private well farther away from the affected area. A few families purchased and installed activated charcoal filters. And some people ignored the health warning and continued to drink contaminated water without boiling it.

It is important to note that the water contamination did not seriously alter the lives or lifestyle of community residents. City water customers only had to boil water for a few weeks until the changeover to river water was made. For residents with contaminated private wells, there was a period of several months in which it was necessary to expend extra effort to procure alternative supplies or to take precautions. Later, filters were provided by the federal Environmental Protection Agency to about twenty homes with TCE in their wells, thus reducing the inconvenience of having to boil or bring in drinking water, although filters do require periodic attention. For residents on the city water system, there was no inconvenience at all after the affected public wells were shut off, and no interruption in service. However, water bills did subsequently increase, in part to cover the higher cost of treatment of river water. Throughout the period, no one in the community had to continue to use drinking water that was known to be contaminated with TCE, and those few households that knowingly continued to drink water with TCE in it did so by their own choice. Additionally, there was no water shortage resulting from the closing of the contaminated city wells or the filtration of contaminated private wells. Everyone had as much water as before. There was no rationing of water, and no need for people to arrange and pay for trucking in of large quantities of water. Water for fire protection and industrial use was in no way restricted. No other aspects of life were seriously affected; no families were forced to move out of their homes.

During the first few months after the contamination was discovered, city and county officials were expecting the problem to be short lived, and felt that they could deal with it locally. Subsequently, they contracted with an engineering firm to investigate the problem. Soon, however, the state Department of Environmental Conservation (DEC) became involved in the local situation because of its jurisdiction over toxic waste problems. When the contamination problem appeared to grow in magnitude and severity, local leaders turned also to the U.S. Environmental Protection Agency (EPA). The wellfield was subsequently designated a federal Superfund site, and the EPA became heavily involved in directing the investigation of the local groundwater contamination problem.

Local Perceptions of the Health Risk

In general, concern with the possible health risk remained low. There was no public demonstration of anger or accusation, or even serious worry. Little concern was publicly voiced over the fact that people may have been exposed to TCE before they knew about it, or that they might, unknowingly, be incurring a health risk now or in the future. Tests for TCE, which had to be done through the state health department, were slow: fully two months elapsed from the time local officials first sent off samples for analysis until they received the report informing them of the presence of TCE above state guidelines. For private wells, with fluctuating concentrations, delays in testing meant people might unknowingly be ingesting high levels of TCE. While this worried some local officials, the public remained calm. Although residents were aware of what had happened at Love Canal, they did not see their own situation as similar, and so neither the events at Love Canal nor the citizen anxiety expressed there seemed to increase concern about the health risk locally. In a typical comment made at a public meeting and reported in the local newspaper, one local resident with an affected well said, "I kept hoping it would clear itself up." She phrased her concern as follows: "I just want to be able to use the ice maker in my refrigerator again." Another woman commented publicly, "I just can't imagine boiling water for the next ten years."

The relatively low level of public concern over the health issue is apparent in the small attendance at the three public meetings that were called by local officials, by the DEC, and by the EPA over the four-year period since the contamination was first discovered. When the situation was new, the public meeting called by a county legislator drew 60 people (quite a few of them public officials). Three and a half years later, only 55 people (again many of them present in an official capacity) attended a well-publicized meeting called by EPA. In between the DEC held a special information session, and despite letters mailed to the 300 households in the potentially affected area, only a handful of people showed up.

The relatively low public concern with the health aspect of the contamination issue is also revealed in the content of these three public meetings. The DEC's information session was held when a new phase of the investigation process was about to begin. The purpose of this forum was to inform the public about the investigation and to answer people's questions. One resident asked, rather vaguely, whether there really was a public health significance. The answer given was that TCE "has been shown to be a cancer causing agent (carcinogen) in mice and it is a suspected carcinogen in people." With such an indefinite and familiar statement about the risk, it seems understandable that people directed their questions and comments to other topics, such as the timing, costs, and technical aspects of the investigation.

The local newspaper during this period ran many articles on the TCE contamination situation. (See Appendix for an analysis of newspaper coverage.) Assuming a responsibility for informing the public, the paper reported frequently on recent water test results, on public meetings, and on the words and actions of government officials, both local people and the state health officials, as well as the DEC and EPA after they became involved. (The newspaper was also reporting, during the same period, on several other nearby problems with water quality, involving petroleum and other chemicals, but mostly not affecting more than a few private wells.) The newspaper was up-to-date and accurate in reporting water test results whenever they came back to the community from the laboratories. But the paper devoted less attention to the health risks resulting from such contamination and sometimes gave mixed messages about health effects. (In part, this unevenness of coverage reflects the contrast between the accuracy of chemical analysis and the uncertainty inherent in toxicological risk assessment.) A headline reporting "startling" levels of TCE found in the latest water tests was soon followed by another headline announcing that the link between this chemical and cancer is "only suspected." The article below the later headline indicated disagreement among toxicology experts, and reported that doubt had been cast on the laboratory experiments that had established a link between TCE and cancer in mice. It also pointed out that carcinogenicity in humans had not been demonstrated. As to the acute health risks from TCE, the newspaper reported very little and provided minimal technical information. After an initial statement of the official version of the acute health risk, as supplied by the state Department of Health, subsequent newspaper reporting generally reverted to mentioning only the cancer risk. And even the cancer risk was soon reduced in newspaper reporting to buzzwords and brief shorthand phrases, such as "suspected carcinogen" or "potential cancer-causing chemical." (These shorthand phrases became a compromise for providing a new reader with sufficient background without boring most readers.)

The newspaper did not carry any stories concerning individual residents encountering health problems that they or others suspected could be linked with TCE. Nor were there letters to the editor from individuals claiming any link between TCE and their own health. Even the quoted comments from public meetings gave little indication of concern over the health risk. Only once was there a quotation of a citizen's public comment indicating anxiety over the health risk. But this comment had no power to stimulate public anxiety because it did not refer to any specific or alleged health complaint and because it was overshadowed by the nonworried comments of other citizens. Furthermore, the person who made the comment was referred to in the newspaper account as "an unidentified individual," whereas other more positive speakers were named and known in the community.

The general public at this time was not actively seeking information on the health risk, and seemed to receive all it wanted from the county health department, either directly or through the local newspaper. There seems to have been no withholding of information, nor any accusation to that effect. But the information available to the health officials, and hence to the public, was neither thorough nor conclusive, and did not address the particular contamination levels actually found locally. The health risk from TCE was continually phrased only in terms of cancer risk, and only at the guideline-level concentration, rather than in terms of the levels actually occurring locally. The effects of inhalation of TCE were not made clear to the public, nor were the effects of absorption through the skin. And there was virtually no public mention of the toxic effects on liver, kidney, heart, and central nervous system that have been documented at these higher levels. The incompleteness of information on health risks may have been another factor muting local response.

Reasons for Limited Public Concern over the Health Risk

Even at the height of citizen interest in the contamination situation, public concern about the health risk was described by local, state, and federal officials as "subdued" or "low-level." Within six months of the discovery of the contamination problem, questions and concerns about health had peaked, without resulting in the creation of any lasting organization or network to represent the public. There was no feeling expressed that the contamination problem was being ignored, mishandled, or covered up by city and county officials. There were no accusatory statements and no attempts to oust anyone from office. In fact, there was apparent and general satisfaction with the activities and competence of both elected officials (city and county leaders and legislators) and appointed officials (health department and water and public works department employees). Additionally, elected officials and the relevant municipal staff were generous in their praise of each other. The community's trust in the way local officials were handling the problem of contaminated drinking water is evidenced in the fact that elected officials were reelected and appointed officials retained their positions throughout the period. No local elections were contested on the basis of the handling of the water situation. The public felt that local officials were doing their job and also keeping the public as informed as it cared to be. Consequently, there has never been a citizen demand for local officials to meet with them to answer their questions; and there has never been a citizen-sponsored public meeting on the problem. Residents seem to have rested assured that their local leaders were taking care of them, that no one in the community was being harmed, and that there was nothing more for citizens to do. This sense of security and trust in their local officials acted as a major factor minimizing community anxiety over the health risk from TCE contamination.

Another important explanation for the limited concern expressed about the health risk is that there was at the time no upsurge of unusual illness or other health effects, such as miscarriages or birth defects. No one publicly reported any health problems that seemed out of the ordinary or appeared linked to the TCE. So citizens went about their business, perceived the whole chemical contamination issue as temporary, and did not pay a great deal of attention to the question of health risks.

Another set of factors minimizing public concern over the health risks has to do with the particular contaminant and its source. The chemical contaminating the groundwater, TCE, was not unknown to local residents. On their jobs in local factories, men and women had handled and known about the degreasers that were used in several of the machining processes. This observation relates to one finding of psychometric research on risk perception: risks that are unfamiliar are more feared, while familiar risks are less feared (Slovic *et al.* 1980: 199). Another aspect of this chemical was that people had not been able to see, smell, or taste it in their drinking water: only expensive, distant, and time-delayed laboratory tests could detect the presence of TCE in the water, even at the higher levels found locally. An additional factor is that the TCE that eventually got into the groundwater had been used by local factories with local ownership or management and local workers. It appears, from our other community studies, as well as from this case, that when the causes or agents of contamination are perceived as members of the community, the risks are less feared than when the causes or agents are perceived as external to the community. This finding seems consistent with observations by cultural anthropologists concerning an American cultural tendency to perceive harm as coming from outside (Hsu 1973; Wolfenstein 1957).

Anxiety or anger over the health risk was probably also minimized because the health risk statements that were given were almost all probabilistic. There were no statements of certainty that any given person would develop cancer or any other adverse effect from consuming the contaminated water — for such statements simply cannot be made. Residents also derived a sense of security from their particular interpretation of the probability statements about the health risk. People reasoned that even *if* TCE could cause one excess cancer per million, it would be unlikely that anyone in this community of under 25,000 people would get cancer from the contaminated water. And so, in the absence of any unexplainable or unusual health problems at this time, the probability statements concerning future health effects did not lead to any overt expression of worry about health effects.

Institutional Response to the TCE Problem

Local officials and the community had initially hoped that the DEC and EPA might be able to supply both technical expertise and funding to help the city find out where the contamination problem lay and get it "fixed." But things have not been as simple as these early expectations indicate, either below ground where the TCE appeared to be "drifting around" in a complex aquifer system, or above ground where state and federal agencies tangled not only with the water problem but with each other and with local officials and leaders. The process of investigation has turned out to be long and slow. Except for some two dozen carbon filters provided and installed as "emergency measures" on private wells with TCE in excess of the guideline, there have been no concrete results of the federal presence, no visible products that local people can identify. And the chemical contamination problem is still a long way from being "fixed."

The institutional complexities of addressing the contamination problem have, in fact, compounded the situation. The EPA has been in charge since the wellfield was declared a federal Superfund site, but through a cooperative agreement, EPA has delegated lead responsibility for carrying out the investigation to the DEC. Both agencies have been heavily involved. Throughout the process of investigation and planning for remediation, both EPA and DEC have been represented locally by several different branches and programs of the parent agency, with each subdivision having different responsibilities and different kinds of expertise. This complex situation of many different institutional actors from both EPA and DEC has made it difficult for local officials to know with whom they are to deal on which aspects of the contamination problem. Lines of communication and responsibility have not always been clear to local officials, much less to the public. Nor has the role of local government bodies and officials been entirely clear. Some local officials have resented the fact that they are often the last to hear about federal or state plans for remediation. They have often felt rebuffed in their attempts to find out what is going on and to participate in decisions concerning the investigation.

Private engineering firms engaged by the government agencies to carry out the actual investigation work have added yet another layer of institutional complexity. Local officials have no say over which firms are awarded the contracts or subcontracts for carrying out the engineering studies and water tests, but they are expected to accept and deal with whichever firms are selected. Yet when local officials have had complaints about the work performed by a consulting firm, they could only address their concerns to the agency overseeing the consultant's work. The fact that the firms engaged to do the hydrological investigations and water sampling and testing mostly come from far away, often out of state, may have further increased the potential for friction and frustration. Local people feel that the investigating firms are as distant as the government agencies, and their trust in them seems to diminish with geographic distance.

Another problem in the interaction between local officials and outside agencies is that there have been differences in perception and disagreements on priorities. At times, the government agencies have undertaken investigations that local people felt were unnecessary — and unnecessarily costly. Sometimes an aspect of the investigation that had seemed minor to a government agency or engineering firm was of considerable importance locally. For example, local officials were particularly anxious to make certain that the privacy and private property rights of local citizens would be protected during such activities as geophysical testing in people's yards and water sampling in their houses. Apparently neither the EPA, the DEC, nor the engineering firms had considered private property rights to be an issue in the investigation, and each had assumed that local officials would "deliver" the necessary signatures and cooperation of residents. The matter of gaining permission for "vehicular access" and installation of "observation wells" on private property had been assumed to be a mere formality. Local officials, however, were concerned about having outsiders trampling uninvited on the lawns (and on the rights) of the local citizens, and felt it their duty to raise these objections with the state and federal agencies overseeing the engineering studies. Such "neglected" issues have generated a large volume of official correspondence and telephone calls, as well as some frustration on the part of local officials. And issues such as this have reinforced the general local feeling that the investigation is being conducted by agencies and firms from far away who know little and care less about the concerns of the local people. Community residents also feel that the personnel, the agencies, and even the laws and policies dealing with chemical contamination are all essentially urban-biased, and not necessarily appropriate for smaller communities in rural areas.

Exacerbating the problems has been the fact that the cast of characters keeps changing — not the local actors, who tend to have long tenure in office, but the personnel of government agencies, particularly on the federal level. As soon as local officials get accustomed to one agency representative, they claim, that person is replaced by another, or else a different branch of the agency becomes involved. When such turnover occurs, it not only impedes smooth interaction, but confirms the local perception that the agencies running the whole investigation are big bureaucracies with no real commitment to the local communities in which they operate. The turnover of institutional personnel may do little to counteract local negative attitudes towards the government agencies as "outsiders."

Community leaders sum up this whole period as a seemingly endless series of "expensive studies," "frequent delays," "numerous inconclusive reports" and "temporary 'emergency' solutions." Local leaders and officials have become increasingly dissatisfied with the whole institutionalized intervention process, and describe themselves publicly as "frustrated."

The Health Risk Issue Becomes Overshadowed

Community residents' perceptions of this entire process have basically mirrored that of its officials and leaders. But the public has not been actively involved, nor has it expressed a desire to become involved. In fact, the public appears only occasionally to pay attention; the rest of the time citizens largely remain outside the whole issue, even bored by it, and content to let local leaders carry on. To the extent that the public does express interests, its concern has not been with the health risk.

At the public meetings, questions and comments from residents have indicated that other interests were uppermost in their minds. For example, at the DEC's information session, people were most concerned that the engineers (who were not local residents) would be coming onto their private property and conducting tests in and around their homes. They were also concerned that the contamination and the publicity about it would lower property values and reduce the salability of their houses.

At another public meeting called a year later by EPA to inform people about the investigation's progress, there was but minimal talk of health effects. Agency personnel said in passing that TCE is a "suspected carcinogen", and that it had been found in the local water supply at levels "exceeding state guidelines for safe drinking water". The audience sought no clarification and asked no questions about these statements. State and federal personnel apparently assumed that any contamination "above acceptable levels" should be thoroughly investigated, and assumed that the local population would share this conviction. But in general the public has remained unconvinced that the health risks and contaminant levels they were told about are sufficiently serious to justify going through a long and costly investigation process.

What has happened, then, is that the public has shifted its attention almost entirely away from the health risk and has come to focus on the investigation process itself. The health risk, which in some communities has had the potential for generating and sustaining wider public interest, has almost been lost as a concern in this community. Residents here simply do not worry out loud or in public about whether the contaminated groundwater has jeopardized their health, whether the substitute river water is any safer than water from the closed wells, or whether there might be similar chemical contamination problems in the future. Both media attention and the attention of local officials have also shifted to the progress — or lack of it — of the investigation and planning for remediation. But this topic seems to have even less power to stir up the residents of this community. (For example, almost no one has requested to see the various reports of the geological investigations and groundwater testing that have been made available in the public library.)

Even the outside agencies conducting the investigation seem to have let the health issue slip from their awareness. The health risk factor gets lost in, or overshadowed by, preoccupation with the process of bureaucratic negotiations and technical investigations, deadlines, and cost overruns. (In fact, health and health risks are not central to the responsibilities or training of most of the personnel of EPA and DEC who have been operating at the local level.) In addition to the relative infrequency of references to the health risk, it should be noted that there have also been some seemingly careless and inaccurate references to the health risk. For example, the consulting firms brought in to do the engineering studies of the aquifer and to test the water may have occasionally overstated the health risk. A few blanket phrases that are part of the language of the Superfund site selection process have slipped into reports written by the engineers: one report labeled the problem "a major health threat," and another said that the situation "could result in a serious public health problem." Even these unqualified statements (by people perhaps not technically qualified to make them because they are not toxicologists or health experts), and some occasional similar phrases by governmental agency personnel speaking at public meetings, have failed to provoke anxiety or to alarm the community. To the contrary, such phrases only add to the frustrated exasperation of local officials who exclaim, "If it really *is* a major health threat, they should *do* something about it — instead of conducting yet another study."

There is a local Citizens' Advisory Committee on the contamination issue, which acts as a liaison between the community and the government agencies overseeing the investigation and remediation. But it is significant that this committee did not come into existence until more than three years after the discovery of the contamination. Furthermore, the committee was not created in response to any rekindling of public interest. At the request of local officials, the federal and state agencies created the Citizens' Advisory Committee and appointed its members, mostly local officials and business leaders who were already involved in the issue in one way or another. There was no clamoring mass of citizens vying for seats on this committee, and few members of the public sit in on its meetings. The committee met several times in its first year and learned a great deal about groundwater in general and the local water situation in particular. But it appears that even for the committee, the TCE problem has been totally redefined away from the health issue. Very little time at these meetings has been devoted to discussion of the TCE contamination as a health risk. Mostly the contamination is discussed as a problem of the state and federal investigation process that is "taking a long time," "costing a great deal of money," and not yielding any "results" that the community can acknowledge as worthwhile.

Because committee members were concerned about the time and expense of the investigation and about the substitute water supply (due to the antiquated filtration plant and to the need for heavy chlorination of river water), the committee has sought ways to hasten or shortcut the investigation process. The committee even briefly discussed whether the community had any right or authority to opt for an exemption from state and federal safe drinking water guidelines. They suggested that the community might be willing to accept a higher level of TCE, with an associated increase in risk to health, as a tradeoff for an earlier return to its groundwater wells. However, the committee soon learned that local governments have no right to set their own contaminant levels above state or federal guidelines or to accept a higher level of risk, even if the local public is willing to assume it. This aspect of risk management has been taken entirely out of local hands by federal and state environmental and health regulations.

As of the present, the public is not really interested in the whole investigation, and seems to have forgotten about the health risk issue which brought it about. Even citizens who are generally well informed about and active in community affairs exhibit relatively little knowledge of what is going on, despite continued newspaper coverage. They remain content to have their elected and appointed officials and unofficial community leaders represent them in the slow state and federal process of investigation and planning for clean-up. Citizens say they are confused by all the technical terms and the plethora of studies, agencies and phases. City water customers know that they are now paying more for their substitute water (in part to cover the cost of the chlorination and filtration required of river water). Town residents on private wells feel that their water situation is far from being settled. Many people suspect that somehow all this investigation and remediation will result in local tax increases sooner or later. They would like to have the whole water situation settled — and soon.

Redefinition of Culprits and Victims

From the beginning, there has been little disagreement over the ultimate cause of the groundwater contamination. Local people have assumed that industrial solvents disposed of by several local manufacturing plants and by the city itself (in its landfills) found their way into the aquifer. But local leaders and residents consider it unfair to blame the manufacturers now, since the disposal practices of earlier decades were both standard and legal at the time, and have subsequently been corrected. There has been no publicly expressed feeling that the companies, which people talk of as "good neighbors" and "responsible citizens," would have knowingly put the local population at risk through faulty disposal practices. And the companies have not attempted to deny that their earlier practices might have contributed to the present groundwater contamination.

Local leaders are not interested in placing blame on any particular company or proving its guilt, but merely in fixing the chemical contaminant problem and returning to good, safe groundwater. The citizenry reflects this focus on solution rather than assignment of blame. At public meetings, citizens have tried to help the investigation along by contributing their own recollections of various dumps and dumping practices of decades past. But their suggestions seem to be offered in a spirit of helping to fix up the problem, rather than pointing a finger at culpable parties.

Local officials had hoped that the EPA Superfund process would pinpoint where the TCE in the aquifer was coming from so that the source of the contamination could be expeditiously cleaned up. By source, community people mean location, the "where,". However, the actions required by the Superfund legislation do not match these local preferences. One expressed purpose of a Superfund investigation is to locate the source in another sense, the "who." Much of the investigation is aimed at assigning liability so that the cost of studying the contamination problem and cleaning up the aquifer can be recovered from the "potentially responsible parties." This search for liable parties has angered community leadership. Community leaders have good reason to think that local companies would cooperate and help pay for cleanup, and they resent the adversarial nature and secrecy of litigation proceedings against the companies.

It appears that there has been a reinterpretation in local minds of the role of the manufacturing plants in the whole situation. In current local thought, the industries are no longer perceived as the *cause* of the water problem, but as the *victim* of the way that the problem is being addressed by "the government." The new perception is that the real problem is not a matter of health risks so much as a problem of state and federal intervention, with its usurpation of local initiative, its slow blame-assessing studies, and its potential harm to the local economy.

Community leaders, in fact, are now particularly concerned with economic repercussions from "all this investigation." It is widely acknowledged that the same manufacturing plants now being legally identified as polluters have been "the backbone of the local economy," the employer of generations of local residents, and the basis of community identity. As local leaders point out, the longer the whole investigation process drags on, the higher will be the costs assessed against the companies. Leaders fear that if specific companies are required to pay for the entire investigation and clean-up, they might be forced to go out of business or to close local factories. Plant closings would be a tremendous blow to the local economy, which has been painfully adjusting to loss of several factories and retrenchment by others over the last few decades. Both the public and their elected and appointed officials seem to feel that these economic risks far outweigh the health risks they have been told about, such as one excess cancer case per million people. However, local people now concede that bringing about an expeditious remedial action to provide safe drinking water is something beyond their control.

Overall, the scenario is reminiscent of "The Sorcerer's Apprentice": in the beginning, the community asked for outside help in dealing with its water problem; but now that the community does not want any more "help," it can find no effective way to say "stop!" Certainly the frustration of local officials over the perceived "unhelpfulness" of the institutionalized risk management process has relegated the health risk itself to a very minor concern.

CONCLUSION: THE DYNAMICS OF RISK PERCEPTION IN COMMUNITY CONTEXT

Several important points about the social construction of risk have emerged from this case study.

1. Risk perception is a dynamic process and has a temporal aspect. Perceptions of a given risk are not fixed and permanent, but may undergo reinterpretation and change through time.

2. Risk perception is an interactive phenomenon. In the period after a health risk is discovered in a community, one or more general public

perceptions of the risk may emerge as the accepted interpretation. The collective perception then acts as a filter through which individual members of the community perceive the risk.

3. Local perceptions of a risk are affected by perceptions of the way the risk is being addressed. In our longitudinal case study of an actual community with an actual risk situation, we have seen that actions undertaken to manage the risk may themselves alter the social interpretation of the risk. In the case presented here, as governmental agencies and consulting firms carried out the process of investigation and remediation, the community's original concern over the health risk was altered and reduced. Residents' attention was drawn away from the health risk and became focussed on the bureaucratic process of investigation and clean-up. In other cases, the way a risk is described and dealt with by state and federal agencies may have the opposite effect: the original local concern over the risk may be increased over time.

4. The social construction of risk at the local level is influenced by factors in the local context in which the risk is embedded. In the particular case presented here, we have identified seven community context factors that seemed most important in shaping local perceptions of the health risk. Specifically, people in this community tended to exhibit relatively little concern over the health risk because: (a) they perceived no adverse health effects; (b) they experienced only minor nuisance or dislocation related to the contamination; (c) people were unable to detect the contaminant by their own senses; (d) many residents were familiar with the chemical that was contaminating the water; (e) people perceived the origin of the contamination as internal to the community; (f) the community's leaders were concerned about other risks, such as further decline of the local manufacturing economy; and (g) citizens trusted their local officials as competent and effective in protecting their health and well being.

Had the community's situation with respect to several of these context factors been different, one might have found a rather different perception of and response to the health risk posed by contaminated groundwater. For example, if residents had not trusted their local government, they might have exhibited more initial alarm over the contamination and more sustained concern over the health risks they had unknowingly incurred. Rather than following their local leaders in being skeptical and unconvinced about the health risk, citizens might have interpreted their leaders' skepticism as a cover-up of the problem. Had this been the case, residents might have been more vocal about the health risk and might have challenged, confronted, and even ousted their local officials. Cases described in the literature support this observation about the role of trust in local government in shaping risk perception. In a New Jersey community with toxic chemicals in groundwater, residents felt unprotected, stonewalled, and deceived by their local government — and were quite vocal in public expressions of concerns about the health risk (Edelstein 1982). In a Massachusetts community with chemical contamination of groundwater, residents aggressively sought expert information about the health risk, and voted some local officials out of office because these officials "were not perceived as having taken the health risks seriously enough" (Berry *et al.* 1984: 46). But here in our study community, trust in local government provided a foundation that minimized anxiety about the health risk.

These four conclusions about risk perception seem to substantiate our premise that the local community is a useful arena in which to study the social construction of risk. The case study approach, using actual situations where health risks occur and are addressed over a period of time, provides rich data for analysis of risk perception. In this longitudinal case study of a real contamination situation, we have found that many factors mediate between the risk itself and the perception of the risk. Indeed, we have found risk perception to be complex, interactive, and dynamic, and have found that it may be significantly affected by perceptions of the risk management process undertaken to address the risk.

This case study was funded by the National Science Foundation — Ethics and Values in Science and Technology (RII-8409912). All views expressed in this chapter are those of the authors and do not necessarily reflect the views of the National Science Foundation. We wish to thank those in the community and agencies who so generously shared their experiences and observations with us.

APPENDIX

ANALYSIS OF LOCAL NEWSPAPER COVERAGE OF THE WATER CONTAMINATION SITUATION

SAMPLE —1 — January and February 1984

Articles dealing with various water quality problems in the city, the surrounding county, and nearby counties, including both public and private well problems. *N*=23 articles (100 percent sample of articles on the topic published in the period).

Of the 23 articles, the seven longest were devoted to the problem of TCE contamination of city and private wells in this community, the case under consideration in this chapter. The remaining 16 articles, all considerably shorter, dealt with a wide range of cases and actions related to contamination of drinking water in and around the community.

SAMPLE —2 — April 1981 through May 1984

Articles on the TCE contamination of city and private wells in this community. *N*=50 articles from the period, selected to be representative of the much larger total number of articles on this subject in this period.

Average length of articles was 15 to 20 column inches, indicating fairly generous coverage of the local TCE contamination story.

The content, mostly reporting the progress (or lack of it) of investigation:

— a report (due to be) released by state agency or consulting firm;
— an investigation begun, or an investigation delayed;
— a new phase of investigation to be undertaken;
— an appearance in town by state or federal officials;
— water tests results received from laboratories;
— a public meeting (either announced or reported);
— the estimated costs of various studies and phases of the investigation;
— negotiations with engineering firms.

The language used to describe contamination of drinking water supply:

— verbs most commonly used in conjunction with TCE: "found," "detected";
— the water was "contaminated by" or "tainted with" TCE;
— TCE was referred to generally as a "contaminant," never as a "poison".

Few mentions of health risk in these 50 articles — breakdown as follows:

Four mentions of a specific health risk, most only three to four words long; mostly "a suspected carcinogen" but not very descriptive mentions.

Three mentions of a general health risk, e.g., "could be some significant health factors to consider."

In contrast, there were:

19 mentions of the legal-regulation aspect of TCE in the water, e.g., "exceeding guidelines," "high levels," "levels above state guidelines";

six mentions of precautionary measures to protect health, e.g., health department advises "boiling before drinking."

Notes

1. For the purposes of our research project, "community" is defined operationally as a locality with a chemical contamination problem in the water supply. Thus in our research, it is not necessary that our sites be "true" communities in terms of geographic and political entities or in terms of social interaction and identification. A "community" for our purposes could be a city, a township or village, or simply a water district; or it could be a composite of several of these. Many of the places we are studying are, in fact, both geopolitical and social communities, although the boundaries of the contamination problem

may not coincide with those of the political or social entity addressing the problem. It is not a relevant research question to determine community boundaries as perceived by residents in each of these locations.

2. Browning, 1965; National Academy of Sciences, 1977; National Academy of Sciences, 1980. The authors wish to thank Bonney F. Hughes for her contribution to this evaluation of the health risks associated with TCE.

3. Douglas (1982), Douglas and Wildavsky (1982), and Rayner (1984, 1985) have raised such seminal questions. They have suggested that answers lie in the hierarchical or egalitarian characteristics of communities, and have developed their "grid-group" theory to explain community differences. Berry and Stoeckle (1985) are conducting comparative studies of community reactions and responses to contaminated groundwater.

References

Berry, David *et al.* *Costs and Benefits of Removing Volatile Organic Compounds from Drinking Water.* Preliminary Draft, 9/28/84. Cambridge, Massachusetts: Abt Associates, Inc., 1984.

Berry, David and J. Andrew Stoeckle. 'Decentralization of Risk Management: The Case of Drinking Water.' Cambridge, Massachusetts: Abt Associates, Inc., 1985.

Brown, H. S., D. R. Bishop, and C. A. Rowan. 'The Role of Skin Absorption as a Route of Exposure for Volatile Organic Compounds (VOCS) in Drinking Water,' in *American Journal of Public Health*, 1984, 74, 479—484.

Browning, E. 'Trichloroethylene,' in *Toxicity and Metabolism of Industrial Solvents.* Elsevier Press, 1965, 189—212.

Douglas, Mary. *Essays in the Sociology of Perception.* London: Routledge and Kegan Paul, 1982.

Douglas, Mary and Aaron Wildavsky. *Risk and Culture: An Essay on the Selection of Technical and Environmental Dangers.* Berkeley: University of California Press, 1982.

Edelstein, Michael R. *The Social and Psychological Impacts of Groundwater Contamination in the Legler Section of Jackson, New Jersey*, 1982. (unpublished ms.)

Fitchen, Janet M. Forthcoming, 'The Importance of Community Context in Effective Risk Management.' Paper delivered at Society for Risk Analysis, Washington, D.C., October 1985. To be publishing in Proceedings.

Fowlkes, Martha R. and Patricia Y. Miller *Love Canal: The Social Construction of Disaster.* Washington, D.C.: Federal Emergency Management Agency, 1982.

Goyer, Robert A. 'Potential Health Effects from Groundwater Pollution,' *EPA Journal*, 1984, 10(6), 22—23.

Hus, Francis L. K. 'American Core Values and National Character,' in Hsu (ed.), *Psychological Anthropology.* Cambridge, Massachusetts: Schenkman, 1973, 241—262.

National Academy of Sciences. *Drinking Water and Health*, Vol. 1. Washington, D.C.: National Academy of Sciences, 1977.

National Academy of Sciences. *Drinking Water and Health*, Vol. 3. Washington, D.C.: National Academy of Sciences, 1980.

National Cancer Institute. 'Carcinogenesis Bioassay of Trichloroethylene.' National Cancer Institute Carcinogenesis Technical Report Series, Number 2, 1976. [H. E. W. Publication number (N. I. H.) 76—802.]

Rayner, Steve. 'A Cultural Analysis of Occupational Risk Perception,' *Royal Anthropological Institute Newsletter*, 1984, 60, 10—12.

Rayner, Steve. 'Predicting Social Acceptance of Future Technologies: Advanced Concepts for Nuclear Reactors,' Lecture presented at Cornell University, 2/7/1985.

Slovic, Paul, Baruch Fischhoff, and Sara Lichtenstein 'Rating the Risks,' *Environment*, 1979, 21, 14—39.

Slovic, Paul, Baruch Fischhoff, and Sara Lichtenstein 'Facts and Fears: Understanding Perceived Risk,' in R. C. Schwing and W. A. Albers, Jr. (eds.), *Social Risk Assessment: How Safe is Safe Enough*? New York: Plenum Press, 1980, 181—214.

Slovic, Paul, Baruch Fischhoff, and Sara Lichtenstein 'Behavioral Decision Theory Perspectives on Risk and Safety', *Acta Psychologica*, 1984, 56, 183—203.

Wolfenstein, Martha. *Disaster: A Psychological Essay*. Glencoe: Free Press, 1957.

Chapter 3

Chemicals and Community at Love Canal

Martha R. Fowlkes and Patricia Y. Miller

Introduction

In August 1978 residents of the neighborhood immediately adjacent to the Love Canal chemical landfill in Niagara Falls, New York, were notified by state health officials of the "grave and imminent peril" posed by exposure to its leaching contents.[1] The uncontested facts of the situation were these: the 21,800 tons of residues and by-products that had been buried by Hooker Chemical Company in the neighborhood's midst contained a number of chemicals known to have adverse effects on human health; this chemical waste had made its way to the surface of the landfill in places and had also leached laterally from its original burial site in the canal; the presence of toxic chemicals had been confirmed in and/or on the property of some specific homeowners. It was far less certain whether these conditions had actually physically harmed or injured residents or had placed them at widespread risk of physical harm or injury.

Although the possibility that life-threatening conditions prevailed in the neighborhood was strongly suggested by officials and clearly recognized by citizens, resident response to this suggestion cannot be understood apart from the ambiguity attendant upon it. Nothing had changed and everything had changed. Family, home and community were clearly intact, but at the center, uncertainty took the place of a once taken-for-granted sense of residential security. From then on and into the present, ambiguity and uncertainty have formed the lens through which policy-makers, scientists and the residents themselves view this community. The play of dangerous forces on human settlements is, of course, not unique to Love Canal and has many antecedents in the record of human disaster. But the multiple paradoxes that characterize this event (if, indeed, it is meaningful to speak of an "event" at all) are exceptional.

The destructive consequences of conventional sudden impact events (either natural or man-made) are apparent and leave no doubt as to the immediate relationship between cause and effect: the earth quakes; a river floods; a dam collapses; a mine explodes; a hurricaine strikes; a building burns and so on. Case studies of such events abound. (See, for example, the 1981 summary of studies of natural disaster events by Kreps, and

B. B. Johnson and V. T. Covello (eds.), The Social and Cultural Construction of Risk, 55—78.

Erikson's 1976 study of the flood at Buffalo Creek induced by mining industry practices.) Situations like these are accessible in commonsense terms, and therefore the definition of the situation and whatever actions follow from it are likely to be spontaneously rather than self-consciously evolved (Scott and Lyman 1968). It is not incumbent on populations affected by such events to identify their consequences, although individuals may exhibit a wide range of responses to them (Wolfenstein 1957). At Love Canal, in contrast, each family found itself in the unusual and difficult position of having to evolve its own definition of the significance of the chemicals. Facing either the possibility or desirability of relocation, families were required to articulate coherent perspectives about the actual or potential implications of the chemicals for their well-being. The consequences of the chemicals were not only less than fully visible, but there was no independent basis — either in the form of visible impact and destruction or definitive official documentation — for the achievement of consensus among residents as to what the consequences actually had been or would be.

Although the public health alarm had been sounded loudly and clearly at Love Canal, the ensuing Emergency Declaration itself, first issued on August 2 1978 by Commissioner Whalen of the New York State Department of Health (NYSDOH), cited only a probable risk to fetal and early childhood development. Thus, the initial Emergency Declaration called only for the temporary evacuation of the 26 residents judged to be most vulnerable to possible health effects — the six pregnant women and 20 children under two living in the 99 homes abutting the canal. These narrowly drawn criteria for evacuation drew strong opposition from many residents who were unwilling to accept that the extent of health risk could be specified with such certainty in the face of the admittedly uncertain boundaries and effects of chemical migration. Within a week, the state yielded to resident pressure and transformed its initial evacuation order into a near mandate for the permanent relocation at state expense of all of the families living in the 239 homes situated in the first two rows of houses adjacent to the canal.

In the weeks and months that followed this evacuation, the residential perimeter of the landfill became synonymous with the official perimeter of risk, and previously indeterminate boundaries of chemical migration were now held to be, if not precisely determinate, lying well inside the boundaries of relocation. This designation can be viewed as simultaneously reasonable and arbitrary — reasonable in the sense that it was plausible, arbitrary in the sense that it lacked empirical verification. Indeed, the boundaries of migration have not yet been and will never, in all likelihood, be ascertained. This is so in part because available techniques for tracking pathways of leachate migration cannot feasibly be applied in the context of the conditions that describe Love Canal. In addition, though, chemical contamination at Love Canal is not self-contained in time or place, but is conjoined with the history of enduring and pervasive chemical contamination of the city of Niagara Falls itself. Since the sources of hazardous waste in the area are so widely diffused, it is virtually futile to attempt to define precisely the boundaries of migration for any one instance of identified contamination. Confronting this dilemma, the state made clear its intent to bound the problem at Love Canal by holding the line on the boundaries of risk.[2]

In the two years following the 1978 relocation, an increasingly antagonistic relationship developed between the state and a core of organized homeowners remaining. While the state insisted that the 1978 boundaries of relocation more than adequately encompassed any risk from chemical exposure, the protesting residents were equally persuaded that chemical risk extended beyond the boundaries set for relocation. Their goal was to expand the boundaries of eligibility for state-funded relocation. The contest, which soon found its way into the national consciousness, was punctuated by congressional hearings, court injunctions, picket lines, aggressive confrontations with officials at every level, massive court-ordered temporary residential relocations at state expense, spates of guerilla theater and intense media coverage of all of these. In the summer of 1980, the state capitulated and the federal government intervened with the promise of funds and enabling legislation to support the voluntary, permanent relocation of homeowners living in some 550 homes in the larger Love Canal area. On each occasion, the official decision to finance and expand the boundaries of permanent relocation went forward as a response to political pressure rather than to recognized disaster victims or to definitive assessments of health effects or risk linked to the chemicals. In other kinds of sudden impact disasters, disaster relief follows from recognition of victims. This was not the case at Love Canal, nor did victim status result from initial or expanded relocation in official eyes.

The Love Canal Community and Its History

In their efforts to understand and resolve the ambiguity confronting them, Love Canal homeowners were divided among and against themselves from the outset. This is not particularly surprising since what passes for social cohesion in almost any contemporary urban neighborhood is a form of integration that is actually rather superficial and tenuous. That is not to say, of course, that individuals resident in urban neighborhoods do not develop highly elaborated and supportive networks, but those networks typically are not founded in residential proximity or commonality (Fischer 1982). Unlike the "mechanical solidarity" of traditional, preindustrial or "folk" societies, the "organic solidarity" of modern local communities is rooted in pluralism in which social "fault lines" inhere

(Durkheim 1893). In fact, there was no Love Canal "neighborhood" as such until the notoriety engendered by the landfill had the effect of renaming the area and redefining it both geographically and symbolically. Before then, it was regarded locally as just a corner of a much larger section of the city of Niagara Falls known as LaSalle.

LaSalle itself, once a free-standing municipality, was annexed by the city in 1927. In the 1930s, Hooker Electrochemical Company (later to become Hooker Chemicals and Plastics Company and recently renamed Oxy-Chemical as a subsidiary of Occidental Petroleum) acquired title to an abandoned canal in the area. The canal had been excavated at the end of the nineteenth century under the direction of William Love with the intent of providing hydroelectric power for his ill-fated scheme to build a planned industrial facility, Model City. During and following World War II, both the city of Niagara Falls and Hooker deposited municipal and chemical waste respectively into the canal. Following a ten-year period of intensive chemical dumping, Hooker covered over the canal with a "cap" of indigenous soil. In 1953, title to the landfill was transferred to the Board of Education of the city of Niagara Falls to provide the site for the construction of a new elementary school.

The electrochemical industry for which Niagara Falls is particularly noted flourished throughout the first half of this century and the city's population grew steadily. In consequence, residential housing gradually gained dominance over earlier agricultural land use in the southeastern LaSalle area of the city. Subsequent federal programs designed to stimulate the construction of moderate income housing, together with the construction of the new school, encouraged even more intensive residential use of the area near the landfill. At the time of the 1978 Emergency Declaration, the Love Canal area of LaSalle was known to local residents as a respectable, predominantly lower-middle class community of predominantly owner-occupied, single-family homes. Low-density, scatter-site public housing set on spacious grounds to the west of the canal site was compatible with the neat, modest residential character of the larger neighborhood. The area was defined and unified less by social ties than geography; an expressway, a major local highway, a creek, and a residential street paralleling the outer edge of the public housing project enclosed the neighborhood and set it off from the larger LaSalle section of the city.

The school was built on the rectangular canal site which stood at the center of the south side of the neighborhood and determined the grid-like pattern of streets built up around it. Except for the school, no other buildings were erected on the landfill itself which served generally as a neighborhood play area, although its conversion into a park as promised by the area's housing developers had never come to pass. Throughout the Love Canal area, much of the housing was of recent vintage, having been constructed in the twenty-five years after the landfill was closed in 1952. In the immediate vicinity of the landfill, many of the homes were small bungalows designed for young families of modest means. The north side of the neighborhood was more suburban in character with somewhat larger ranch-style houses set on winding streets. Despite variations in the settings and styles of houses in the Love Canal neighborhood, homeownership itself was of paramount importance for all residents both as a locus of autonomy not generally available to them occupationally (Sennett and Cobb 1972) and as a major economic asset (cf. Halle 1984).

While the young, blue-collar family with dependent children so favored in media coverage of Love Canal was certainly well represented in the neighborhood, the overall population of the area was actually far more disparate with respect to occupational status, family status, and age. There was a substantial proportion of white-collar families, post-parental, and elderly couples, especially on the north side of the neighborhood. Religiously, the neighborhood was part of a wider area served by a Catholic parish, and two Protestant churches (Methodist and Church of God) were among the very few non-residential structures in the immediate Love Canal area.

Such differences as these, of course, are typical of present-day residential communities, and the social and occupational mix represented in the Love Canal neighborhood is very like that of the residential settings described by Halle (1984) in his study of New Jersey chemical factory workers. Typically such differences among neighbors and across the neighborhood lie dormant and are of little or no consequence to the undisturbed community. However, these stand to become lines of fission and controversy when neighbors are called upon to speak for the gravity of a situation because the situation does not speak for itself. Of necessity, then, the process of formulating understandings and explanations by Love Canal homeowners drew substantially from the interests, experiences, values, and attitudes that characterized the different social worlds they inhabited with all the potential for division and polarity that those contained.

Sociocultural Factors and the Process of Definition

Over time, the views of Love Canal residents coalesced into two major, and essentially opposing, categories of perception and response to their situation. Just what this process entailed serves to illustrate Perrow's observation (1984) that social and cultural rationality organize thinking and decisions about risk. One-third of the families we interviewed in a random sample of homeowners are best described as "minimalists."[3] They are of the opinion that chemical contamination from the landfill was probably limited in scope with little, if any, serious consequences for health. Not surprisingly, these families chose either to remain in their Love Canal homes or moved only reluctantly in response to the initial relocation, or to the fear or experience of declining property values and a deteriorating quality of life in the area generally. On the other hand, two-thirds of the families arrived at what we will term a "maximalist" stance regarding the situation. They are disposed to believe that chemical migration extended throughout and beyond the entire area that was eventually declared eligible for relocation and that serious health risks in all likelihood paralleled the migration of the chemicals. Overwhelmingly these families relocated out of the area, and many of them were instrumental in pressing government officials to widen the initial boundaries of relocation.

In their social construction of risk at Love Canal, residents came to hold beliefs about the chemicals based on what they knew from what they saw in the evidence available to them. In this process, they drew on their perceptions of the presence of chemicals in their own homes and the homes of others, their perceptions of chemical effects on their own health and the health of others, test results and epidemiological reports, media coverage, official statements, the "off the record" remarks of officials, and declarations issued by grassroots neighborhood and other community organizations. The credibility assigned by families to these different sources of information and, indeed, the inclination to seek information varied with the social and structural realities of the household.

Among demographic factors, age and household composition are most strongly associated with the formation of one or another perspective on risk from the canal. Overwhelmingly, the minimalists are in or near retirement and have no children living at home. Thus, the most pressing concern in their lives is not the welfare of their children but their own economic welfare in old age. For them, their Love Canal homes were their permanent and final homes, and bespoke both residential and financial security. For minimalists, the meaning of home was shaped primarily, if not entirely, by material considerations. In most instances these homeowners had paid off their mortgages and any costs related to home improvements and looked forward to living comfortably on small, fixed retirement incomes in a respectable neighborhood. For persons in this situation, the state's set purchase price for their homes fell woefully short of the amount of money required to purchase a comparable home elsewhere in the city without incurring a mortgage. The elderly who did relocate typically moved to small apartments or mobile home parks with the result that their current standards of living are considerably lower than what they had enjoyed in their Love Canal homes.

Younger homeowners, of course, faced a different set of problems owing to their concerns as parents. Although they are not immune to worries about financial or residential security, the welfare of their children understandably takes priority over such other worries. Parents are generally unwilling to tolerate the presence of gratuitous risk in their children's environment and tend, therefore, to be conservative in their assessments of the probability of the presence of risk. In the context of the ambiguity of the situation at Love Canal, parental conservatism was typically expressed in a maximalist definition of the boundaries of chemical migration and its attendant hazard. Of course, not all maximalists are parents of young children, but virtually all parents of young children are maximalists. At the center of this definition is the conviction that chemical risk could very well be widespread since it has not been proven to be limited.

The strength of the life-cycle factors — age and household composition — is that they can be understood as a shorthand reference to what families have at stake in their lives. Older people can ill-afford to believe that their emotional and financial investments in their homes are in jeopardy. Younger people, most especially parents of dependent children, can ill-afford to believe that their children (including their children and grandchildren yet unborn) are in jeopardy. The economic investment in homeownership is no less important to younger than to older homeowners. For young families, though, this is potentially in competition with and stands to be overriden by a commitment to family-based autonomy. Such autonomy extends beyond the fact of homeownership itself to the quality of life and well-being inside the home. Either vantage point can be seen as the result of an eminently rational calculus aimed at achieving resolution in a highly ambiguous context. Clearly a family's beliefs concerning the chemicals are not evolved independently of its own sense of what it stands to gain and lose in the circumstances. And, as will be discussed more fully later, neither are beliefs formed independently of a family's sense of what might have already been lost, particularly with reference to health.

The connection is an obvious one between specific life-cycle factors and resident views on the risks posed by exposure to toxic wastes at Love Canal. Nonetheless, such factors in and of themselves do not "cause" beliefs so much as influence the *perspective* from which individuals collect information and evaluate experience as the evidential basis for forming beliefs and definitions. By the same token, experience and information do not lead "automatically" or predictably to a given conclusion or point of view. The various life situations of the different families undoubtedly affected in several ways both the process of gathering evidence and the process of interpretation out of which their views were formed.

Structural factors encourage the desire for evidence in the first place. Thus we would expect families raising small children in the Love Canal neighborhood to be more interested than older couples or childless couples in learning about the effects of chemicals on childhood development. Then, too, the presence of children implies certain things about the quality of neighboring ties and the kinds of things parents speak about in conversation with other parents. The demands of the parenting role itself, and the greater opportunity it affords for interaction with other parents, suggest greater access to information and greater attentiveness to it on the part of parents compared to others. While factors such as these constitute the framework within which definitions are constructed, the process of constructing the definitions themselves is an interpretive one. Here we do well to bear in mind Blumer's observation that, "There is a process of definition intervening between the events of experience presupposed by the independent variable and the formed behavior represented by the dependent variable. . . . This intervening interpretation is essential to the outcome" (1956: 687).

For both minimalists and maximalists this "process of definition" is embedded in a more general set of cultural values and social attitudes which are consistent with their perspectives on Love Canal in particular. It is apparent that prior to the introduction of crisis in the community, there were marked differences among Love Canal homeowners in terms of the kinds of ties they had established to their neighbors and neighborhood and to the meaning of being "at home" there generally. For the most part, those who emerged as minimalists lived encapsulated in the highly privatized worlds of their individual homes. Although they were typically long-term residents of the area, their lives there over the years were not characterized by strong social or friendship bonds with neighbors except in instances were neighbors were also kin. Their idea of a "good" neighborhood is one where people are fastidious in the up-keep of their homes and yards and maintain a cordial distance from one another, i.e., where neighbors "mind their own business."

In addition, minimalists not uncommonly are company people who have long histories of employment with local chemical and related industries. By virtue of their occupational commitments, they are sympathetically predisposed toward a view of the world in which "chemicals are our friends," both as a manufacturing enterprise which is essential to the city's economic viability and as a set of products that make a substantial contribution to the convenience and comfort of daily living. In their view, Niagara Falls is so pervasively a "chemical city" that Love Canal represents nothing special or different to worry about compared to the rest of the city. Although chemicals represent risk, that risk is generally felt to be greater in the workplace than at home. In either case risk is considered to be intrinsic to the business of living and working in modern society, and any costs attached to chemical exposure are outweighed by the benefits represented by chemicals themselves (see Lowrance 1976 for a formal exposition of this perspective).

In keeping both with their social isolationism and occupational values, minimalists place little emphasis on — indeed, do not trust — the notions of the collective welfare or of privately or publicly based social responsibility directed toward the common good. *Laissez-faire* values and assumptions organize their thinking concerning the desired behavior of individuals as well as of organized corporate and government entities. The best interests of everyone are served by maximizing everyone's self-interest, and toxic exposure is regardless an individual rather than a collective or social problem. Following this reasoning, minimalists hold that any Love Canal family that wanted to move should have borne that burden on its own.

> If I'd lived over there by the canal, I'd have just moved. . . . But why ruin the whole city? But I wouldn't make any noise about it. I'd just sell the house and move. . . . No, I wouldn't have worried about having somebody else live there. I would have sold it. That's their business. If you don't look after yourself, nobody's going to look after you.

The minimalists recognize that the "bigness" of the interests represented by government and business means that they are inherently at odds with the "little" interests of individual people. On the other hand, individual interests are seen as expendable in the face of the higher order of interests represented by corporate activity and its importance to the nation's general economic health. The assertion of individual claims to environmental protection can only serve to subvert and erode the strength of industry.

> I don't feel Hooker has any responsibility for it at all. . . . I mean, Hooker does do it's share as far as industrial waste, but my god, you've got to have it if you want a plant there. You've got to pay the consequences. . . . If it ever came out that Love Canal was severely polluted and really destroyed the area in which these people were residing, and these physical problems were caused by the chemicals, I mean, there's no saying where this could ever end. In a way [if this were true], I think it would probably be best if they never got it out, because there might be two people who really have a problem and a situation that was caused by the chemicals, but you're going to have another 300 that are going to accuse the government. . . . I think the financial consequences could be just devastating.

This general value system embodies a pronounced cynicism about "human nature" which predisposed minimalists to respond reactively rather than proactively to the initial Love Canal Emergency Declaration. This response, in turn, encouraged and contributed to the development of the minimalist perspective itself. As events unfolded at Love Canal in the form of media publicity, informational meetings and citizen protest, minimalists experienced and resented these as intrusive in their lives. At the same time, by their own accounts, they were noticeably and — in many cases — intentionally uninformed about these events. Their descriptions of "what happened" are vague and contain few inquisitively based details of the "who, what, where, when" variety, thus substantiating their own professed inattentiveness to relevant details and central actors in the situation. ("I couldn't even tell you what year it was. I never looked into it or got that much involved that I would be able to say that there could really be a problem.")

For the most part the families that espouse and adhere to a minimalist definition of risk at Love Canal do not so much marshal evidence in support of their position as they discredit those who stand against it. The maximalists are the "wrong" kinds of people and, on that account, the maximalist perspective is the "wrong" kind of perspective. Where minimalists had come into contact with residents who attributed serious health problems to the effects of toxic exposure, these claims were discounted as expressions of mistaken causality and/or avarice. Maximalists are persistently stigmatized as liars, cheats, crazies and radicals — opportunistic ne'er-do-wells who were out to get something for nothing.

> I think people invented sicknesses, people that didn't have them, just to leave. . . . Half of them were sick when they moved in. . . . Everybody there was crying and hollering about the canal. They claimed they had health problems. Whether they did or not, I don't know. It could be an inherited thing, too. . . . Those families that organized weren't such nice families over there. I didn't know them very well. I mean, I wouldn't associate with them. I'm not too picky, but I can't stand too boisterous people.

or

> I'm not saying some of the health problems might not have been caused by the chemicals, but I'm sure the liquor had a lot to do with it. When you drink a pint or a quart a day, it's got to have an effect. . . . These people, as far as the residents, one of them lived near me and I know he was the worst. I think he was the type that would sue anybody. He was always out for the free buck, wanted something for nothing out of life, trying to get compensations. You know this type of person; wherever you can get a free handout, he would take it. . . . Living in the neighborhood, you know some of the people and they were the worst ones. . . . You found out that the people that were unemployed, people that were on welfare were the ones who were really causing the most trouble.

In this manner the identities of the maximalists have been constructed as illegitimate, thereby rendering their opinions illegitimate as well through the process described by Goffman (1963: 2—5):

> While the stranger is present before us, evidence can arise of his possessing an attribute that makes him different from others in the category of persons available for him to be, and of a less desirable kind. . . . He is thus reduced in our minds from a whole and usual person to a tainted, discounted one. . . . [T]hose who do not depart negatively from the particular expectations at issue [are] the *normals*. . . . By definition, of course, we believe the person with a stigma is not quite human.

The minimalists, then, see themselves as the owners of "normal" identities, in this sense, and were considerably reinforced in this view by state and local officials who themselves advanced a minimalist definition to "hold the line" on relocation. The maximalists, on the other hand, their identities as such and the roles they variously assumed in consequence, are perceived by minimalists as a significant departure from the conventional (or normal) and this departure became the rationale for the imputation of stigma. Our research does not indicate that the imputation of stigma was reciprocal. This may be a function of timing. At the time of our interviewing the maximalists had prevailed in the matter of relocation and the views of the minimalists are of little consequence to them now. At an earlier point in time, it is certainly likely that stigma was attached to minimalists as a part of the process by which maximalists marshalled support for the relocation effort.

The maximalists — those who hold that contamination was widespread with serious implications for health — share a noteworthy unity of perspective on the events in their neighborhood and the meaning of those. The majority of these families had dependent children at home, a fact that undoubtedly motivated them in their search for evidence about the situation that confronted them. While there was certainly no conclusive evidence available to homeowners, all residents did share access to some basic information. Unlike the minimalists, the maximalists took a great deal of initiative in seeking out information. They were typically regular attendants at meetings called by both officials and residents and engaged in persistent questioning of relevant authorities regarding the management and meaning of toxic exposure in their neighborhood. They were alert to opportunities for health and environmental testing and insistent in seeking clarification of test results from doctors and other experts. Eventually, most of these families came to possess their own Love Canal "documentaries" built from first-hand engagement with events as they unfolded. The following comments, spoken by a resident who came to a maximalist view of the situation, illustrate the kinds of considerations that lay behind initial resident engagement or disengagement with the Love Canal "problem" and foretell the divergent definitional perspectives that arose from each.

> It wasn't until the summer of '78 that I was really aware there was a problem. When the canal thing hit, August 1st, I went over to my neighbor and said, "Hey, I don't know what this whole canal thing is about, but you can bet it's going to affect our property value.
>
> There's a meeting tonight, why don't you go?" He said, "We lived here for 40 years. Don't get worried. There's no problem." I said, "I don't know if you folks are aware but we're expecting another baby and our older one has a lot of birth defects. I don't like the sound of them moving pregnant mothers out four blocks away. We're going to get involved.

For many of the younger maximalist families their Love Canal homes were their first homes. Some of these newer arrivals to the neighborhood planned to move up and out of the neighborhood as and when greater financial stability and family size combined made it possible and necessary to do so. For this group any decline in property values which would impact on the potential resale value of their homes would obviously be of major concern. However, the majority of those families more recently settled in the area did plan to stay. In contrast to the minimalists, the families who became maximalists typically lived in more sociable and geographically dispersed worlds within the general Love Canal neighborhood. Many formed at least casual ties through their children to a large number of area residents. Still others built ongoing, adult-centered friendships encompassing some degree of intimacy, shared leisure and interdependency with their neighbors. For some, neighborhood ties were synonymous with childhood and extended kin ties.

This general pattern of sociability appears to have been influential in shaping the maximalist definition and its expression. Not only did these families both know (or know of) a great many more residents in the area than the families who came to be minimalists, but they were inclined toward a sympathetic view of others and their accounts of experiences related to the chemicals. In this fashion, the collective experience, as perceived and interpreted by individual families, formed part of the base of evidence adduced by maximalists in support of their position. Thus, their accounts of "what happened" to them as individual families at Love Canal are often simultaneously their accounts of what happened to other families as well.

> I don't remember anything about [how it all started]. When it started getting in cellars, that's when we heard. The houses right close to the canal would get the goop coming up through their sump pumps. . . . We felt sorry for them over there, so we went to meetings anyway and still had our blood tested and so on. But I think the immediate concern of ours was people over there, not so much for over here. . . . But when you stop and think of the people that you knew that died of cancer, it's excessive. Died young. There've been a lot of people on this street who've had cancer. . . . And then you got to thinking about the babies that died at birth, or the cancer and all the rest of it, and you start putting everything together and it became your concern. . . . All of our children were born in the area. All the children have had [health] problems, all of them.

Science and Medicine

Perspectives on the present are shaped not only with reference to background factors and value systems laid down in the past as those are applied to the present, but with reference to situated encounters in the present itself (Becker and Geer 1958). In this connection, resident encounters with scientific and medical management of the Love Canal situation were important in different and complicated ways to the adoption of both the minimalist and maximalist definitions. Insofar as the maximalist perspective is substantially based in reactions to and interpretations of these encounters, the minimalist perspective also received reinforcement from them. Lodged in the differences between the two definitional systems and the evidence appealed to in each are fundamental and profound differences in the *incentives* of families to believe that chemical migration was either widespread or narrowly circumscribed. The quality of perceived family health experience is of particular significance in this connection.

Although informal self-reports of family illness stand at a far remove from the kind of data required for conventional epidemiological analysis, they are nonetheless of considerable interest sociologically. In comparing the accounts of family health given by both minimalist and maximalist families, it is apparent that more is at issue than a simple one-to-one correspondence between self-reports of serious illness in the family and the conviction that the chemicals constituted a serious health risk. It is true that, overall, minimalists report fewer health problems in the family than maximalists, but the differences of particular interest between them are more qualitative than quantitative. First, in the view of many older residents, the risks of advancing age exceed any risk from the chemicals, and their own years to date constitute *prima facie* evidence for their assertions of minimal risk and, therefore, for the minimalist position itself.

Cripes, we plant tomatoes out here, and cucumbers, and eat them every year. I'm not dead yet. I might die of old age. . . . There's nothing wrong with my kids. They were raised here. . . .

The kinds of health problems reported by minimalists are typically of the sort that traditional medical expertise is able to label and treat. The disorders they describe are both conventional and age appropriate conditions — heart disease, gall bladder disease, high blood pressure, diabetes, and even cancers. The gravity of some of these conditions notwithstanding, they nonetheless conform to prevailing medical paradigms of what constitutes legitimate illness for which there is a well-developed medical response. Insofar as minimalists experience their health problems as medically intelligible and manageable, they have little incentive to resort to the view that the situation in the community was out of control environmentally.

Well, there was a series of problems with us . . . problems this past year — malignant, female. . . . Yes, I'm getting through it anyway. I had the surgery done in the winter and then, after I get done with my treatments, I wound up with infections and other problems. . . . Then I had to have surgery again. . . . I'm coming along a little bit at a time. . . . It's possible it could be related to the chemicals. I never gave it much thought, 'cause really what happened to us could happen anywhere in the country, right?'

Not all maximalists identified personal or family health problems attributed to chemical exposure, but overwhelmingly they find the claims of others in this regard to be credible and sympathetic. Some of them, especially parents, were very concerned because the absence of family health problems to date did not preclude the possibility of their emergence in the future. They shared this preoccupation with the majority of maximalists, who portrayed a constellation of health problems for which traditional primary care medicine frequently has neither name nor specifically effective control or treatment over time. Many of the conditions they described — uncontrollable shaking; disabling, prolonged headaches;

recurring, localized skin eruptions; recurring respiratory disorders; persistent vomitting; stomach pains and bloating; severe, sudden and heavy bleeding — are the sort that often elude definitive medical diagnosis while entailing considerable erosion of regular routines, energy, money, and the sense of personal security. Despite the fact that the knowledge base of professionalism is never wholly complete and always harbors uncertainty as well as certainty (Fox 1951), patients whose symptoms are not amenable to routinely available professional expertise are commonly dealt with either as though they have emotional problems or are themselves to blame for their problems, as was the case for many maximalist families.

I just worry about my son all the time. . . . Nerves is another thing doctors don't seem to believe in. Oh yes, you feel like you're lost, making it up. The doctors look at you as though you're making it up too. This one doctor that he went to for a physical, he wanted him in the hospital. He wanted to check out everything. He did check out everything. This doctor wouldn't talk about Love Canal either. None of them would. I don't know why they won't talk about it. Yes, and you go to a doctor and you try to tell him what your problem is, and it's hard to tell him because you really hardly know yourself. They look at you strange when you tell them how you feel.

When he was a pre-schooler, finally when he could talk, he would tell you, "Mommy, my tummy hurts." He'd come in from playing, he'd be white as a ghost, doubling over with pain, and his stomach would bloat up like a balloon. The doctor said it didn't have anything to do with his kidneys. He sent us to the gastroenterology clinic. They did so many tests on him. Finally they said that he was just doing this to get attention. I just hit the roof. I said, "A kid cannot turn pale to get attention. He does not bloat up to get attention. He's an only child. I do not work. He is not lacking for attention. There's something wrong with this boy." They put him back in. He was in every two or three weeks. He was doubling up and waking up crying that his stomach hurt. . . . He seemed to be the sickest kid around. We were always broke with doctor's bills. We thought all kids were sick. . . .

There had to be something going on. [My husband] carried a lot of headaches in that place. And it gave him an ulcer before it all broke. He was nervous a lot, too. . . . My oldest kid was an asthmatic [and] very hyper. She was on medicine for hyperactive kids. . . . I was a very hyper person over there. I was on medication. I thought at times I really was cracking up. I had a headache that lasted seven days. . . . They had done all kinds of tests. They thought I had a brain tumor. . . . Five years ago, before it all broke, I had a lot of hemorrhaging; for two years I hemorrhaged. I wound up getting a hysterectomy after we moved out. . . . My son had a lot of ear problems. One time, his ears ran like a runny nose for six weeks. I kept him on medicine. He was in and out of the hospital quite a bit with it. . . . My other daughter had nose bleeds all the time. She was hemorrhaging all the time from the nose. It would be all of a sudden she'd just wake up in the middle of the night in a pool of blood. She had a lot of kidney and bladder problems, too. She was loaded with blood in the urine . . . they couldn't believe how much blood. . . . My kid would come home crying every day. She'd be nauseated. She complained of headaches and that. . . . When I filled out the form about medical problems, I just kept going, page after page after page, with the kids. I took a shoebox full with all my doctors' bills, all my prescriptions, for my whole family. They couldn't believe it. The guy [from NYSDOH] was shocked. I said to him, "How would you like to pay all of that out?". . . .

For families like this, prevailing medical conceptualizations of and response to illness offer few answers and little consolation. They had few options but to resign themselves on an individual basis to coping with health problems that were bewildering at best and terrifying at worst. Their sense of vulnerability was additionally heightened where the health of their children was affected.

The dissemination of information about the toxic contents of the Love Canal landfill and the potential for a variety of insidious effects on human health these families' altered drastically and understandably interpretations of their situations. The presence of the chemicals offered an efficient and plausible explanation for the kinds of illnesses they had endured; they learned that they were not alone in their experiences of health problems. Viewed from the perspective of traditional medicine, the health problems of these families were a collection of incoherent anomalies. Viewed with reference to the chemicals, their problems began to make sense. The chemicals provided a framework for understanding their health experiences that accorded them a legitimacy that professional medicine was unable or unwilling to give them. Sociologically it matters very little whether these health problems are "real" by epidemiological criteria. What matters, in keeping with the classic observation of W. I. Thomas (1923), is that situations which are defined as real are real in their consequences. The maximalist definition, accordingly, owes much to experientially based perceptions of many residents of what was real in their own lives as it concerned their health and medical response to it.

Not surprisingly, families whose health histories gave them the most cause to worry about the magnitude of chemical hazard were the ones most eager for expert information that would definitively confirm or disconfirm their fears related to health. Armed with the official identification of environmental risk in the community generally, they sought a heretofore elusive validation of their specific ailments as well as diagnostic and prognostic information regarding the linkage between their health and the presence of environmental toxins. From the vantage point of these residents, the quest for such information was eminently reasonable in that the questions emerged directly and obviously from the situation in which they found themselves. Appropriately enough, residents looked first to their own physicians and ultimately to the state's public health officials to address the questions of concern to them.

As it turned out, the management of health and environmental testing of Love Canal families and homes furthered communication and comparison among neighbors in ways that encouraged and strengthened the maximalist perspective as a group perspective. Families who wanted blood-testing done waited together in long lines at the school where the work was conducted. Health histories of residents were gathered by means of self-administered survey questionnaires distributed to homeowners who quite naturally turned to one another for discussion and clarification of specific questions. Results of testing for chemical levels in the basement air of individual resident homes were sent to families without interpretation as a series of numbers recorded on computer printouts. Families resorted to consulting one another in their efforts to understand the information at hand, and as a result it was inevitable that they would come to consider the situation in terms of the comparative conditions prevailing in the wider neighborhood: "The only thing we could do with the readings we got was we showed our totals to everyone else and saw whose was worse."

In the end, the maximalist perspective came to be conjoined with distrust of official science primarily because of the lack of fit between scientific inquiry and the kinds of experientially and collectively based questions about health and safety that loomed largest for residents. Maximalists began as maximal information-gatherers. Although initially they extended "full faith and credit" to state health experts, residents repeatedly understood official explanations of their problems as thinly disguised attempts to explain them away. This construction was based in part on the many lapses that occurred around testing and data collection and the frequently contradictory results of those, often communicated in a bureaucratic and impersonal style.

They drilled a hole right here in front of the lot, and then they drilled a hole over there and they found chemicals in both of them. They found chemicals in the sump and in the air in the basement. . . . We had stuff raise up in our basement. One day I went down there and there was little black spots all over the floor, kind of like an oily substance had come right through the cement. It looked like someone had taken a paint brush and just flicked it. It was all over down there. Then it disappeared. Then when they took the reading in our sump pump, we had such a high reading I went to a meeting and discussed it with them. They tried to blame it on, they said it probably came from your plastic, the vinyl, you know, the plastic sump pump. I said, "I don't have one. I have all brass." They said, "It has to be from your lines." I said, "I have all galvanized ones." They said, "Well then we'll test it again." So they tested it and they said, "Oil must have seeped through your sump from someplace." I wish it was oil. . . .

They came in and took air tests in the basement one time, and two or three of the chemicals were quite high. Then they came back another time and wanted to re-do it and they had it marked down. . . . They never sent us a report. They said if you wanted to know any more about it to call a certain number in Albany and ask for so-and-so. So I did that and they wanted to know what I wanted. I said that I had a letter stating the chemical content of my cellar and that if I wanted to know more about it I should call the girl there and ask her about it. They said, "Well, she's not here right now, and we'll have her get back to you." So when she called back she wanted to know what I wanted to know. And I told her the same thing, and she said, "Well, all I can tell you is I wouldn't spend over two hours down in that cellar and I wouldn't sleep there."

Perhaps more important than instances of "bungling" and ineffective and insubstantial communications in the evolution of distrust of science was the appearance of the politicization of the scientific endeavor itself.

Despite the early promises and proposals of the state, its agencies were slow to undertake formal epidemiological and environmental studies. When funding was cancelled for a study begun by the Centers for Disease Control to answer questions about the health impact of residential exposure to chemicals at Love Canal, this signalled for residents an indifference of the government to their welfare in the present which left them alone with their fears for the future.

After the first relocation official attention shifted from an emphasis on new data collection to remediation at the landfill. The emphasis on construction that promised to contain the migration of leachate implied that health hazards in the neighborhood belonged to the past, not to the present or future. Since the developmental well-being of children was of such overriding importance to those inclined toward a maximalist view, the remedial approach was not a reassuring response to concerns about past exposure in relationship to health effects that might emerge in the future. Instead it left residents apprehensive about the possible risks associated with the remedial work and frustrated at what they saw as the inertia of the state in documenting the toxicity of the wider area.

NYSDOH had at the outset used its own scientific studies as the basis for declaring Love Canal a "public health time bomb" (New York State Department of Public Health 1978). However, such study findings as were issued by the state after the August 1978 relocation consistently asserted "no significant health effects" which themselves were asserted as the rationale for preserving the extant boundaries of permanent relocation and alleged chemical migration. The studies on which these findings were based were not available for peer review or criticism by the wider scientific community in an open forum (Levine 1982).

At the same time, the work of independent scientists — particularly that of Paigen and Picciano, whose studies suggested an association between chemical exposure and damage to health — was persistently and publicly discredited, largely because their data *were* available for review and criticism. Paigen's work was important for its identification of the correspondence between patterns of certain health effects exhibited by area residents and patterns of surface water and groundwater flow away from the canal site. Since Paigen had enlisted the assistance of the Love Canal Homeowners' Association in her data-gathering, interested residents were well aware of the findings of her study and of the professional and personal harrassment to which she was subjected in response to her research (Paigen 1982). The Picciano study, released in May 1980, indicated a high level of chromosome abnormalities in a small test group of Love Canal residents. Although the study and its findings were quickly overwhelmed by debate (Kolata 1980; Picciano 1980), it constituted decisive evidence for the maximalist point of view and became the catalyst for the mobilization of protest around it.

Chemicals and Community

Ultimately maximalists carried the day in effecting the eligibility for voluntary, permanent relocation of residents remaining in some 550 homes in the larger Love Canal area. State health officials, though, have consistently questioned the validity of the expanded boundaries for relocation on the grounds that they are based more in popular protest than in informed risk assessment. The problem of Love Canal, it is asserted, has been misconstrued: it is properly a scientific rather than a political problem.[4]

But the failure to recognize the interrelationship of the political and the scientific *is* the problem of science in general (Lewontin 1983). The science most likely to be credited as "good" scientific research is more readily mobilized by elites (in both the public and private sectors) than by nonelites, in part because it is an expensive endeavor. Then, too, the results of "normal science" are biased toward minimal definitions of harm, because the method itself is inherently conservative in its assumptions and procedures (Kuhn 1970) and, therefore, in its readiness to make assertions about cause and effect relationships. And in the field of public health a conception has evolved of individual responsibility for health that places environmental factors at a politically acceptable remove from risk (Starr 1982). The problem of science at Love Canal is the problem of scientific risk assessment, in particular as Douglas and Wildavsky (1982: 73) have described it:

> Thinking about how to choose between risks, subjective values must take priority. It is a travesty of rational thought to pretend that it is best to take value-free decisions in matters of life and death. . . . One salient difference between experts and the lay public is that the latter, when assessing risks, do not conceal their moral commitments but put them into the argument, explicitly and prominently. . . . The private person does not isolate the risk elements to address them directly. . . . The ordinary individual admits that his loyalties and moral obligations are largely the matter at stake, but the risk expert claims to depoliticize an inherently political problem.

As Short (1984) has argued, the risk assessment model assumes that "cultural values, political wisdom, and judgments as to what is worth preserving are related solely, or in some linear fashion, to technical knowledge." For residents of exposed communities, however, the reality of what is involved in living with and responding to risks is far more socially than technically driven. In the final analysis, then, both the maximalist and minimalist perspectives on risk at Love Canal must be understood as *equally and simultaneously* subjective and rational in the context of the respective considerations and concerns that gave rise to each perspective. The maximalists began as residents whose age, household composition, ties to group life in the neighborhood, and family health experiences quite naturally encouraged them to interpret the situation in high risk terms.

The failure of official science to place and address their concerns in context (and, indeed, in the context of the concerns expressed by state officials themselves at the outset) served only to exacerbate this and to put science itself in a context that was decisive in crystallizing the maximalist definition and its activist expression. Maximalist resident distrust of science built not in reaction to the intrinsic adequacy of any given inquiry but in reaction to what was and was not studied, by whom, under what auspices, with what results and with what ensuing dialogue.

The minimalists, on the other hand, had less at stake in assigning legitimacy to questions concerning the safety of the area by virtue either of their own life-cycle concerns or their general value system. In fact, they had a great deal at stake in deflecting attention away from such questions altogether and therefore from an aggressive search for answers that could undermine their own economic and residential security. For these reasons, minimalists were implicitly allied with official science, which confirmed not only what they were prepared to believe about the "facts" of the situation at Love Canal but about the kinds of people who disagreed with those.

In contrast to sudden impact events, the problem for residents at Love Canal was not the problem of response to *what happened*, but the problem of constructing and responding to the *meaning* of what happened. Accordingly, patterns in the assignment of blame and responsibility paralleled the patterns by which residents evolved perspectives on risk (see Drabeck and Quarantelli 1967). If the situation at Love Canal got out of hand, as many minimalists claim it did, then in their view the maximalists, not the chemicals, are to blame. For the maximalists, blame and responsibility are lodged primarily with the state's Department of Public Health. While Hooker's accountability for the creation of the chemical landfill is unequivocally recognized, the company's responsibility is typically mitigated for both minimalists and maximalists by reference to "state of the art technology" and corporate guilelessness in much the same way that Hooker itself has used these to explain its actions (Hooker 1980a).

The maximalist indictment of the state is not rooted in the belief that NYSDOH was incompetent to handle the situation. To be sure, maximalists frequently noted instances of apparent bungling, but not without sympathy for the unprecedented challenge confronting officials. Neither does it typically spring from an exaggerated or naive "faith" in the power of science, since the legacy of the Love Canal experience for many residents is a rather sophisticated appreciation of the problems posed by, for example, the relative absence of standards for uncontrolled exposure to toxic chemicals in combination. Rather, the sources of disillusionment lie elsewhere and are twofold. First, the scope of official scientific inquiry was not broadly enough drawn to address and respond to particular kinds of commonsense questions about the well-being of home, family and neighborhood that evolved logically from resident experience and concerns. Then too, the particular patterns of controversy and dissent that emerged around the crediting and discrediting of science at Love Canal fostered a perception that scientific independence was compromised in the service of political and economic interests. Consequently, blame is reserved for those who are held responsible for the public health and, therefore, for exploring fully the magnitude of chemical exposure and related health effects. And a sizeable number of maximalists suggest that the limited scope of scientific inquiry is linked to a general unwillingness to set politically and economically costly precedents that would both identify and protect other communities at risk (cf. Bucher 1957).

Although the maximalist definition of the impact of chemical exposure on the exposed population eventually prevailed politically and compelled national sympathy, it was and is a definition at odds with prevailing medical, scientific and other official definitions of the situation. Their departure from and challenge to established expert judgment has left maximalists vulnerable to the assignment of deviant and stigmatized identities by minimalists with whom they once coexisted as neighbors. What represents gain for the adherents of one perspective stands, of course, as loss for the adherents of the other. Minimalists are the symbolic winners and material losers with respect to the particular meaning and value they attached to their Love Canal homes and homeownership. Although many maximalists have also sustained material loses in the process of relocating to other homes, they have regained their futures as those relate to the more important concerns of family-based autonomy. Minimalists, on the other hand, whether they relocated or remained in the area, see themselves as having been unreasonably displaced from their futures as those relate to their primary concerns for residential security. In the end there are no winners.

What is clear from the Love Canal experience is that the destructive toll of residential exposure to environmental contaminants is not limited to health effects alone. Where potential and/or actual risk is unknown and unbounded, the identity of victims, or whether there are victims at all cannot be definitively determined. As residents attempt to understand and resolve the ambiguity confronting them, they divide and polarize in ways and for reasons that are inevitably and irreparably damaging to the social fabric of the modern urban neigborhood — knit together by pluralism in the good times and rent apart by it in the bad.

Notes

1. The summary of events surrounding the toxic waste emergency at Love Canal was compiled and reconciled from the following documents: "Love Canal Chronological

Report, April 1978 to January 1980" (Love Canal Homeowners' Association n.d.); "History of Disaster at Love Canal: Chronology of Events" (in Ecumenical Task Force, Addendum to *The Love Canal Disaster: An Interfaith Response* 1981); *Love Canal: Public Health Time Bomb*, (New York State Department of Health 1978); "Love Canal Chronology" (in New York State Department of Health, *Love Canal: A Special Report to the Governor and Legislature*, 46—52: 1981); newspaper coverage appearing in the *Niagara Gazette*, the *Buffalo Evening New*, and the *Buffalo Courier-Express*, 1975—date; *Love Canal: Science, Politics and People* (Levine 1982); *Laying Waste: The Poisoning of America by Toxic Chemicals* (Brown 1979); Love Canal: My Story (Gibbs 1982); and "Factline Hooker" Nos. 11 and 12 (Hooker Chemical Company 1980a and 1980b).

The following references informed our summary of the social and economic history of the area: "Chemical Industry on the Niagara Frontier" (R. B. MacMullin *et al.* 1940); *Salt and Water, Power and People: A Short History of Hooker Electrochemical Company* (Thomas 1955); *The Rise of the American Electrochemicals Industry, 1880—1910: Studies in the American Technological Environment* (Trescott 1981); "Niagara Falls: Community Renewal Program" (Niagara Coalition *et al.* 1971). To cite all of the relevant documents pertaining to any given assertion would burden the narrative with excessive interruptions; selective citations would misrepresent the research actually undertaken. We have elected the reasonable alternative of omitting all citations in the narrative except those pertaining to quoted material.

2. There were numerous and varied incentives for the state to bound the problem at Love Canal by preserving the extant boundaries of relocation and alleged chemical migration. The state of New York had already assumed an enormous financial burden in committing funds for buy-out and remediation. In the weeks and months that followed the Emergency Declaration, some 50 active and closed dump sites were identified in Niagara County alone; in the intervening seven years, 252 have been documented in Niagara and nearby Erie Counties. New York State was found to contain about 600 hazardous waste sites (Interagency Task Force on Hazardous Waste 1979); and estimates of the number in the United States reached 50,000 (Hart 1978).

In addition to the potential magnitude of the toxic waste "problem" in New York State, the city of Niagara Falls was facing its own rather pressing problems. Since 1950, its population had declined by 30 percent, a correlate of an eroding employment base. This is accounted for in part by the national trend in industrial migration out of the Northeast in this period. For Niagara Falls, industrial decline was further accelerated by the destruction of the local hydroelectric power plant in 1956. Nevertheless, manufacturing remains the dominant industry in the area and more than half of the area's labor force is employed in chemical and related industries. Clearly, the economic well-being of the city is heavily dependent on the continued vitality of its chemical industry. The historical record reveals a century of local government accommodation to the industry and its toxic waste by-products. This tradition of accommodation and harmony had been increasingly strained in the years following World War II by the growing number of legislative and enforcement initiatives undertaken by federal and international bodies. The situation at Love Canal promised both increased scrutiny and legislation regarding the conditions of chemical manufacture, and Hooker — the largest chemical plant in the city — stood in financial jeopardy, its corporate image in shambles. Thus, the bleak economic picture that described Niagara Falls by the mid-1960s stood to become even bleaker in 1978 following the toxic waste emergency at Love Canal.

Moreover, in response to its declining industrial base, the city has retreated to an economic re-emphasis of its long-neglected tourist industry on the American side of the Falls. Local officials were understandably reluctant to attract negative attention to Niagara Falls in the midst of its efforts to rebuild tourism. As Love Canal gained increasing notoriety in the media, the Mayor of Niagara Falls turned repeatedly to the media to assert the symbolic distance between the unaltered tourist appeal of the city proper and the 16 acre landfill tucked away in its southeastern corner (Fowlkes and Miller 1985).

3. Discussions of the Love Canal community and quotations from residents are drawn from a study of remaining and relocated homeowner families (Fowlkes and Miller 1982). Structured, open-ended interviews were conducted with 63 families selected by stratified random sampling techniques.

4. In 1984, for example, a group of scientists was assembled to serve as consultants to a technical review group established to determine criteria for the eventual rehabitation of the Love Canal emergency declaration area. The initial mandate to this group of consultants (of which the authors are members) as represented by the Director of the Centers for Disease Control (CDC) and the Commissioner of Health of New York State is as interesting for what it *excludes* as it is for what it *includes*. Clearly the operative assumption is that knowledge and information relevant to risk assessment and to any determination of eventual habitability are limited to and, indeed, are one and the same with technical and scientific knowledge and information. Social, cultural, moral, and political information are notably not encompassed in the definition of what constitutes knowledge in this case, although our participation has served to broaden the definition of what constitutes relevant knowledge.

> CDC and NYSDOH request your technical expertise and advice in our efforts to help establish criteria for habitability in the area. . . . You will be asked to consider various options for determining habitability, given your knowledge of the circumstances at the Love Canal neighborhood, including: (1) the results of human health studies and animal exposure experiments already done; (2) the environmental testing of soils, water, residences, drains, air, etc.; and (3) the existing hydrogeologic data for the area. . . . [Y]ou will be asked to recommend, in writing, the criteria for habitability based on your best scientific judgment (Mason and Axelrod, letter to the authors, February 9, 1984).

References

Becker, Howard S. and Blanche Geer. 'The Fate of Idealism in Medical School,' *American Sociological Review* 1985, 23, 50—56.

Blumer, Herbert. 'Sociological Analysis and the "Variables,"' *American Sociological Review*, 1956, 21, 683—690.

Brown, Michael. *Laying Waste: The Poisoning of America by Toxic Chemicals*. New York: Pantheon, 1979.

Bucher, Rue. 'Blame and Hostility in Disaster,' *American Journal of Sociology*, 1957, 62, 467—475.

Douglas, Mary and Aaron Wildavsky. *Risk and Culture*. Berkeley: University of California Press, 1982.

Drabeck, Thomas E. and Enrico L. Quarantelli. 'Scapegoats, Villains, and Disasters,' *Transaction*, 1967, 4(4), 12—17.

Durkheim, Emile. *De la division du travail social*. Paris: Felix Alcan, 1893. Translated by George Simpson, under the title *The Division of Labor in Society*. New York: Macmillan, 1933.

Ecumenical Task Force of the Niagara Frontier, Inc. *The Love Canal Disaster: An Interfaith Response*, 1981.

Erikson, Kai T. *Everything in Its Path: Destruction of Community in the Buffalo Creek Flood*. New York: Simon and Schuster, 1976.

Fischer, Claude S. *To Dwell Among Friends: Personal Networks in Town and City.* Chicago: University of Chicago Press, 1982.
Fowlkes, Martha R. and Patricia Y. Miller. *Love Canal: The Social Construction of Disaster.* Washington, D.C.: Federal Emergency Management Agency, 1982.
Fowlkes, Martha R. and Patricia Y. Miller. 'Toward a Sociology of Unnatural Disaster: The Case of Love Canal' Paper presented at the annual meetings of the American Sociological Association, Washington, D.C., 1985.
Fox, Renee C. 'Training for Uncertainty,' pp. 207—41 in Robert K. Merton *et al.* (eds.), *The Student Physician.* Cambridge, Massachusetts: Harvard University Press, 1957.
Gibbs, Lois. *Love Canal: My Story.* New York: Grove Press, 1982.
Goffman, Erving. *Stigma: Notes on the Management of Spoiled Identity.* Englewood Cliffs, New Jersey: Prentice-Hall, 1963.
Halle, David. *America's Working Man: Work, Home and Politics among Blue-Collar Property Owners.* Chicago: University of Chicago, 1984.
Hart, F. C. Associates. *Task IV — Economic Analysis.* U.S. Environmental Protection Agency Contract 68014895, 1978.
Hooker Chemical Company. 'Love Canal: The Facts (1892—1980),' *Factline Hooker,* 1980a, 11.
Hooker Chemical Company. 'No Demonstrated Health Effects From Love Canal,' *Factline Hooker,* 1980b, 12.
Interagency Task Force on Hazardous Waste. *Toxic Substances in New York's Environment.* Interim Report, May 1979 (as cited in Paigen 1982, note 7).
Kolata, G. B. 'Love Canal: False Alarm Caused by Botched Study,' *Science,* 1980, 208, 1239—1242.
Kreps, G. 'The Work of the NAS-NRC (1951—63) and DRC (1963—present): Studies of Individual and Social Response to Disasters,' pp. 91—121, in J. D. Wright and P. H. Rossi (eds.), *Social Science and Natural Hazards.* Cambridge, Massachusetts: Abt, 1981.
Kuhn, Thomas S. *The Structure of Scientific Revolutions, International Encyclopedia of Unified Science,* Vol. 2(2). Chicago: University of Chicago Press, 1970.
Levine, Adeline G. *Love Canal: Science, Politics and People.* Lexington, Massachusetts Lexington Books, D. C. Heath, 1982.
Lewontin, R. C. 'Science as a Social Weapon,' pp. 13—28 in Occasional Papers I. Amherst, Massachusetts: Institute for Advanced Study in the Humanities, 1983.
Love Canal Homeowners' Association. Love Canal Chronological Report, April 1978 to January 1980.
Lowrance, W. W. *Of Acceptable Risk: Science and the Determination of Safety.* Los Altos, California: William Kaufmann, Inc., 1976.
Mason, James O. and David A. Axelrod. Letter to the authors, 1984.
MacMullin, R. B., F. L. Koethen and C. N. Richardson. 'Chemical Industry on the Niagara Frontier,' *Transactions of American Institute of Chemical Engineers,* 1940, 36(2), 295—324.
New York State Department of Public Health. *Love Canal: Public Health Time Bomb.* Albany, New York, 1978.
New York State Department of Public Health. *Love Canal: A Special Report to the Governor and Legislature.* Albany, New York, 1981.
Niagara Coalition and Schwartz, Fichtner and Bick, Planning Associates. *Niagara Falls: Community Renewal Program,* 1971.
Paigen, Beverly. 'Controversy at Love Canal,' *The Hastings Center Report,* 1982 12(3), 29—37.
Perrow, Charles. *Normal Accidents: Living With High-Risk Technologies.* New York: Basic, 1984.
Picciano, D. Letter on Love Canal Chromosome Study, *Science,* 1980, 209, 754—756.

Scott, Marvin B. and Stanford M. Lyman. 'Accounts,' *American Sociological Review*, 1968, 33, 46–62.

Sennett, Richard and Jonathan Cobb. *The Hidden Injuries of Class*. New York: Vintage Books, 1972.

Short, James F., Jr. 'Toward the Social Transformation of Risk Analysis,' *American Sociological Review*, 1984, 49(6), 711–725.

Starr, Paul. *The Social Transformation of American Medicine*. New York: Basic, 1982.

Thomas, R. E. *Salt and Water, Power and People: A Short History of Hooker Electrochemical Company*. Niagara Falls, New York: Hooker Electrochemical Company, 1955.

Thomas, W. I. *The Unadjusted Girl*. New York: Little, Brown, 1923.

Trescott, Martha M. *The Rise of the American Electrochemicals Industry, 1880–1910: Studies in the American Technological Environment*. Westport, Connecticut: Greenwood Press, 1981.

Wolfenstein, Martha. *Disaster: A Psychological Essay*. Glencoe, Illinois: Free Press, 1957.

PART III

Part III

Environmental Protest Movements, Citizen Groups, and the Social Construction of Risk

In recent years, risk assessment and risk management activities have become increasingly politicized. Virtually every major health, safety, and environmental decision is subject to intense scrutiny and lobbying by a vast array of environmental, consumer, and other citizen groups (Covello and Mumpower 1985). Not only has the number and size of such groups increased, but also their level of scientific expertise and political sophistication. These developments have contributed to at least two others. First, it has become increasingly necessary for decision-makers to consult representatives from citizen groups on virtually every major health, safety, or environmental decision. Second, the dissemination of competing risk analyses by government, industry, and citizen groups has contributed to public confusion about a wide array of risk issues — ranging from the risks of nuclear power to the risks of hazardous wastes and toxic chemicals (Covello and Mumpower 1985).

Some of the most intriguing questions related to these developments are why people form environmental groups, why people join them, and how they operate. For purposes of this book, a critical issue is how and why these groups identify and select specific risks for attention and political action. These and related questions are the subject of the chapters in this section of the book.

The first chapter, by Edward Walsh, focuses on the action of citizen groups in the controversy over the restart of the second, undamaged nuclear power reactor at Three Mile Island (TMI). Walsh pays particular attention to the variety of factors that contributed to the survival and longevity of these groups. First, local citizen groups received assistance from external sources. This included assistance from outside experts who provided data about restart effects. Second, the original TMI accident left a lingering distrust of utility executives and federal regulators. This facilitated the task of mobilizing parts of the local population (Soderstrom *et al.* 1984). Third, the Governor of Pennsylvania and the Nuclear Regulatory Commission disagreed over the restart of the reactor. This undermined the solidarity of elite groups in opposing actions by citizen groups. Finally, the discovery of "target vulnerabilities," including revelations that plant operators were cheating on tests, provided local groups

B. B. Johnson and V. T. Covello (eds.), The Social and Cultural Construction of Risk, 81—83.

with additional ammunition. According to Walsh, these four factors played a major role in the survival of citizens' groups for some seven years after the original TMI accident. These factors also provide an important supplement to explanations based solely on the risk perceptions and environmentalist beliefs of group members.

The second chapter in this section, by Luther Gerlach, focuses on the influence of five factors — organizational networks, recruitment, commitment, ideology, and opposition — in the formation and expansion of environmental organizations and protest movements. First, organizational networks provide additional resources and flexibility for individual organizations. They also partially compensate for weaknesses in the protest movement due to organizational fragmentation. Second, recruitment into environmental organizations often occurs on a face-to-face basis with family, friends, and fellow workers. As a result, social networks and protest movement networks tend to overlap and reinforce one another. Third, organizational and ideological commitment is promoted through a process of personal transformation and radicalization, which binds the member to the group and provides a new vision of the world. Fourth, ideology functions to motivate, persuade, legitimate, and integrate the protest movement and its members. Finally, opposition — especially attempts to suppress the protest movement — only serve to underline the truth of the movement's ideology to group members. Gerlach illustrates the functioning of these factors through several case studies, including protests by farmers against high-voltage electrical transmission lines and group protests at hazardous waste sites.

The final chapter in this section, by Branden Johnson, provides a critique of Douglas and Wildavsky's analysis of environmentalism as presented in their book *Risk and Culture* (Douglas and Wildavsky, 1982). Specifically, Johnson takes issue with Douglas and Wildavsky's argument — derived largely from grid/group analysis and Olson's theory of group formation (Olson 1965) — that environmentalist ideology stems from the inability of environmental groups to maintain themselves without postulating a global threat to nature from external invidious forces. Johnson argues that this thesis does violence both to grid/group analysis and to the realities of the environmentalist movement. In concluding his critique, Johnson offers several suggestions for refining and testing grid/group models and for integrating such models with other theories of social movements.

The chapters in this section make several important contributions to the literature on the social construction of risk. For example, they extend our understanding of the degree to which the public statements and concerns of environmental groups may be derived from sources and motivations other than environmentalist beliefs. The chapters also highlight the importance of values and symbols in debates about health and environmental risks. Despite these and other contributions, however, the chapters are only partially successful in explaining (1) why different environmentalist values arise and (2) why specific health or environmental issues are selected for attention and concern (Leahy and Mazur 1978). Adequate answers for these questions are still lacking.

References

Covello. Vincent T. and Jeryl Mumpower. 'Risk Analysis and Risk Management: An Historical Perspective,' *Risk Analysis*, June 1985, 5(2), 103—120.

Douglas, Mary and Aaron Wildavsky. *Risk and Culture*. Berkeley: University of California Press, 1982.

Leahy, Peter and Allan Mazur. 'A Comparison of Movements Opposed to Nuclear Power, Fluoridation, and Abortion,' in Louis Kriesberg (ed.), *Research in Social Movements, Conflicts and Change*. Greenwich, Connecticut: JAI Press, 1978, 143—154.

Olson, Mancur. *The Logic of Collective Action: Public Goods and the Theory of Groups*. Cambridge: Harvard University Press, 1965.

Soderstrom, E. Jonathan, John Sorenson, Emily Copenhaver, and Sam Carnes. 'Risk Perception in an Interest Group Context: An Examination of the TMI Restart Issue,' *Risk Analysis*, September 1984, 4(3), 231—244.

Chapter 4

Challenging Official Risk Assessments via Protest Mobilization: The TMI Case

Edward J. Walsh

Introduction

Prior to the Three Mile Island (TMI) accident, the handful of local opponents at the licensing hearings for the Unit 1 and Unit 2 reactors in the early 1970s were not permitted to raise questions about citizen evacuation in the event of a serious accident because such an event was defined by mainstream risk assessors and hearing officials as virtually impossible. The partial meltdown of Unit 2 in 1979, however, undermined the credibility of the organizations responsible for such probability estimates while also serving as a major catalyst in transforming a previously docile and trusting local population into antinuclear activists. This paper provides a brief analytic summary of the tandem development of protest groups and risk disputes in the wake of the most serious nuclear accident in U.S. history.

Powerful elites have always had their risk assessors, ranging across a wide spectrum of occupations from astrologers to shamans, and the historical record reveals few successful challenges to their decisions by the less resourceful majority. Scientists and engineers have replaced the shamans of yesteryear as advisors to decision makers in contemporary society where technological centralization greatly increases the importance of risk assessment (see Perrow 1984). Organizational elites, rather than the rulers and property owners of previous generations, now selectively absorb the recommendations of such experts and then routinely define and decide societal risks. In modern society, however, the risk assessments and related decisions of the technological elites are being increasingly challenged by an aroused segment of the larger population, resulting in legitimacy crises. Public trust, in brief, has become a critical political variable in our society (Short 1984; Perrow 1984).

Even where technological elites confront a skeptical public, however, a successful challenge by citizen groups is problematic. The Santa Barbarans, for example, were not able to prevent the resumption of oil drilling in the wake of a massive and destructive offshore spill in 1969 (see Molotch

B. B. Johnson and V. T. Covello (eds.), The Social and Cultural Construction of Risk, 85–101.

1970). A significant erosion of public trust in the nuclear establishment also resulted from the TMI accident, but in this case citizen protest groups emerged to channel the discontent into protracted political and legal challenges to the official risk assessments. They were not able to accomplish this on their own, however, but only with help from resourceful outsiders. Vulnerabilities of the nuclear establishment itself, especially the TMI utility and the Nuclear Regulatory Commission (NRC), were also instrumental in this struggle. The following analysis will draw upon concepts and insights from the contemporary literature in social movements.

Social Movement Perspectives

During the past 15 years, the focus has shifted from psychological and social psychological variables allegedly explaining why people become involved in movements for social change to an emphasis on the organizational and societal factors promoting and inhibiting such collective efforts (Oberschall 1973; McCarthy and Zald 1973). Fragmentation among societal elites (Ash-Garner 1977), outside organizational support (Gamson 1975; Walsh 1978), and the avoidance of radical displacement goals by challenging groups (Gamson 1975; Goldstone 1980), for example, have been found to correlate positively with protest movement success. Each of these variables will also be important in the TMI case.

The central thrust of these new emphases, sometimes loosely referred to as "resource mobilization" perspectives, is that collective goals rather than individual members' motives should be the analyst's main concern. Collective actors are viewed as drawing upon available resources in attempting to influence political decisions (Tilly *et al.* 1975). The relationship of protest organizations to the mass media, political authorities, and other third parties is considered critical (McCarthy and Zald 1977; Zald 1979; Zald and McCarthy 1979).

Some writers, however, have gone to extremes in their emphasis on structural factors. McCarthy and Zald (1977), for example, ignore grievance fluctuations in the population as they focus on societal supports and constraints of movement phenomena. Goldstone (1980) goes even further in challenging the relevance of all organizational and tactical parameters of social movements while suggesting that the critical variable explaining their success is the occurrence of national crises.

More recent studies have accentuated the role of grassroots discontent and other social psychological variables as integral components of any comprehensive analysis of protest mobilization processes (Useem 1980; Snow *et al.* 1980; Gamson 1982; Turner 1981; Walsh and Warland 1983; Klandermans 1984; Killian 1984). While individual grievance levels were indeed important factors in this conflict (Walsh 1981), they operated indirectly in challenging official risk assessments. This report will briefly summarize quantitative measures of accident-generated discontent in both the general public and among those citizens who became most active in the struggle against the utility before turning to the decisive structural context within which these sentiments were situated.

While drawing upon data from the individual, organizational and societal levels, this analysis of the TMI conflict reveals the special importance of three structural variables, two of which are variants of ones recently accentuated in the literature and a third which is commonly overlooked by social movement analysts. Specifically, a critical type of outside support for the TMI citizens' groups was the counter-expert whose testimony might neutralize that of pronuclear experts (see Mazur 1981). Fragmentation of elites, and especially the rift between the governor and the federal regulators of nuclear power, was also crucial. The third variable I call emergent target vulnerabilities, and it refers to the technical as well as managerial weaknesses of institutional targets. Although it has not previously been isolated for special consideration in the social movement literature, its importance in this conflict will become obvious.

Sources

This paper relies on data from systematic surveys, informant interviews, observation, document analysis, and a six-year file of newspapers and newsletters.

Survey data were collected from 149 activists in the four most politically dynamic protest groups as well as from random samples of 288 of their less involved, but generally sympathetic, neighbors. Access to the most active citizens came via the protest organizations themselves while the four random samples of sympathizers were contacted by a telephone survey organization. More complete details on these survey data are available elsewhere (Walsh and Warland 1983).

Informant interviewing and participant observation has been carried out by the author over the six-year period since the accident (see Walsh 1981, 1985).

A six-year file of newspaper articles and protest group newsletters was also drawn upon for the following analysis (Walsh, 1983, 1985; Walsh and Cable 1985).

Background and Findings

TMI lies a few hundred yards off the east shore of the Susquehanna River, ten miles south of Harrisburg, in the middle of a partially industrialized region of south-central Pennsylvania. In the late 1960s, when the Atomic Energy Commission (AEC) approved the construction of two nuclear power plants on the island, there were an estimated 620,000 people living within twenty miles of the site. The AEC regulatory staff's safety evaluation reports for the Unit 1 and Unit 2 reactors, issued on May 18, 1968 and November 4, 1969, respectively, stated that there was "reasonable assurance" that the proposed plants could operate "without undue risk to the health and safety of the public" (Ford 1981: 35). The possibility of an accident serious enough to warrant evacuation of surrounding populations was officially regarded as negligible. The Unit 1 reactor began producing electricity in 1974, and Unit 2 followed four years later in December 1978.

Within three months of its start, however, the partial meltdown occurred at Unit 2, forcing more than 150,000 residents to evacuate (Flynn 1979; Walsh 1981). At the time of the accident, Unit 1 just happened to be out of service for routine refueling operations.

Numerous local protest groups formed in the weeks after this initial emergency period as residents challenged various Unit 2 cleanup and Unit 1 restart proposals of both the industry and its federal regulators (see Walsh 1981). Four of these groups are of particular relevance. Middletown and Newberry Township, both within five miles of the island, had their own anti-TMI citizen organizations on the east and west shores of the Susquehanna River, respectively. A bit further out from the TMI reactors, Harrisburg and Lancaster — within fifteen- and twenty-mile zones, respectively — also had their anti-TMI groups. Although these various organizations eventually formed loosely-knit anti-TMI coalitions, they nevertheless continued to retain their own individual identities.

Had local residents not mobilized to challenge the utility and the NRC, Unit 1 would have restarted within a few months of the Unit 2 accident and cleanup procedures at the latter reactor would have been carried out with much less attention to public safety concerns (Walsh 1981, 1983, 1985). In fact, Unit 1 restarted in October 1985 — six and one half years after the accident — and the heavily damaged and lethally radioactive core of Unit 2 has yet to be removed from the reactor at this writing (January 1986).

After a summary overview of public opinion in the TMI area on restart and cleanup issues as well as a more detailed look at attitudes on key questions of risk, the paper focuses on the dynamics of critical disputes deriving from the accident itself, the Unit 2 cleanup, and the Unit 1 restart controversies. Because the survey was conducted approximately two years after the accident when conflicts over both cleanup and restart issues were still in process, a certain amount of chronological foreshadowing is unavoidable in the following topical analysis. Questions on the disposal of radioactive water and krypton from Unit 2, for example, were included on the survey, which is discussed before these cleanup issues are taken up in detail in the subsequent section. The utility's attempt to restart Unit 1 in the midst of a dangerous and unprecedented cleanup of Unit 2 also meant that activists would draw upon perceived management shortcomings in cleanup procedures to argue against permitting a Unit 1 restart.

Public Opinion

Random samples of adults in the four TMI communities with the most active citizen protest groups were interviewed by a professional telephone survey firm approximately two years after the accident (Walsh and Warland 1983: 768ff). The results, displayed in Table I, give the reader an overview of public opinion on the critical issues. Outspoken opponents of a Unit 1 restart and/or the dumping of the hundreds of thousands of gallons of filtered radioactive water from Unit 2 into the Susquehanna River ranged from 30 percent in Middletown to 72 percent in Lancaster, the only community with a drinking water intake station downstream from the heavily damaged reactor. Although at first glance it may seem strange that Middletown, the community closest to the reactor by land access, had the lowest opposition rate of the four, the statistic becomes more understandable when one remembers that it also had the highest percentage of TMI workers, their families, and friends. The 27 percent refusal rate for Middletown, much higher than that for any other community, is also an empirical indicator of the ambivalence many of its citizens experienced in being asked to decide between a risky plant and their neighbors' jobs.

TABLE I Telephone survey results for four TMI communities approximately two years after accident (1981)

	Middletown (N= 399)	Newberry (N= 144)	Harrisburg (N= 219)	Lancaster (N= 79)	Total (N= 841)
Against Unit 1 restart and/or dumping of Unit 2 water	30%	51%	37%	72%	39%
For Unit 1 restart and/or dumping of Unit 2 water	38	35	42	23	37
Undecided	5	3	5	0	4
Refusals	27	11	16	5	20
	100%	100%	100%	100%	100%

The low percentages of undecided respondents reveal the salience of the TMI issues in the area. The two communities closest to the TMI plant entrances on the eastern shore, Middletown and Harrisburg, had more TMI supporters than the others because they contained more workers, suppliers, and service establishments with business interests linked to nuclear power station. Regardless of such neutralizing forces, however, these data show the strong opposition which persisted in the vicinity two years after the accident.

Opponents' Attitudes

While public opinion polls may have little influence on authorities deciding questions of risk, organized publics have better chances of making their voices heard. As mentioned earlier, activists from the four focal communities were also given questionnaires, some of the items from which are relevant for our discussion. Table II compares the responses of these

TABLE II Comparisons of activists and randomly selected TMI opponents[a]

	Activists ($N = 149$)	Unorganized opponents ($N = 288$)
A. Accident responses		
Perceived seriousness of threat[b]	3.8	3.3
Evacuated during initial emergency	87%	56%
B. Radioactive water		
Opposition to dumping[c]	5.0	4.8
C. Radioactive gas		
Opposition to krypton venting in June 1980[d]	4.7	4.0
Evacuated during krypton venting	39%	10%
D. Attitudes toward Unit 1 restart		
Level of opposition to restart[e]	5.0	4.4
Would consider moving in event of restart	52%	26%

[a] All differences, except 2.B, significant at 0.01 level or beyond, using chi-square or two-tailed t-tests (see Walsh and Warland 1983: 775).
[b] Response categories ranged from (1) no threat to (4) very serious threat (Walsh and Warland 1983: 770).
[c] Item was only asked of Lancaster residents where *N*'s are thus reduced.
[d] Response categories ranged from (1) strongly favor to (5) strongly oppose (Walsh and Warland 1983: 770).
[e] Response categories ranged from (1) strongly favor to (5) strongly oppose (Walsh and Warland 1983: 770).

citizen activists in the four TMI communities with their unorganized neighbors, in the top row of Table I, who also opposed a Unit 1 restart, certain Unit 2 cleanup procedures, or both. The reader will notice that the intensity of the organized activists' attitudes and behavior is stronger on each of the items. The activists, for example, were more likely to have evacuated during the initial accident, to oppose the dumping of the radioactive water and the venting of the krypton gas, and to oppose any restart of the Unit 1 reactor. They worked against the utility on each of these issues. Such organized preferences, however, were only one factor in the charged political atmosphere of decision making following the accident. These protest organizations found themselves confronted with numerous political, legal and technical issues necessitating outside assistance from various sources. The accident and its immediate aftermath provides an initial example of their need for outside help.

The Accident and Collective Response

Despite warnings from the growing national antinuclear movement about the possibility of serious plant accidents, the vast majority of residents in the TMI area trusted the word of the nuclear industry prior to the radiation emergency which started in late March 1979, at TMI. The accident, and especially the behavior of the utility as well as the NRC during the emergency period, was critical in alienating significant segments of the local population. Upon their return from the forced evacuation, hundreds of residents joined existing antinuclear organizations or formed new ones in their own communities. The details of these early mobilization ventures have been analyzed elsewhere (see Walsh 1981).

The organized citizen groups in Middletown, Newberry Township, Harrisburg, Lancaster, and other area communities opposed the restart of Unit 1 as well as the dumping of hundreds of thousands of gallons of filtered radioactive accident water from Unit 2 into the Susquehanna River. The NRC, however, ignored these citizen demands and was preparing to allow both the dumping and the restart by early summer of 1979. Local residents were incensed, and some planning for widespread civil disobedience took place, but critical interventions by outside actors forced the NRC to delay both the dumping and the restart.

Only after the city of Lancaster filed its own court suit against the dumping of the filtered radioactive water did the NRC order an environmental assessment prior to its disposal. This confrontation between local and federal officials was critical for the Lancaster activists and will be discussed at more length in the following section.

Outside intervention also forced the NRC to prevent the immediate restart of Unit 1. In this case, it was Governor Richard Thornburgh who intervened with a threat to file a lawsuit against the federal agency unless it delayed the restart. This restart issue will also be discussed in a later section.

The point to be emphasized in both cases is that the success of the citizen groups depended upon interventions by city and state authorities. These officials, in turn, were themselves significantly influenced by the maturing national antinuclear movement. This movement was ready for the TMI accident, as evidenced by the massive national rally held in

Washington, D.C., on May 6, 1979, less than one month after many local residents had returned from their forced evacuations. The national movement, in turn, relied upon the local groups to channel widespread discontent into politically efficacious behavior as efforts were made to persuade elected officials, the courts, and other third parties to challenge risk assessments of both industry and the NRC.

The assistance from local and state officials was regarded by most activists as critical but also quite fickle. In the case of the cleanup of Unit 2, for example, the dumping of the water was blocked while the venting of the trapped krypton gas into the surrounding atmosphere was supported by the Thornburgh administration. The Unit 1 restart was only temporarily thwarted by the NRC's lifting of the utility's license and holding of hearings. A closer examination of both processes will reveal the importance of organized counter experts and emergent target vulnerabilities in these protracted political struggles over risk assessments.

Unit 2 Cleanup Issues

The Lancaster citizens' group became singlemindedly opposed to the dumping of the water only after they were alerted by John Gofman, a Berkeley professor of medical physics and outspoken critic of the nuclear industry, to the possibly harmful effects of allowing it even in filtered and diluted form to be mixed with the city's drinking water. In May 1979, Gofman was invited by the Lancaster activists to speak in that city. While there, he also visited with the mayor and other elected officials, convincing them of the health risks involved from the Unit 2 water. The city subsequently filed its own court suit against the proposed dumping after Gofman convinced them that industry officials and the NRC were underestimating the risks involved.

If the role of risk assessors is to transform political issues into technical ones, then the challenge for citizen groups is to peel away this technical veneer while exposing the underlying political nature of such assessments and mobilizing available resources for a struggle. Lancaster's activists pressured the mayor to listen to Gofman's pessimistic risk assessment, and then went both to the streets and the courts to challenge the dumping.

After the city filed its suit, the NRC rescinded the utility's dumping permit and, in August 1979, ordered an environmental assessment. The citizens's suit was dismissed by a federal district court — on the grounds that the protest group should not have circumvented the NRC — but appealed. In March 1980, the appeals court ruled in the Lancaster citizens' group's favor, and the U.S. Supreme Court eventually supported that ruling upon an appeal by the utility. A more detailed analysis of this litigation process is available elsewhere (see Walsh and Cable 1985).

The accident also caused significant quantities of radiocative krypton-

85 gas to be trapped inside the Unit 2 containment building, and this had to be disposed of in some way in the course of the cleanup processes. The most expeditious method, and that favored by both utility and NRC officials, was atmospheric venting, but the citizen groups, publicy expressing doubts about the scientific credibility and moral integrity of the utility and NRC, strongly opposed this method. Rancorous public meetings occurred over the issue during the early months of 1980, and the NRC even introduced its own novel approach whereby the federal agency trained ordinary citizens to read the radiation monitors in order to increase levels of public confidence. Skeptical opponents insisted that this was like the blind leading the blind.

In a letter to NRC Chairman Ahearne on April 12, 1980, one of the Middletown protest leaders, James Hurst, asked for a listing of the studies and other scientific information supporting the argument for the venting. He never received a response.

As a result of citizen group pressure on this issue, Pennsylvania's Governor Thornburgh turned to the Union of Concerned Scientists (UCS), in the spring of 1980, for an independent assessment of the likely effects of any krypton venting. This professional organization includes technical and legal experts who, while not absolutely opposed to nuclear power in the abstract, insist that existing reactors in the U.S. and elsewhere are threats to public health and safety. Although generally sympathetic with the citizens' groups, the final UCS report was careful to distinguish between physical and psychological effects of the venting, concluding that even though there was no scientific evidence that the krypton might be physically threatening there was reason to believe it would cause additional psychological stress for local residents.

Thornburgh ignored the UCS suggestions about possible psychological effects and supported this method of disposal. Weeks before the scheduled venting in late June 1980, another group of scientists from Heidelberg, West Germany, issued a much more negative assessment of its possible effects. Many area residents were understandably confused by these claims and counterclaims from presumed experts. As shown in Table 2, above, approximately 40 percent of the most active opponents in the four communities mentioned earlier evacuated during the krypton venting.

The dynamics of this venting case dramatically illustrate the complex relationships among the three variables of technical expertise, public trust, and political power in risk assessments. The NRC, for example, had its political power diluted significantly by Thornburgh's active intervention in the process. The governor, in turn, was pushed to intervene by the widespread citizen activism resulting from public perceptions of NRC incompetence and pronuclear bias. Unwilling to rely on NRC technical experts, Thornburgh turned to UCS, but then ignored the group's policy recommendations against the venting because they were based on psychological considerations presumably outside UCS's area of expertise. The Heidelberg scientists, with no invitation to enter the controversy from any established political authority such as the NRC or the state governor, found their technical report ignored by all but the protest groups in the area.

Although numerous events and processes associated with the Unit 2 accident are inextricably related to questions of a Unit 1 restart at TMI, two general risk categories are noteworthy for the purpose of this paper. The first derives from technical and organizational issues common to the nuclear industry, and the second is more specific to the TMI facility.

The industry underestimated both the probability of the occurrence of accidents as serious as TMI and, according to some experts, the impact of radiation releases resulting from such accidents on the public. In July 1982, for example, the NRC radically revised its own earlier estimates of accident probabilities, admitting that accidents as serious as TMI should have been expected every 10 to 15 years rather than every 200 years, as previously predicted in its Rasmussen Report (Ford 1982; Walsh 1983). Adequate radiation monitors were not in place during the early days of the accident, and thus it is impossible to give reliable estimates of population radiation doses released (Beyea 1984). Personal injury claims against the utility by area residents are mounting, however, and TMI's insurers settled 73 of them for $4 million dollars, in early 1985, out of court. The *Harrisburg Patriot* (3/28/85) commented on this settlement:

> Could it be that the accident was more serious than the industry cares to admit? Over and over again the people of this area have been assured that there were no health implications associated with the accident. . . . Yet, the industry — through its insurers — chose to avoid what in American society is the ultimate test of any claim of blamelessness — placing their case before the judgment of a court. The excuse that this tactic was taken solely to avoid the expense of a trial is either false or unbelievably stupid. This was a golden opportunity to put to rest the "mythology" that the accident produced a local epidemic of cancer.

There are a growing number of personal injury claims against the utility for birth defects, cancers, and related health problems allegedly deriving from radiation releases during the emergency period. More than 1550 additional suits were filed in Dauphin County Court by people with cancer and other ailments throughout 1985, or by the estates of people who have died since the accident (*Harrisburg Patriot* 12/18/85).

The second risk category derives from a growing public distrust of the owners and managers of the TMI facility throughout the course of the Unit 2 cleanup processes. For example, some high-ranking employees of the utility charged in the early months of 1983 that the company was taking safety risks in its effort to cut cleanup costs, and such whistle-blowing cost them their jobs at TMI. These employees insisted that they tried to persuade their own superiors before going public. The *Philadelphia Inquirer* also ran a three-day front page series (2/10/85 to 2/12/85) summarizing a wide range of risks to which workers and the public have been exposed in the course of the Unit 2 cleanup procedures. The utility was concerned enough with the public impact of that series to pay for double page ads in newspapers throughout the state attempting to challenge the reliability of the *Inquirer* series. This is an example of a social movement target's attempting to reduce the public's estimates of the risks it imposes.

The estimated costs of the cleanup will exceed $1 billion, pushing the utility to the brink of bankruptcy. Many citizen activists are convinced that the company will do anything it can to avoid further costs, including the taking of serious risks with the health and safety of its own workers and the general public in the course of the cleanup. The lethally radioactive core is still awaiting removal almost seven years after the 1979 accident. The utility insists that inadequate accident insurance explains its delays. It has not convinced Governor Thornburgh, however, who has scored the nuclear industry's refusal to help in the cleanup as a major factor in his own opposition to a Unit 1 restart.

The Unit 1 Restart Case

As noted earlier, the utility intended to ignore the public opposition to a Unit 1 restart in the wake of the accident and evacuation. Neither did the NRC reveal any intention of intervening to prevent the utility from bringing Unit 1, down for routine refueling at the time of the accident, back on line. Citizen groups were even preparing for massive civil disobedience in the late spring of 1979.

The rapid transformation of thousands of previously conservative middle-class citizens into outspoken challengers of a Unit 1 restart would probably not have occurred in the absence of a fragmented elite. Had the President's Commission on the Accident, commonly called the Kemeny Commission after its Chairman John Kemeny, been less critical of both the industry and the NRC, for example, the partial meltdown and frightening evacuation might have been publicly perceived as an unfortunate series of unlikely events with little probability of recurrence. This is not to suggest that the proceedings of the Kemeny Commission were the critical variables because they were only one example of elite fragmentation. The existing antinuclear movement's political strength, in fact, put constraints on President Carter's appointments to this post-accident commission. Had Carter excluded all critics of the nuclear industry from his selections, the commission's public credibility would have been challenged and seriously undermined.

This politically charged atmosphere of distrust precipitated Governor Dick Thornburgh's intervention, in June 1979, to demand a postponement of any Unit 1 restart until the questions about public health and safety had been answered to his satisfaction. Confronted with the threat of a legal suit by the State of Pennsylvania, the NRC suspended the utility's license and ordered that hearings be held. This was the first time any commissioned reactor's operations had been interrupted by federal regulators for further safety hearings. The interruption turned out to be much longer than anyone expected, due especially to the emergence of a series of target vulnerabilities which were used by opponents to mobilize strategic elites.

Citizens' groups found themselves confronted with the challenge of creating legal arguments against a Unit 1 restart in preparation for the NRC hearings. The Lancaster group remained on the sidelines offering support to the other communities, but preoccupied in their own legal struggles over the Unit 2 water disposal issues. The Harrisburg citizens' group focused its attention on the utility's management capability and financial qualifications. Newberry Township concentrated on emergency planning which had been so poorly handled during the accident. Middletown's activists decided to argue against any Unit 1 restart on the grounds that it would precipitate additional psychological stress for large numbers of people who had already been through enough. The preparations required for the coordination of legal efforts delayed the opening of the restart hearings until October 1980, approximately 18 months after the accident.

A new and critical series of target vulnerabilities — in addition to those associated with the initial emergency period during the accident — began in July 1981 at the conclusion of the NRC's Unit 1 restart hearings. Evidence surfaced of cheating by the utility's operators scheduled to work at Unit 1. Additional hearings were ordered, and all operators were retested by the NRC. Twelve of the 33 tested failed. The Special Master appointed to oversee the reopened restart hearings urged prosecution of the utility for the cheating, but the NRC's own Atomic Safety and Licensing Board (ASLB) reversed his recommendations in July 1982, and urged the commissioners to permit a Unit 1 restart. Citizen activists had abandoned any hope of gaining a favorable ruling from the ASLB because of their perceptions of pronuclear bias on the part of its head, Judge Ivan Smith, throughout the restart hearings. Most restart opponents also expected the majority of the five NRC commissioners, who had said they would make the final decision themselves, to vote in the utility's favor. These challengers' strategy was to appeal such a decision immediately to the courts, and the Lancaster group's earlier success in the Unit 2 water appeal was a source of hope.

By February 1982, the utility had discovered more than 15,000 leaky steam tubes in its Unit 1 reactor. After months of repair work, and an expensive public relations restart campaign, the utility again found the repaired tubes leaking in July 1984, and then again in March 1985. It assured the public, however, that these multiple difficulties were unrelated and did not signify any serious structural weaknesses in the steam tubing. An even more serious problem for the utility, however, was its low credibility with both the public and elected representatives. Critics insisted that the utility would significantly underestimate the seriousness of all technical difficulties in its attempt to get Unit 1 back into service and, especially, back into the rate base.

Risk assessments based on technical factors are of relatively minor importance in the TMI controversies compared to those deriving from what have come to be called "management integrity" issues. Critics of the utility insist that the operator cheating scandal was only the tip of a massive iceberg of corporate corruption. In support of this assessment, they cite the utility's alleged shortcuts in the cleanup process and, especially, its "guilty" or "no contest" pleas, in February 1984, to seven of eleven grand jury charges involving the falsification of leak rates prior to the Unit 2 accident. After first claiming complete innocence, the utility changed its plea rather than allow the case to go to trial, as opponents had hoped it would. The NRC fine amounted to $1,045,000, and, apparently as a concession to critics, the president of the utility's nuclear operations at TMI was removed. Citizen activists argued that they and the rest of the public know very little of the truth about record falsifications involving both the Unit 2 and Unit 1 reactors, but the NRC terminated adjudicatory hearings on those issues.

The various phases of the Unit 1 restart hearings have also convinced opponents that the NRC is suspect as a judge of issues involving nuclear power (see Walsh 1983, 1984, 1985). Such charges received strong support from Governor Thornburgh, in January 1985, when he called for the NRC's removal of the agency's ASLB head, Judge Ivan Smith, whom Thornburgh claimed "has gone out of his way to prove himself incapable of fair and impartial decision-making in this complex and difficult matter." Other elected officials joined the governor in his demand for Smith's removal. The five NRC commissioners, however, voted two months later to keep Smith at the helm.

The final NRC decision on the Unit 1 restart issue occurred on May 29, 1985. The commissioners voted 4-1 in favor of allowing Unit 1 to return to service, despite demands by Governor Thornburgh, U.S. Senators Heinz and Specter, and numerous other elected officials that further hearings be held before that vote was taken. In casting the lone dissenting vote on this restart issue, Commissioner James Asselstine stood with the local protest groups against the nuclear establishment. He publicly charged his peers on the commission with ignoring or discounting important unresolved questions about the integrity of the utility's management.

The Harrisburg protest group led the civil court appeals of this NRC ruling. The group was joined by the Commonwealth of Pennsylvania, the Union of Concerned Scientists, and a Coatsville couple. On June 7, the

Third Circuit Court of Appeals in Philadelphia issued a stay of the NRC restart order, agreeing to decide on the necessity for further hearings. Negative decisions by a three-judge panel in August (2-1) and then by the full court (10-2) in September left only the U.S. Supreme Court as opponents' last resort. On October 2, the High Court ruled 8-1 against reconsidering this case, and lifted the stay. The utility restarted the reactor on October 3, 1985.

Conclusion

The TMI struggle illustrates the centrality of political power in risk assessment. "Ultimately," writes Perrow (1984: 306), "the issue is not risk, but power; the power to impose risks on the many for the benefit of the few." Successfully organizing "the many" to challenge existing risk assessments and impositions by entrenched institutions, however, depends on the convergence of various structural factors supporting the challenges of citizen groups. There which were shown to be particularly important in this struggle were the availability of counter-experts, divided political elites, and emergent target vulnerabilities.

Despite widespread public opposition and protest mobilization among citizens in surrounding communities, the NRC was prepared to allow the utility to resume nuclear operations at Unit 1 within a few months of the Unit 2 accident. The national antinuclear movement, however, had helped mobilize counter-experts whose technical and legal expertise in matters nuclear was drawn upon by citizen protest groups as well as by local and state officials challenging federal risk assessors. Three important examples noted in this paper are the John Gofman testimony against the dumping of the radioactive water, the Heidelberg scientists' report on the krypton venting, and the UCS involvement in both the venting and restart issues. The national movements's strength was also important, as has been noted, in assuring a critical presence on the Kemeny Commission. Under different conditions, even a seriously disruptive accident (the Santa Barbara oil blowout of 1969) was insufficient to guarantee any critical members on the investigative commission (see Molotch 1970; Walsh 1984).

State and local political officials depended on the technical advice of some of the counter-experts in their own evolution to increasingly anti-TMI ideologies. Thornburgh's selective use of UCS recommendations, however, would challenge the conclusion that political decision-makers became mere pawns of technical experts. The importance of local and, especially, state political intervention on the challengers' behalf was also emphasized. Thornburgh's opposition was critical in forcing the NRC to suspend the utility's license in 1979, and his refusal to oppose the krypton venting was commonly considered the key factor in that 1980 controversy in which opponents' arguments were overruled. Yet neither the strategies and tactics of the protest groups nor the arguments of the counter-experts would have been sufficient, by themselves, to sustain the opposition of Governor Thornburgh, Senators Heinz and Specter, and other political leaders. The continuing emergence of utility weaknesses was also necessary.

Both the counter-experts and political authorities seized upon emergent target vulnerabilities throughout the course of this conflict in developing their anti-TMI positions. Had such things as the operator cheating evidence not surfaced at the conclusion of the first phase of the restart hearings in the summer of 1981, the thousands of leaky steam tubes not been discovered the following winter, the utility's criminal conviction for Unit 2 leak-rate falsification not taken place in early 1984, and evidence of the NRC's increasing pro-industry bias not continued to emerge, these elected officials would have been more willing to trust the confident risk assessments of the utility and the NRC.

The Unit 1 restart decision was imposed upon the citizens of central Pennsylvania and their elected representatives by the federal government. Despite opponents' demands for a more thorough investigation of the utility management's integrity and related issues, the NRC arbitrarily terminated its own hearings on these issues. Commissioner James Asselstine, the lone NRC dissenter in the 4-1 vote to permit restart, called it "the single biggest failure of the agency," and candidly acknowledged that "the NRC has viewed the public as the enemy" (*Harrisburg Patriot* 9/20/85). Asselstine also warned listeners that his fellow commissioners intend to circumvent the Sunshine Act and further eliminate the public from critical decisions involving nuclear power. He said that a possible reason for his being somewhat out of step with his colleagues might be that they are scientists or engineers while he is the only lawyer of the five, and that he cares more about procedural issues such as the opportunity for public involvement than the others.

Most opponents insisted that no possible benefits could outweigh the risks involved in the proposed restart. Some were convinced, for example, that serious harm had already come to many residents who happened to be in the path of the radioactive plume during the Unit 2 emergency in 1979. Their concerns about this and other physical threats from the reactors, however, are secondary to their distrust of the utility's management. Many feel that the utility executives will do anything to prevent the public from knowing about further risks associated with the operation of the plant.

This case involved two very different types of risk. On the one hand, the utility and indeed the whole nuclear industry had considerable economic interests dependent on a Unit 1 restart. In the eyes of many proponents, it was not only the future of this particular utility but that of the entire nuclear industry which was at risk. How could an already ailing industry hope for more nuclear reactor orders from utility executives if a completed one could be shut down by local opponents? On the other hand, the citizen groups and their supporters were concerned primarily with the threats to public health and safety occasioned by both the technology and by the suspect managers of the TMI plant. The restart took place because of the superior political power of the nuclear establishment, including both the NRC and the industry, which emphasized the primary importance of reducing its own economic risks rather than local health and safety risks.

Acknowledgements

Special thanks to the following for their various forms of assistance in the organizing and analysis of the data drawn upon for this report: Sherry Cable, Betsy Graham, Merlin Gehrke, Rosemary Walsh, Eileen McCaffery, James Hurst, Beverly Hess, and Kay Pickering. Grants from the National Science Foundation are also gratefully acknowledged.

References

Ash-Garner, Roberta. *Social Movements in America*. Chicago: Rand McNally, 1977.

Beyea, Jan. *A Review of Dose Assessments at Three Mile Island*. Philadelphia: TMI Public Health Fund, 1984.

Fingrutd, Meryl Anne. 'The Three Mile Island Commission and the Language of Legitimacy.' Doctoral dissertation. Department of Sociology. State University of New York at Stony Brook, 1984.

Flynn, Cynthia. *Three Mile Island Telephone Survey: Preliminary Report on Procedures and Findings*. Seattle, Washington: Social Impact, Inc., 1979.

Ford, Daniel. *Cult of the Atom: Secret Papers of the Atomic Energy Commission*. New York: Simon and Schuster, 1982.

Gamson, William. *The Strategy of Social Protest*. Homewood, Illinois: Dorsey, 1975.

Goldstone, Jack. 'The Weakness of Organization: A New Look at Gamson's *The Strategy of Social Protest*,' *American Journal of Sociology*, 1980, 85, 1017—1043.

Killian, Lewis M. 'Organization, Rationality and Spontaneity in the Civil Rights Movement,' *American Sociological Review*, 1984, 49, 770—84.

Klandermans, Bert. 'Mobilization and Participation: Social Psychological Expansions of Resource Mobilization Theory,' *American Sociological Review*, 1984, 49, 583—600.

Mazur, Allan. *The Dynamics of Technical Controversy*. Washington, D.C.: Communications Press, Inc., 1981.

McCarthy, John and Mayer Zald. 'Resource Mobilization and Social Movements: A Partial Theory,' *American Journal of Sociology*, 1977, 82, 1212—1241.

Molotch, Harvey. 'Oil in Santa Barbara and Power in America,' *Sociological Inquiry*, 1970, 40, 131—144.

Oberschall, Anthony. *Social Conflict and Social Movements*. Englewood Cliffs, New Jersey: Prentice-Hall, Inc., 1973.

Perrow, Charles. *Normal Accidents: Living with High-Risk Technologies*. New York: Basic Books, Inc., 1984.

Short, James F. 'Toward the Social Transformation of Risk Analysis,' *American Sociological Review*, 1984, 49, 711—725.

Snow, David, Louis Zurcher, and Sheldon Ekland-Olson. 'Social Networks and Social Movements: A Microstructural Approach to Differential Recruitment,' *American Sociological Review*, 1980, 45, 787—801.

Tilly, Charles, Louise Tilly, and Richard Tilly. *The Rebellious Century: 1830—1930.* Cambridge: Harvard, 1975.

Turner, Ralph. 'Collective Behavior and Resource Mobilization as Approaches to Social Movements: Issues and Continuities,' pp. 1—24 in L. Kriesberg (ed.), *Research in Social Movements, Conflicts and Change*, Vol. 4. Greenwich, Connecticut: JAI Press, 1981.

Useem, Bert. 'Solidarity Model, Breakdown Model, and the Boston Anti-busing Movement,' *American Sociological Review*, 1980, 45, 357—369.

Walsh, Edward. 'Mobilization Theory *vis-à-vis* a Mobilization Process: The Case of the United Farm Workers' Movement,' pp. 155—177 in Louis Kriesberg (ed.), *Research in Social Movements, Conflicts and Change*, Vol. 1. Greenwich, Connecticut: JAI Press, 1978.

Walsh, Edward. 'Resource Mobilization and Citizen Protest in Communities Around Three Mile Island,' *Social Problems*, 1981, 29, 1—21.

Walsh, Edward. 'Three Mile Island: Meltdown of Democracy?' *Bulletin of the Atomic Scientists*, 1983, 39, 57—60.

Walsh, Edward. 'Local Community vs. National Industry: The TMI and Santa Barbara Protests Compared,' *International Journal of Mass Emergencies and Disasters*, 1984, 2, 147—163.

Walsh, Edward. 'Three Mile Island Revisited: The Battle of Unit 1,' *Bulletin of the Atomic Scientists*, 1985, 41(5), 30—31.

Walsh, Edward and Sherry Cable. 'Litigation and Citizen Protest after the Three Mile Island Accident.' Working Paper, Pennsylvania State University, 1985.

Walsh, Edward and Rex Warland. 'Social Movement Involvement in the Wake of a Nuclear Accident: Activists and Free Riders in the TMI Area,' *American Sociological Review*, 1983, 48, 764—780.

Zald, Mayer N. 'Macro Issues in the Theory of Social Movements.' University of Michigan, Center for Research on Social Organization (Working Paper no. 204), 1979.

Zald, Mayer N. and J. D. McCarthy (eds.). *The Dynamics of Social Movements*. Cambridge, Massachusetts Winthrop, 1979.

Chapter 5

Protest Movements and the Construction of Risk

Luther P. Gerlach

Introduction

Public disputes between established authorities and interest groups over the development, deployment and use of a broad variety of technologies have become common and often dramatic events in Western society.

These events are seen as problems calling for explanation and solution. There have been many attempts to provide this. They have come from the disputants themselves (claiming the power of experience), from scholars and policy analysts (some times claiming objectivity), and from consultants (offering plans for successful action). These have reflected different schools of thought, different objectives, biases, and premises. None has produced a dominant and generally accepted paradigm of explanation or resolution. Rather, these various analytical attempts have often themselves become part of these events. That is, they provide ideas which enter into the strategic ideological struggle characteristic of these events, and which provide the means to achieve tactical advantage. People interpret the events in ways which enable them to apply their special capabilities to the situation, or which enable them to claim that "right" is on their side according to some intrinsic standard above and beyond their selfish gain. I suggest that there are five major interpretations:

- The events constitute problems in risk assessment and management.
- The events constitute problems in decision-making procedure.
- The events are part of or a result of larger sociocultural changes associated with post-industrialization.
- The events are driven by social movements, and it is the structure, function, and operation of such movements which is the problematic to be understood.
- The events produce the public debates by which contemporary ways of life are criticised, defended, and changed.

I shall argue the merits of the last two of these, and focus upon them. I shall examine social movements as a phenomenon in their own right, and then consider how they work to drive technological disputes to control both risk/benefit assessment and procedural adaptation, and to generate

B. B. Johnson and V. T. Covello (eds.), The Social and Cultural Construction of Risk, 103—145.

the public debates. To conclude, I shall briefly advance some ideas about how these debates might promote or counter sociocultural change. I will argue that a rhetoric of risk is constructed and used in the context of movement growth and related public debate, and is hence more product than producer of this growth.

The Events, the Worries, and the Hopes

It is a matter of common knowledge that many important technologies and projects for their use have been militantly protested in recent years by groups of people, some mobilized in social movements. These protesters have argued that the technologies and uses are unfair in their distribution of costs and benefits, often unnecessary, and above all, unsafe. The protested technologies range widely in type: from nuclear and fossil fuel energy systems to recombinant DNA technique, from pesticides and herbicides to pharmaceuticals and infant formula; from SST's to jetports and highways; from technologies which produce hazardous wastes to technologies to manage such wastes. Sometimes that which is protested is the technology itself, often it is the proposed use of the technology in a particular community or, as in the case of infant formula, its marketing among a certain population. Usually the leading disputants are those directly affected by the proposed land use or technology. Yet, as in the case of infant formula, major protesters can come from outside the affected population, but claim to speak not only for those impacted, but for the public interest.

Once protest is underway the technologies or uses can be neatly labeled "controversial" as if this controversy is their inherent characteristic. In Minnesota a report by the Citizens League, a non-partisan public affairs research and education organization, thus stigmatizes "power plants, power lines, landfills," but also "trails, and wild and scenic rivers" and "controversial facilities" (Committee on Facility Siting. 1980). Yet, when people begin to challenge such technologies or uses they find that their first task is often to convince others that these are, indeed, something which should not be routinely accepted but instead contested, made an object of controversy. And it is the manner in which this contest is conducted which also becomes controversial.

A technical (technological) controversy, according to one analyst (Mazur 1981: 7—8), has three important factors. First, its focus is some product or process of science or technology. Second, some of the main participants in the controversy must qualify as experts in technology or science. Third, there must be experts on opposing sides of the controversy who disagree over relevant scientific arguments which are too complex for most laymen to follow. These are the characteristics which, he says, roughly set technical controversies apart from other controversies. I assume he means controversies in the public arena.

This is a useful introduction to the topic. Yet, most of Mazur's essay deals with how the scientific and technological dimensions of technical controversy are used and shaped by the kinds of social forces including protest movements which characterize so many other public controversies. They become, as physicist Alvin Weinberg (1985), has observed, "trans-scientific." Thus, another characteristic of public technical controversy (in the U.S.A. at any rate[1]). is that members of the scientific and technical establishment do not control the dispute. They are not able to contain its conduct within the framework of scientific discourse. Some lay activists are not hesitant to enter the scientific-technical debate, no matter how complex. When criticized for this by officials, they are more likely to seek to improve their command of the subject than to leave it to the specialists. A related characteristic is that such disputes are not contained within the normal, mundane, political process of representative democracy. Just as people consider the issue too important to be left to scientific technological elites, so do they consider it too immediately important to be left to elected or appointed officials. They volunteer, and recruit others into groups which challenge not only the discussion but the authority of public officials. Similarly, they contest the actions and question the capabilities or integrity of executives in the corporation promoting the technology.

Doing this, protesters often identify themselves as being "grassroots," representing ordinary citizens and communities rather than being of "the system," the people in official power, the "established order." In any particular dispute they consider themselves to stand in opposition to this established order with its conventional wisdom, even though in other matters they may well flow with this mainstream. The protesters wage their struggle using a variety of means, many of which also have been considered unconventional. These range from participating in official decision-making public hearings to holding confrontations in the fields and streets; from suing or defending in the courts to conducting boycotts and stockholder resolutions against actual or symbolic target corporations; from getting prime time coverage to spreading the word in advertisements, pamphlets, newsletters or convocations. The protesters are often considered to be a single constituency with a single issue — for example, the farmers protesting a power line over their land, as was the case in west central Minnesota from 1974 to the early 1980s. But protesters characteristically search widely for allies. As they expand their network of allies, so do they share the concerns of these allies and search for an overarching and integrating ideology. Thus those who joined in protesting the power line were first the landholding farmers, then their rural townsfolk neighbors including local church leaders, then counter-culture activists from the cities of St. Cloud and Minneapolis-St. Paul, and many more. These developed not only an ideology of stopping the line, but also of protecting the family farm and rural life, of promoting alternative energy technologies, of challenging big business, and — as women began to lead in the protest — of advancing women's liberation. As their protest continues some of them become increasingly unwilling to make trade offs and negotiate a compromise settlement. "The bottom line," agreed protesters of the high voltage direct current transmission line across west central Minnesota, "is no line at all" (Gerlach 1978).

Thus, protest movements do not interact with government and industry in ways which produce the mutual adjustments and consensus approvingly attributed to the pluralistic American political process by some political analysts (Bentley 1949; Truman 1951; Almond and Verba 1965) who, as Ornstein and Elder (1978) observe, may have been influenced by particular political and economic stability and prosperity of their particular eras to be "interest group liberals." Instead protest movement actions would confirm or exceed the fears of critics (Schattschneider 1961; Lowi 1969) of this liberal interpretation of the role of interest groups. They fear that interest groups have come to undermine the authority of government, its legitimacy, and ability to plan and manage in the public interest. Indeed, they say, interest group actions tend to weaken democratic institutions as they replace formal procedure with informal bargaining. How much more critical would they have been of protest groups which used direct action and civil disobedience such as sabotaging protested power lines — as happened in west central Minnesota.

Certainly both the private and public proponents of the challenged technologies have been disturbed by the challenge of protest groups. They have understood this as a challenge not only to their projects but also to their legitimacy, their claim of serving the public interest, their ability to plan and manage within the certainties of formal procedure. This is evidenced in the way they have gone to the public with often elaborate campaigns of their own to counter these challenges and to influence government decisions. For example, the two area associations of the National Rural Electric Cooperative Administration (NRECA) which were building the protested Direct Current line in west central Minnesota ran a series of advertisements in the late 1970s warning the public that the protesters toppling pylons and shooting out insulators were "destroying *your* line," and with this increasing *your* costs and jeopardizing *your* law and order.

Both proponents and protesters of technologies and their use have gone to the broader public with their claims and counter-claims of technical and related economic, political, social — even religious — costs, benefits, implications. Amplified by the media, overlapping with other movements, protest events do produce public debates about technology and much, much more. In some cases, such as when energy technology has been contested, the whole shape and future of Western society and culture can be put to question and called to change. Many others have been drawn into these conflicts, some to report or analyse them, some to police or manage them, and many to ponder and often to worry.

It is a worry if technology is out of control (Winner 1977), if technologies are hazardous and the proponent corporations wrong (Lovins, 1977). It is a worry if modes of making and implementing decisions about these technologies are wrong (Commoner 1972). It is a worry if core technologies cannot be effectively used and if their proponents and their investors and their employees are put into jeopardy (Tucker 1982). It is a worry if demonstrations of protest and response force expenditure of much money and time in legal and administrative action.[2] It is even more of a worry if these disrupt the social peace and rightly or wrongly jeopardize trust in established institutions, in government, or, for that matter, in grassroots participation.

There is all this worry, and more. But, on the other hand, all of this can be interpreted positively. It has been proposed that, through this, Western society is trying to bring technology under better control or at least actively debating its cost and risks as well as its benefits (Mazur 1981). It has been proposed that this is a part of the ongoing evolution of participative democracy in which ordinary people, from the "bottom up," act in self-help groups to question the experts and make change (Gerlach and Hine 1973; Theobald 1972).

That there are these protests, responses, worries and yet some hope seems clear enough. What is not clear is why this should be, what is does mean, what kind of a problem or promise is it? Among the many interpretations of these events we can identify five as particularly significant.

Interpretations

The disputants generally argue over the risks and benefits of the technology and the distribution of these costs and gains. Hence a logical interpretation of the dispute is that it is a problem in such risk/benefit analysis and management (Hadden 1984; Lowrance 1980). But disputants also argue about how decisions are made, and about the representativeness, capabilities and fairness of the decision makers. Thus, this to some is basically a problem in procedure, in designing decision or mediation procedures which will be open to the affected yet move efficiently and expertly to conclusion (Kunreuther *et al.* 1982; Advisory Commission on Intergovernmental Relations, n.d.). Such definitions of the problem as one of procedure or of risk/benefit management suggest that solutions can be reached through culturally quite manageable changes in technique. These changes call for more input from lawyers and policy researchers and analysts, from technical specialists and from public relations people. They can be expected to keep defining the problem as one which can be solved using their techniques, and they are opinion leaders. Putting them to such tasks meets the need of legislators, administrators, and corporate executives to do something positive, but not drastic, about the disputes. Funding research on these approaches to the problem also gives foundation trustees and directors an appropriate way to act on the issues.

But, of course, the disputes continue. For this and for more personal reasons people look for deeper causes. For some the cause lies in particular psychological characteristics of the disputants (Slovic *et al.* 1980) or in basic decision-making approaches (Self 1975; Wynne 1982). For some it is a product of existing social and cultural factors; such as long standing interpersonal or intergroup rivalries (Gulliver 1979), or the disputants' articulation to social institutions which differ markedly in the way they assess risks and produce agreements (Douglas and Wildavsky 1982: Rayner 1984; Gross and Rayner 1985).

For yet others the presence of disputes of this number and magnitute and intractibility are best explained as a kind of by-product of larger social and cultural changes associated with the post-industrial transformation of the Western world (Bell 1973; Benjamin 1980; Huntington 1974). As part of this transformation people have come to expect not only the benefits of technology and development but relief from the disbenefits or negative externalities of such material advance. As part of this they have come to expect not only more individual and local freedom but also more comprehensive and integrated planning and resource management by government (Gerlach and Radcliffe 1979). When these contradictory expectations cannot be met (Bell 1975; Huntington 1974) people mobilize into collectivities to fight for their competing interests.

Explanations of these types clearly will have quite partisan constituencies and themselves become a focus for dispute. An explanation which says that a farmer protested the construction of a high-voltage power line over his land because of his particular social affiliations, psychological characteristics, or unrealistically high expectations implies that the protest is socially invented rather than the result of natural aversion to real biophysical risk or to fundamental injustice. Conversely, an explanation which attributes protest to such intrinsic threats will help legitimate resistence to the technology. In Western industrial culture, social "facts" simply do not carry the clout of those considered biophysical or according to "natural law."

Yet, it is always the action and interaction of the disputants and other concerned parties which play the critical role in the choice and use of problem explanations. It would seem reasonable to examine this action and interaction not only to understand problem explanation but more basically to understand the problem. The approach to the problem which I have taken and found most attractive examines technological disputes as a problem in the structure, function, and operation of social movements, and in the interaction of such movements with the established orders or institutions which they seek to challenge and change. It is the approach which I shall advance in this essay. Building on past studies (Gerlach and Hine 1971, 1973) I shall endeavour to describe and analyze how a movement works without taking a position on whether it is right or wrong, a risk or a benefit; or whether or not the protested technologies are right or wrong. I shall argue that it is the movement which launches and drives the dispute, which generates concern about risk and the distribution of costs and benefits. It is the movement which raises concern about how decisions are made, and which leads authorities to change procedures to permit more participation from the public. It is the movement which frustrates decision-making and negotiation.

It is a useful approach. But it is not enough. It leaves people with the question: So what? So you explain how a social movement works to drive dispute but also to frustrate negotiation and settlement. But of what use is this information? What can be done about the dispute? Project protesters can worry that such an analysis again attacks the legitimacy of their fight by indicating that it is not the result of natural reaction to real risk and/or unjust decision-making, but instead is the result of social process. In some ways this seems even more critical of their action than approaches which attribute dispute to ongoing and underlying social and cultural context. Process suggests impure manipulation rather than pure response. Project defenders or dispute mediators can feel that such an analysis still does not enable them to get at the cause of dispute — and to eliminate it.

A way to move beyond this is to propose that dispute over technology is not simply a problem to be identified, defined and solved, but rather a recurrent and patterned event which has a larger meaning and impact. This larger argument is that these dispute events, driven by social movements, generate public debate and through this make change. That is, they move from being arguments about specific impacts to being debates about technological risk and benefits and their distribution; about procedure of making fair and acceptable decisions; about the positive and negative features of our present way of living and of using technology and managing resources; and about how we should change. It is an argument which this essay will explore. But its exploration must be launched from a base of understanding about social movements as the driving force. We turn now to social movements — movements of protest and change — as our unit of study.

Movements of Protest and Change as the Unit of Study

In his study of the dynamics of technical controversies Allan Mazur

(1981) says that the time has come for a comparative study of these as a social phenomenon, lifted above the specifics of any one case. This is needed, he says, to provide a general theoretical understanding and to guide further research with testable hypotheses. His work provides a good example of the "protest as a phenomenon" approach, which I find so useful. I would urge that a comprehensive study examine a range of cases extending beyond those dealing specifically with technology.

That is, we know that just during the past 25 years much more than technology has been at issue in major protest events: civil and human rights; animal rights; the military draft; policy and war in Vietnam and Central America; abortion; anti-abortion; the management of water, forest, wilderness and other natural resources; investments and marketing strategies and decisions of industry; urban renewal projects; farm foreclosures and farm prices; fuel shortages and price gouging; religious belief and practice and its application in public education, public morality, politics and other secular domains; the nuclear freeze/war issues; and much, much more. A materialist bent on showing that technology is the prime mover might seek to trace these other disputes to technology. But in our examination of protest as the unit of study, it would seem far more useful to consider these as disputes about other big issues and to ask in what ways these are like or unlike disputes directly concerned with technology. Technological disputes are thus a subset of the larger class of major disputes about decisions affecting the major aspects of our way of life or sociocultural system; economic, societal, political, religious — and technological. All of these disputes are driven by the ways people mobilize in collective action to advance or defend interests or positions. So this becomes our unit of study,

Collective action can take many forms. The simplest form which concerns us is that of people organized in a local group for a single, specific, short-lived purpose; for example to protest a particular use of neighborhood land. The entities involved in such bounded (Mazur 1981) protest have been termed and indeed deprecated as "Not in my Backyard" (NIMBY) organizations (O'Hare 1977) or as "single issue" groups. Their critics complain that such groups are narrow and selfish. The designers of 1980 legislation in Minnesota to site hazardous waste management facilities, recognized that people would organize in local interest groups to protest such facility siting in their particular neighborhoods (Todd, n.d.). The designers hoped to create a siting process which would produce a decision through compromise following competition among these rival local interests, and their recognition of trade-offs. In this they seem to have followed the philosophy that in Western democracies, decisions in the public interest are produced through such interest group interaction. Robert Dunn, Chairman of the Waste Management Board, charged with directing this process, began to despair that it would work because "people just do not want to make the hard decisions anymore" (Gerlach 1987). They have become too self-serving.

But protest usually does not stay bounded in space and time and focus; groups do not remain with single issues or fight in isolation. The ideas of protest, the organization of protest and the commitment and perspective of protesters all evolve as the protester groups act, and interact with the established order. In this, the established order also changes, or evolves. This co-evolution is also marked by increasing schism or divergence between protesters and their established opposition, rather than by negotiation toward convergence, compromise, and consensus.

Protester organization evolves to become an entity which is often called a movement, not only by scholarly analysts but by participants, the media, the public, and the established opposition. For example, the first three movements which my students and I studied, beginning in 1965, were known by us and their participants — for a time — as the Neo-Pentecostal Movement, the Civil Rights Movement, and the Ecology Movement. I say for a time, because movements often undergo many name changes during their history. These changes reflect both the noninstitutional, irregular nature of movements and their dynamism. Thus, what was Neo-Pentecostalism in the 1960s became the Charismatic Renewal in the 1970s and some of those who call themselves Charismatics may also accept the popular label, "Born Again." Thus, Black Power evolved from "Negro" Civil Rights as new, more militant "Black" activist groups emerged in the 1960s. The term Ecology Movement was used interchangeably with Environmentalism and both emerged in the late 1960's from what was called the "New" Conservation, an upstaging in the mid-1960s of traditional conservationism.

Movements often defy easy naming. There really is no simple name for most of the events in which people protest technologies and their use. For example, when in 1974 farmers and rural townsfolk in west central Minnesota began to form local interest groups to protest the construction over their land of a high voltage (± kV d.c.) transmission line this was conveniently labeled "power line protest." Yet as it grew by the late 1970s into a broad movement challenging even the way Americans get and use energy and make decisions about this, this term seemed too limited. Had this become part of the Ecology Movement? Such identification was made by a few television reporters. But the protesting farmers did not use it. Rather, one of their arguments was that it was the "environmentalists" who forced the state to site the line over farm land because 1970s environmental legislation excluded parks and wildlife areas from consideration as transmission corridors. Yet, there were ways in which the organization and ideology of the power line protest did contribute to the overall push of what has generally been called environmentalism. It is probably just as "environmental" as much of what has been called the "antinuclear (energy) movement." Some of the people who protested the construction of this d.c. power line came to regard their effort as part of a struggle waged by many other groups across the state and country against domination by big government, big industry, and urban managerial elites. They sent out representatives to urge people in neighboring areas and states to mobilize against the siting and operation of other power lines, and to cooperate with people protesting fossil and nuclear fueled power plants. They formed temporary alliances with groups of "ordinary people" elsewhere in Minnesota who were fighting efforts by governmental agencies and established environmental organizations to designate new areas as wild or scenic public lands or trails or to impose new restrictions curbing the use of snowmobiles and motorized boats in the Boundary Waters Canoe Area of Minnesota. Some also formed alliances with essentially urban "citizen" groups organized to protest high fuel prices. Later, some of these same people or neighbors stimulated by their action formed the initial groups which organized to resist efforts by the Minnesota Waste Management Board (WMB) to carry out the provisions of the 1980 waste management act and locate urban and rural sites for hazardous waste management facilities. Some saw this as an extension of their resistance of the same kind of threats and enemies; some saw it as a new activity requiring their particular resistance and organization. Quite apart from these people and groups, many others organized in local groups across the state also to challenge the siting of waste management facilities in their communities. During the evolution of this protest these groups followed the usual protest pattern and developed a broad ideology about the forces in our society and culture which lead to waste management problems and environmental risks. They came to see themselves as more environmentally conscious than the waste management board. At least some WMB staff and board members, on the other hand, regarded facility resisters as being the NIMBY-type, single issue protesters who wanted the environmental benefits of waste management but who did not want to accept responsibility for solution.

What name can we give to these activities which will help us refer to them as comprising a set of events sharing some common characteristics, not only in the minds of their participants but also according to the models to be presented, below? Researchers (Casper and Wellstone 1981; Gerlach 1979) have referred to the energy facility protesters as "Energy Wars," but obviously this is not broad enough to encompass all the other disputes. Nothing simple really works well, and I am not aware of any term used by activists themselves. But for this essay let us somewhat awkwardly lump these as "Technology and Resource Management Issues," or TARMI. They may begin as NIMBY, but they become TARMI, following the usual evolution of interest group action into movements of protest and change. What name can we give to the various Christian activist groups protesting secular humanism, evolutionism, pornography, homosexuality, abortion, the ERA, and a host of other things which they regard not only as religious sins but as dangerous social and political ills, which put America at grave risk. These groups have sometimes been lumped together as Moral Majority, but the Moral Majority is but one part or segment of this whole movement. They have been called Fundamentalist, but not all share the same traditional fundamentalist religious views. They are often evangelical and "Born Again," but not all. And not all are Protestant. Their participants often simply refer to themselves as "Christian," but usually use this to mean only people who share their own type of Christian belief. I have elsewhere referred to these groups as Conservative Christian Activists, or CCA (Gerlach and Radcliffe 1983), and this seems as useful a compromise title as any for our purposes in this essay.

Names are important to help us know when we are talking about the same phenomenon. But they are not enough. We require a definition and the elaboration of a set of operational characteristics. I define a movement as *a group of people who are organized for, ideologically impelled by and committed to a purpose which implements some form of personal or social-cultural change; and who are actively engaged in the recruitment of others, and whose influence is spreading in opposition to the established order within which it originated.* This definition identifies five key factors or characteristics as critical, namely *organization, ideology, recruitment, commitment*, and *opposition.* It is as these factors take certain form and interact that a collectivity moves along the continuum, from interest group to become a full-fledged movement. The nature and development of these factors shows strong consistency or regularity across the many different cases of the movement phenomenon which we have studied since 1965. That is, the characteristics and patterns which we found and delineated in our 1970 and 1971 (Gerlach and Hine) analyses of the Black Power, Neo-Pentecostal/Charismatic Renewal and Ecology movements we found to be generally repeated in newer movements and protests; for example, the various Technology and Resource Management Issues, (TARMI) and in Conservative Christian Activists (CCA). I have used our model of these factors to predict stages and events in the evolution of two major TARMI, in Minnesota, namely the protest of the $\pm$ 400 kV d.c. power line (Gerlach 1979), and the Hazardous Waste Facility Siting Controversy (Gerlach and Meiller 1986). This predictable consistency is one good indication that the phenomenon of protest and movement is the one which drives technological disputes and that approaching them as this kind of problem is most useful. I also find it useful to examine movements and their development as products of the adaptive strategies of capable people. This is an approach somewhat like that taken by scholars advancing theories of movements as products of resource mobilization (Zald and

McCarthy 1979) or rational calculus (Oberschall 1973). It is thus markedly different from approaches once so traditional in academia and still popular in the general public which theorized that since movements were abnormal their participants were themselves abnormal or troubled by abnormal events so that they joined movements in collective response (Turner and Killian 1957) to construct "a more satisfying culture" (Wallace 1956).

For instance, much of this traditional and popular literature concentrates on determining the causes of movements. As a pathology, they can be treated or prevented if only their cause is known. Introducing her anthology of essays on the movements of the 1960s and 1970s, Jo Freeman (1983) observes that people have been able to speculate quite freely about the causes of movements because movements have "inconspicuous beginnings." With few facts to encumber them, people could imagine anything and everything as cause: from external plots and agitators to spontaneous eruptions of the deprived, the disorganized, the defective. We (Gerlach and Hine 1970; Hine 1974) have attributed such creative speculation less to an ignorance of origins than to dominant paradigms in the social and behavioral sciences favoring equilibrium or economic motive, or simply to self-interest. It comforts authorities and the larger society to believe that protesters are defective. Movements seem less threatening to their targets if they can be explained (away) as pathologies. On the other hand, movement activists or supporters find it useful to attribute their movement and its growth to flaws in the established order, for instance in TARMI to use hazardous technologies or inadequate public participation in decision-making. Each side in a movement interprets the movement in ways which legitimize its position. Explanations of cause are thus constructed in the process of dispute. They are a product of movement action and its interaction with established orders and the larger society.

I argue that the search for movement cause will prove as fruitless as it has been beguiling if by "cause" is meant necessary and sufficient conditions. Yet movements do have backgrounds. They come to life in conditions which facilitate rather than directly cause. And almost from the beginning of this life the factors of organization, recruitment, commitment, opposition, and ideology interact dynamically with these conditions and increasingly control outcomes.

Let us now turn to these factors since they hold the key to movement success. These constitute a system so that it is often difficult to analyze any in isolation. We could begin our exposition with any factor, anywhere in the system. But let us start with organization.

Organization

Movements begin to take shape as people begin to organize to express their concerns and to identify and advance shared interests. In the various movements I have studied in the U.S.A. their organization has regularly taken a particular form. It is different from conventional centralized bureaucracy with a distinct chain of command. But it also is something far more coherent than implied by the terms so often given to it of decentralized, amorphous, loosely organized.

It is an organization best summarized as segmented, polycentric and reticulate or networked. By *segmentary* I mean that it is cellular, composed of many different groups of varying size, scope, mission and capability, and of varying life span. By *polycentric* I mean that it has many different leaders or centers of direction, evangelical persuasion and decision-making or consensus-building. Leaders are often no more than "first among equals" and some act chiefly as traveling evangelists, criss-crossing the movement network. Groups and leaders come and go, and rise and fall in influence and power both in the bourgeoning movement and in society as a whole. By *networked* I mean that the segments and the leaders are integrated into reticulated systems or networks through various structural, personal, and ideological ties. Networks are usually unbounded and expanding. They do not have one command center, but may have different nodes of special influence, where a number of segments overlap.

I have found it useful to abbreviate this organization as SPIN, that is an organization which is a segmented, polycentric, ideologically integrated network. This acronym helps us picture this organization as a fluid, dynamic expanding one, spinning out into mainstream society. SPIN does not match the popular image that movements consist of a single charismatic leader commanding a mass of blindly following true believers (Moorhead 1960; Hoffer 1965). It also does not conform to the model of the centralized bureaucracy, which is supposed to prevail in effective institutions, including conventional voluntary associations. There is in Western society a powerful presumption that bureaucratic organization, with a pyramidal chain of command, is efficient, rational, proper. Its presence is considered evidence that the organization is mature and effectively able to mobilize its members, and identify and accomplish its objectives. A collectivity which does not have such a central structure has classically been considered disorganized, or, at best, as having "an organization in embryo," and constituting an "emergent institution" (Kopytoff 1964). Some observers have attributed lack of formal bureaucracy in collectivities and movements to limitations peculiar to the participants, for example, to their lack of experience, their immaturity, their emotionalism, their gender, their stubbornness, or to their social marginality and related egalitarian ideology (Douglas and Wildavsky 1982). Early in our research we found that activists in movements, even those with egalitarian and "do your own thing" ideology, shared this bias against segmentation and for centralization (Kahn 1969). But there has been change in this perception. Interviewed by Minnesota Public Radio during the July 1985 meeting of the Minnesota Women's Consortium, a speaker explained that the Consortium, like other women's movement groups, has been successful because it has been "small, flexible, fast moving." It needs to "stay this way," while also "getting ever bigger and bigger." How can it do both at once? By building "coalitions," by "networking."

In her own way she was describing the formation of SPIN organization. And in their own ways so many other groups have built SPINs by seeking to maintain the flexibility, entrepreneurship, commitment, independence and innovation characteristic of small-group action, while gaining the clout accorded large organizations.

In our way, we can delineate some of the many ways SPIN does help movements adapt to their circumstances, advance against established orders, and generate change.

1. Such organization counters suppression by its opposition. Multiplicity of groups and leadership and lack of central control helps ensure the survival of the movement. It is difficult to arrest the movement by removing its leader because others will remain or new ones will come forward. Autonomy and self-sufficiency of local groups make it difficult for its opposition to gather intelligence information about it and to know what all of its groups will do.

2. Factionalism and schism facilitate penetration of the movement throughout a society. The variety of ideological emphases, group arrangements and tactics has something for most everyone, no matter what his or her preference in goals or methods might be. Yet, all of these can come together by integrating ideology and a common foe.

3. The multiplicity of group types maximizes adaptive variation during a time of threat or challenge. As each group does "its own thing" in its own way, each contributes synergistically to the success of the whole. In such a system moderates as well as radicals have a mutually enhancing role.

4. The presence of so many different groups often competing among themselves (but uniting in the face of a common enemy) produces an escalation of effort. As one segment of a movement emerges to take militant action and thereby to attract public attention, some other segments are motivated to upstage it. Thus, demands or concepts which are once viewed as outrageous, soon appear as relatively moderate and reasonable. And thus, today's radical becomes tomorrow's moderate. Establishments threatened by the newest radicals will seek to negotiate with the newest moderates, who can at least claim to speak for most people.

The dynamics of small group structure also contribute to the escalation of effort. Most movement segments are small enough to permit face-to-face interaction among participants in it. Participants can observe, evaluate, praise, or condemn the contributions of other participants to the operations of their small group. Participants can also observe how their own activities help or hinder these operations. This contributes to the striving of each segment, which, in turn, carries the whole movement forward as segments and leaders compete. Our findings here are in accord with the proposition (Olson 1965; Hardin 1979), that small groups are more effective in mobilizing the energies of members to achieve shared objectives than large collectivities within which individual efforts or dodges can escape notice or control. Networking can then integrate the smaller efforts to serve the larger purposes.

5. And, of course, another characteristic of such organization is that it does frustrate those who wish to negotiate settlements with it. Just as there is no center which can be identified for purposes of intelligence gathering or counterattack, so is there no center which can make agreements which are binding upon the various segments. One feature of this is that the movement as a whole does not admit defeat. Someone, some group, will always keep on fighting. While this frustrates established orders and blocks attempts to bring peace through procedure, it does keep the movement alive.

6. This organizational form contributes to the overall reliability of the movement. The presence of so many different groups each trying to act in its own way means both that many different tasks are done and that some tasks are done over and over again. While such duplication is often seen as inefficient by exponents of centralization, this in fact helps assure that if one group fails another will do the job. This is a system with many backups. Furthermore, since the various groups are not connected through a central command, the failure of one does not jeopardize the entire system. Competing groups will experiment with new tactics or ideas. If these work, they will be copied. If they fail they will be dropped and the movement disavow the errant group as too radical — or too moderate — or honor its martydom.

7. Because of all of this, SPIN promotes experimentation and innovation on the design and implementation of various kinds of sociocultural change, depending upon the particular forms and ideologies of the movement. This is a function of SPIN which can be broadly adaptive.

Recruitment

For new organizations to grow they must gain new adherents. Our term for getting these is recruitment, though movement participants often say it differently. Charismatic Renewalists, for example, rather speak about witnessing to others, or "bringing them to the Lord." Power line protesters talked in various ways about getting people out, lined-up, active, involved. How is this done? Popular beliefs have been that people become involved in movements through exposure to mass media, because they get caught up in mass hysteria, because they are swayed by a charismatic leader, or because they are drawn to it by their agreement with the ideology of the movement. Eric Hoffer (1965) projected the image that participants in mass movements are somehow weak-willed, searching for something, so that they are sucked up into the movement as if they are like pieces of chaff sucked up by a vacuum cleaner. In his famous model of revitalization movements, Anthony Wallace (1956) essentially ignores the process of recruitment, but implies that people flock to the movement as they recognize that their culture no longer satisfies various needs, and that it lacks vitality, order, clarity, power. It is a combination of the push of cultural disintegration and the pull of the revitalization message which makes the movement grow. In our study we found that activists like to explain that their movement is growing because people are just naturally being pushed into it by intrinsic flaws in the system and threats to their welfare. But they do not rely upon such push factors to gain adherents, and they do not rely upon the mass media, or mass hysteria, or mass persuasion. Instead, they actively beat the bushes for new participants. And it is "the bushes" which they know that they work on, and which provide the most recruits.

Individuals do not join movements as such: they join a specific local group or cell that is but one node in the total network. And, people are recruited primarily through face-to-face contact with a participant whom they have known before and who has been significant in some way, or at the very least can claim some personal ties. Members recruit others along lines of pre-existing relationships from families, friends, neighbors, associates, or colleagues. Recruiters use the capabilities, emotions, rights and duties already existing in the social relationships to influence other to join their cause. The movement grows exponentially as each new recruit becomes in turn a recruiter of others through his or her network of personal ties. Interestingly, this importance of pre-existing networks was recognized when officials of the rural electric cooperatives building the west central Minnesota power line explained that protest of the line was abetted by the "clannishness" of the German Catholics living near its route in Stearns County. These protesters used their ethnic and church networks to organize and recruit. The power line builders regarded this as a troublesome social fact which clouded objective assessment of the technological and economic virtues of the line.

Even when whole communities are exposed through the media to news of the impending construction of a facility which they could be expected to dislike, such as a hazardous waste disposal center, activists in emerging protest organizations still feel they must personally persuade people from the population to assure their participation. Similarly, when movement groups wish to draw crowds to a rally featuring the appearance of one of their big name evangelists, they do more than rely on mass media announcements and the fame of the evangelist. They encourage attendance by spreading word through the grapevine of local and regional segments of the movement. When Jerry Falwell visited Minneapolis for a big rally in 1980 it was such personal networking and recruiting which brought the people out to support him, to contribute money to projects he endorsed and in some cases to join for the first time the Moral Majority organization. It was also this type of networked communication and persuasion which brought people and groups out to demonstrate against the stand of his supporters on abortion, women's rights, homosexual rights, defense, and other issues.

Big jumps in organization size come when people in one group or movement persuade those in another that they share a common cause and should participate in each other's activities. Activists as well as researchers (Oberschall 1973) have remarked on this. And again, it is the personal touch which builds these interrelationships. One of the factors which binds groups together in networks is the participation by one or more individuals in both groups. These social bonds facilitate the development and sharing of ideological bonds and perception of common enemies, as we shall see.

Commitment

It is important for organizations to have adherents. But it is not enough. The adherents can still "let George do it." Successful recruitment into small groups still does not overcome the "free rider problem" so often bemoaned by collective action theorists (Olson 1965, 1971; Hardin 1982) as well as movement activists. It is also necessary to motivate people to act and to keep acting. What we shall discuss below as ideology contributes to this. But this is still not enough. People in a movement talk about commitment. This is as good a term as any for the necessary ingredient for action.

We have found that in any of the different types of movements we have studied there is what we term a process of commitment. For many participants in many movements this is an unplanned, spontaneous process which can be analyzed only in retrospect, and which goes largely unnamed. In other cases people can be consciously led through the process by the already-committed, who tend to interpret it in terms of movement ideology. This process of commitment has three important components:

The first is that of experience. Individuals undergoing commitment have one or more subjective, highly emotional experiences after which they feel themselves to be different and to understand their relationships with others differently. There is a change, sometimes a significant one, in sense of identity.

The second component is an act, or series of acts, which burn the doers' bridges to past patterns of behavior, which cuts them off in some significant way from the conventional social order or their previous role in it, and identifies them with other committed participants.

The third component is the social context of these acts and experiences which supports their interpretation as some kind of separation and transformation. People are committed not only by risky acts and identity altering experience, but by the support of "soul brothers and sisters," who join them in defining their action in positive terms, according to movement ideology.

We found that the most well-constructed and conceptualized forms of this process were in the religious movements we studied. Scholars have referred to this process as conversion. Activists in the movement we studied usually talked about it as finding the Lord, as being "Saved," or being Born Again, or being Baptized in the Spirit, or simply becoming Christian. A commonplace event in charismatic groups is that those newly awakened and saved will witness to others of their act and experience, and receive strong verbal and often physical support for this from their "brothers and sisters in Christ."

The process has also been recognized more or less concretely in various contemporary social activist movements, including those over technology and resource management. In these the process is often called radicalization. Farmers who faced arrest when they stood fast on "the Hill in front of Jenk's place" to stop the power line surveyors explained later that now they knew what anti-war protesters meant by being radicalized. It had happened to them when they were threatened with incarceration in the town jail — the "slammer" — by officials. So they called the place "Slammer Hill," until some said that they were going to be arrested simply for fighting for their rights, so they renamed it Constitution Hill. Symbolic interpretation and commitment go hand in hand.

In the technology protest and the social issue movements we have studied, people enter confrontations with established orders both to make their statement to the world, through the media, and to take members through the commitment process, perhaps for the first time, perhaps as a renewal of dedication. Some of these confrontations seem to reflect a breakdown of orderly decision-making procedures, but it is recognized by many that confrontations have become very stereotyped and ritualistic. In July 1985 the Minneapolis/St. Paul affiliate of NBC presented as a part of evening news a short special about the evolution of confrontation tactics from the anti-war to the nuclear freeze movement. The news editors suggested that there is now a cadre of professional protesters in the Twin Cities, who are active in many different movements, and that they have indeed made confrontation into a ritual. Protesters interviewed agreed that the police have also learned how to deal with protest and have joined the protesters in making it a ritual. In this, for example, agreements are made beforehand about what will be done to get arrested and how many people will be taken in arrest. Apparently some protesters now feel that they must find new ways to demonstrate which again exceed the expectations of the police — and again put the demonstrators at risk.

Public hearings and meetings designed to head off or mediate disputes and ensuing confrontations can themselves be construed by protesters to provide the experiences which commit. People who have never spoken in public before now, suddenly are called upon by their fellows to get up and speak before official decision-making authorities. In preparation for this they meet with their protester fellows to lay plans and tactics. Preparations frequently also prepare them to see authorities as people who have already made up their minds; who will try to "put down" the citizens, particularly if these are women. People may agree "not to be emotional," but to present the facts, since this will have a better effect on the authorities and the observing media. But at the event itself emotion does often run high. People are encouraged with applause, and vocal outbursts by their fellows when they make telling — or emotional — statements, just as the authorities are interrupted with negative comments. This often provokes authorities to call for order, or to seek to assert control by cutting off discussion. All of this can make of ostensibly simple procedure an event of drama and commitment process.

The commitment process exists; the question remains; what does it do? An important question is unresolved. From our field research we have argued that it leads to changes in personal identity, and with this to changes in the behavior in accord with movement goals. Some scholars and many activists argue in similar direction.

For example, anthrolopogist Anthony F.C. Wallace (1956, 1966) describes a process like what we call commitment by which people can be led to make radical changes in orientation. He calls it a ritual learning process, and he considers it basically different from the traditional learning process. In it people learn through the restructuring of cognitive and affective elements, and they become changed persons — at least as long as they are reinforced in this by others. Anthropologist Ward Goodenough (1963) also explains how people in traditional societies experience an alteration in self-image and cultural identity through ritualized process and suggests that from such personal change people will be active in modernizing social and cultural change.

Activists in the Charismatic Renewal and in what we are calling CCA say that those who are brought to Christ through their particular conversion processes are changed persons and do contribute more to the mission of the movement, and do behave in ways which, by their example alone, shows others the way and act as a leaven to keep God from turning his back on the country (LaHaye 1980; Schaeffer 1976). Liberal social activists criticized the Charismatics and Evangelicals in the 1960s and 1970s for not being concerned with the social and economic problems of the world. The response of the Charismatics has always been that once people are personally transformed they can then and will then work in the secular realm to make life better (Stott 1975). Analysts of Pentecostal movements in the developing world say that converts to these do tend to move ahead economically, and do so chiefly because of the values instilled by their new faith, much as did Protestants in the sixteenth and seventeenth centuries in Europe (Willems 1964, 1966). Historian Terence Ranger reported in a colloquium at the University of Minnesota in Spring 1985, that a forthcoming publication by anthropologist Jean Comaroff will argue that Pentecostalism among the Tswana of South Africa has served them as a vehicle for political rebellion as well as economic gain.[3]

I find that a study conducted for Resources for the Future (RFF) by Cameron Mitchell suggests the importance of commitment as I theorize it. Mitchell analyzes how people across the U.S.A. showing environmental interests and membership in environmental groups responded to direct mail appeals for financial aid from national environmental lobbies. The lobbies asked for contributions to protect environmental resources from various risks of pollution and development. Mitchell says they contributed in excess of what would be predicted by Mancur Olson's (1965, 1971) collective action theory. The national lobbies are organizationally too removed from people on their mailing lists to reward contributors, for instance by praising them before their fellows, or to punish noncontributors, for instance by calling group attention to their omission. Mitchell explains the apparent Olsonian illogic of their action as actually rational response to their perception that they will experience costs or "bads," if they do not contribute, namely the loss of these resources. He considers this to be more important than motivations of guilt or self-esteem, though he suggests how these also can work. For instance, before reading an appeal from an environmental lobby a responding individual may already have felt a responsibility to "increasing environmental goods," and experience remorse that he had not yet done so." In addition, an individual may be positively reinforced by feelings of heightened self-esteem if he does contribute.

Based on his own research (Tillock and Morrison 1979), Denton Morrison suggests (1979) an addition to Mitchell's analysis. Morrison feels that Mitchell's alternative to Olson's pessimistic assessment is "on the right track," but could be improved by distinguishing among types of "battles" waged for "new goods," "lost goods," and "threatened goods." Environmental goods are "threatened goods," that is neither new nor yet lost. Fighting for lost or new goods is much more difficult, he suggests, than for those threatened. The social movements which fight for the lost or new will have much more trouble securing active support from collectivities than the latter. This approach implies that it is important that environmentalists do see environmental goods as being threatened, but still protectable.

In his critique of Mitchell, Russell Hardin (1979) defends Olson's theory as verified rather than disproven by Mitchell's data and analysis. He seems to admit that some (Sierra Club) environmentalists will behave "extra-rationally" because of their particular ecology orientation. But he does not find this an anomaly requiring much explanation. A theory of rational calculus will accurately predict most collective behavior, he argues, and do so from efficiently small amounts of data. It is not necessary, he says, to "wallow" in the vast quantities of data required by most social scientists to make analyses and predictions using Olson's elegant economic theory.

I suggest that Mitchell, Hardin, and Morrison are each trying to interpret rationality according to narrow criteria of cost and benefit. Mitchell may be closest to the mark in proposing that environmentalists perceive environmental goods and bads in novel ways. But perhaps this is what Hardin means by extra-rationality.

My models of movement organization and commitment (but also ideology) provide explanation to elaborate Mitchell's analysis (and to clarify the "extra-" in Hardin's "extra-rational"). I suspect that many of the people who contributed to the national lobbies had been involved in local environmental groups — segments of the environmentalist SPIN — and that it was this local involvement which did give them reinforcement for environmentalist actions. I also suspect that these people had been committed to environmental causes through participation in these local groups, and this commitment carried them to "extra-rationality," or rather that it changed their calculus of cost and benefit. A person committed to environmentalism will have perceptions of the environment and obligations to protect it which are different from those held by others. Our exploration of movement ideology, below, will amplify this. Movement ideology shapes the identity changes which come with commitment.

Ideology

Participants in a movement may not like to use the word ideology for their idea system. It connotes, at least in our society, something contrived, maybe distorting, maybe even devious. Participants will rather talk about getting their facts straight, *the* facts straight; and about digging at what causes the problem. They will talk about "getting the message out," and informing, persuading, motivating, even "sensitizing." Or understanding, believing, and knowing the facts, the truth.

But to us, as scholars, the term ideology is useful. In this study it is a system of ideas, a system itself and part of the system of the movement, growing at it grows, helping it grow. One of its characteristics is that it informs the two major levels of a movement, namely the segment or local group, and the movement as a whole. It is "split level." At a micro-level, each group develops its particular ideas which both stem from and guide its particular endeavors. Furthermore, groups or factions come into being over division in ideology. It is this which allows ideology to be adapted to local tasks and also which perhaps most sparks the creation of new, different "wild" ideas. Yet, at a macro-level, all groups share in central tenets or concepts, and it is this conceptual unity which helps glue the various groups together.

Movement ideology, particularly that at the macro-level, deals in part with the state of society, culture, and environment in the past, present, and future. It provides a vision or master concept of the ways things should be and can be in the future, even as it identifies flaws in the fabric of contemporary ways of living and believing. It also makes specific claims about particular problems and prospects facing local groups. In technology and resource management disputes, for example, technology itself is attacked as a Frankenstein-like force "out of control" which has come to dominate humans instead of serving them (Winner 1977). Specific technologies are protested for their particular risks. Farmers warn of the "as yet unstudied ways" electromagnetic fields from high-voltage transmission lines threaten the health not only of their livestock but of their families (Rebuffoni 1986).

In Conservative Christian Action, the world is pictured as threatened by "secular humanism," which does the bidding of Satan by deluding humans into believing that "Man is the master of all he surveys" instead of knowing that God is Lord. A particular group of activists seeks to force its school to remove from the curriculum cross-cultural case studies of family life, because these show that "anything goes," and delude their children into believing that "there are no absolutes."

Such ideas and models are often present in easily communicated terms, as metaphors, slogan, jargon. Basic themes are repeated, many times word-for-word. Outside observers find it easy to disparage these as "mere rhetoric" or the "party line." For movement participants, however, they are often a codification of common and deeply felt experiences. This codification facilitates communication to new recruits, helping them to interpret their own personal experiences in terms of movement ideology. It also provides patterned answers to questions raised by outsiders, reinforcing ideological solidarity among participants and strengthening their defenses against criticism. Converts have ready-made and time-tested answers to most objections. It is not surprising that the objections raised to a movement often become just as patterned as the "party-line" statements, and that some of these established objections are repeats of traditional cultural themes. Thus, we should not be surprised that protesters of particular technologies attack these as "hard," and claim that there is no "need" for them if we "stop worshipping growth for growth's sake" or find "soft" and "appropriate" "alternatives," or that proponents find all of them "unrealistic,' and even that "you can't stop progress," and that protesters "cannot believe that people still say this."

Anthropologists have elucidated and argued the importance of metaphor, ritual expression and other ways of codifying and communicating fundamental cultural models (Firth 1973; Lewis 1977; Sapir and Crocker 1977). Some have stressed the role of ritual and symbolism in maintaining established culture and society (Turner 1967). When we shift focus to the phenomenon of protest and movement we recognize that symbolic, ritualized expression also plays an important role in challenging and changing ways of life.

People and groups use ideology to challenge and change established ways essentially by directing ideology to some major purposes, including those of empowering and communicating. One function of ideology is, as we have indicated, to give participants the conceptual tools to become strong in their beliefs and positive in their opinion about their role in the movement, and about the role of the movement in society. A related function is to gain support and positive evaluation from the concerned public, the general public and the media as well as other opinion-shapers, and then to sensitize and shape the opinion of these audiences. In the case of the Black Power Movement, the concept of "Black Power," "Black Pride," that "Black was Beautiful," helped to instill a positive image among a broad black population (Barbour 1968; Carmichael and Hamilton, 1969) and to sensitize whites to the evils of racism. Along with other aspects of this movement, it surely helped spark the rise of other rights movements among other "minorities."

Generally, movements, or rather media reports of them, in the U.S.A. have reinforced each other by making people think more acceptingly about the right and the need to protest. The Ecology Movement certainly gave so many people some understanding of ideas like ecology, interdependence, pollution, the environment is fragile, "you can't change just one thing," the unanticipated consequences of seemingly innocuous development projects, limits to growth . . . . Thus, it is not surprising that in fact actions more bounded to immediate neighborhood protection should by extension become identified as environmentalism. People, including media reporters, learned to use or simply to mimic ecology terms to get on this thematic bandwagon. It often helped them, just as, for a time, it helped spread the environmental ideology. When the media termed the d.c. power line protest an environmentalist struggle, they were doing what had become commonplace.

Motivating and Persuading Ideology

Movement participants want to do more than stimulate general support and gain a more positive image. They want to recruit new participants to join them, not only to swell the size of the group and its war chest, but to become committed enough to strive and drive for the cause. Ideology is developed and used to persuade and motivate in these processes of recruitment and commitment. Let us summarize this as motivational ideology.

This returns us in part to the distinction made between ideology at the micro- and macro-levels. Motivating people to become involved and to stay involved requires the application of ideology from both levels. Those charged with recruiting newcomers to participate in the activities of a protest or movement activity characteristically say that it is necessary to appeal first to very specific, immediate, selfish interests. Only this, it is argued, will get people out, away from their other pressing concerns. This is particularly so in technology and resource disputes, when people in a community are suddenly faced with a projected impact. Micro-level ideology, the particular ideology of the local groups, is tailored to do this job. Those doing the recruiting in collective action may themselves need to be reminded of how this concerns them, immediately. Often, at this time and level, people will know little about the proposed project or its impact. Our studies and personal experience shows that at this point the activists and recruiters go straight for threats to the pocketbook, the customary qualities of the community, health, and welfare. They also bring up the recent history of other attempts to bring development or management to the community, and warn that the same officials and conniving interests are at it again, and must be stopped. This assault with warnings of risk and malfeasance may have the additional effect of delaying decision-making on the topic while further studies are conducted. This gives the protesters time to put together a better effort. It is with this goal of delay in mind, indeed, that they may enter public hearings. Yet, while these particularistic ideas and warnings are important at first, and continue useful throughout a dispute event, they are soon seen as not enough. So, people turn to macro-ideology.

People search for better ways to understand the problems they are facing. They look for explanations which place their struggles in a larger class of events, and are motivated by these. For example, a traveling evangelist in the network of Conservative Christian Activists explained how she now goes out to give people in local communities the tools to understand many local problems of school and community as manifestations of secular humanism. This motivates them from a larger perspective to be involved and *stay* involved. For example, whenever there are questions about school course materials, about the teaching of evolutionism, or the operation of family planning and sexual counseling in the school or community, or of the opening of an abortion clinic . . . Also, protesters of the d.c. power line found renewed motivation to keep on fighting by placing their effort in the context of the struggle to keep family farming alive. The line, they warned each other, would not only threaten their health with its emission it would interfere in many ways with their operation of equipment and irrigation systems on their fields and take land out of production. The primary purpose of the line, they came to agree, was to put power into the midwest area power pool grid, from whence it would serve urban and suburban interests. They had a duty to protect rural life and agriculture; after all, they were feeding the world. And with this idea, they could find common cause not only with other power line and power plant protesters, but also with the American Agriculture Movement.

Legitimating and *Leveraging Ideology*

Movement ideology also serves legitimating and leveraging functions. It declares not only that the movement cause is legitimate, but that the means used by its participants are legitimate, even though they are irregular and are criticized by the established order as being anything from "too emotional" to "absolutely illegal"; certainly "not according to the rules."

Movement participants claim that they are justified in using these means because their cause is so just, because they are acting under a higher law, and serving the broadest public or human interest. They also may claim that the monopolization of power by the established order forces them to use unconventional tactics. Counter-attacking, they declare that it is the established order which is behaving illegitimately and illegally. The more that movement ideology develops an alternative view of the world, and what culture and society should be, the more that it can declare that it is the established order which is blocking progress by its adherence to old ways and means. Activists may break the rules of decision-making because they are unaware of them, but it is more likely that they recognize that these rules help the powers-that-be to be powerful. For some movements, such as the student anti-war, anti-draft protest, rule-breaking became the rule, not only as a tactic, but as a strategy to expose the artificiality of the rules.

In a similar vein, movement ideology helps support movement tactics and strategy by accomplishing various "leveraging" functions. Movements, especially in their formative stages, are objectively much weaker in power and authority than their established opponents.

"We are like the flea fighting the elephant," explained a leader of the Infant Formula Action Coalition mobilized in the 1970s to protest the infant formula marketing methods used by Nestle and other giant multinational food and drug conglomerates (Gerlach 1980). How can the anti-Nestle fleas hope to win? How can David resist Goliath, the guerrilla overcome the superpower? It is usually not by fighting according to conventional rules of engagement, since these favor the established force. It is by using a sociopolitical form of the principle of judo. It is by turning the power and weight of the mighty foe to the foe's disadvantage. It is by leveraging the foe out of balance on the fulcrum of upsetting ideas. A favorite kind of fulcrum is "system" guilt, often the guilt associated with institutionalized hypocrisy. Under the probe of movement action, "the universal gap between principle and practice appears in all its nakedness" (Kenniston 1968). It is important but not enough that protesters believe that it is the establishment, the system, their opposition which is guilty. It is also important that those who are being protested feel this. One function of ideology, then, is to force this awareness. During Civil Rights and Black Power days' "whitey" especially the white power structure, was made to admit of institutional and often of personal racism. Blacks and whites alike spoke of the way white guilt then was used to lever change.

In technology disputes protesters claim that the established order should be "ashamed" that they are making people be guinea pigs to "untried" technology. But proponents of the technology are likely to feel equally right. Much is made of risk probability assessments, and what these really mean, and how they are understood by different publics. Rayner (1984) is surely right when he says that most people, like himself, can't really grasp the difference between a probability of 10^{-7} and 10^{-8}, or the significance of these measurements. But when people are mobilizing against a project, any probability is construed by them as unacceptable, as obscene. Again, in the d.c. power line case, a succession of Minnesota governors agreed that they themselves would not want the line in their backyard, but that is was sadly necessary that someone carry this for the public good. This, of course, helps make "need" and its determination a vital concern in the legitimization struggle.

Perhaps some movements begin because their leaders have in mind a master design for reshaping the world. But in many of the ones which we have seen develop in the U.S.A in recent years, the design has been a product of action and selective adaptation. For example, as protests evolve to become movements the quest for legitimization contributes to production of alternative cultural models. We see this in technology and resource management disputes. They may begin as ordinary people wish simply to stop or change a specific project which will affect them in the pursuit of traditional cultural goals. But it is common in such disputes for protesters to say that they must do more than simply find fault and say no. Reflecting their articulation to traditional America culture, they recognize that it is not enough to "be negative." It makes them feel uncomfortable. They are challenged in their small group to invent something better. And also they know that this is necessary if they are to keep challenging the establishment and win the good opinion of the public.

Under these conditions and pressures people do search for and propose alternatives to the technologies, projects, or management programs which they protest. Sometimes they innovate alternatives more or less from the bottom up, engaging in what has been called "lateral thinking" (DeBono 1968) since they view problems and possible solutions in new ways apart from the conventional mode or "mode lock" (Gerlach and Hine 1973). Often it is a matter of building rather freely on ideas generated elsewhere, perhaps diffused more as stimuli than as concrete models. For example, during the d.c. power line protest farmer protesters learned about Nicola Tesla's idea of broadcasting electrical energy without wires. It was suggested to them by a participant in the range of urban-based, counter-culture, alternative lifestyle activities of the day. A few farmers then agreed to collaborate to build such a wireless transmission system with participants in a Minneapolis alternative lifestyle collective. Some members of the collective had grown up in west central Minnesota, on farms. Many members worked in various food cooperatives in the city and wanted to build closer ties with farmers as potential suppliers of organically grown produce. The farmers and the collectivity participants did work together and did build "a Tesla receiver" to receive electricity which would be used to power an irrigation system. The electricity was to be sent from people in Timmons, Ontario, who were interested in the project mostly because they were Tesla buffs (and had the notion that Tesla's most advanced ideas had been suppressed by a conspiracy financed by the copper industry barons — who wanted to sell copper transmission wire). With this project the farmers and their allies sought to show that there were technologies — high technologies — which the established Rural Electric Cooperative Associations had not even considered. The project was completed, but did not work. The builders did not consider it a failure. And yes, it was possibly more risky than the official d.c. power line, and yes, through building it a farmer's field was disturbed maybe as much as it later was when the DC line was erected. But to the Tesla builders their effort symbolized not only their opposition to the d.c. line. but their claim that this resistance was not simply a result of their unreasoned fear of modern technology. To them, it was the d.c. line proponents who were the backward ones. This was one more reason to claim that their resistance to the line was certainly legitimate — even when their demonstrations and blockades led the governor to deploy hundred of state troopers to enforce the construction order, and even when the Rural Electric Association asked that the line be placed under the protections of the FBI to stop persisting sabotage of the towers even after the line had been completed.

It is not surprising that these line protesters joined with others protesting lines and energy facilities elsewhere to invite Amory Lovins to tell them more about his proposition that there were alternative "soft energy paths" to follow rather than the conventional "hard paths." It is also not surprising that Lovins referred to these protests and other "energy wars" to show that the conventional paths were indeed "hard." Similarly, other researchers have argued that resistance to energy facility production is evidence that the Western industrial world has reached some sociopolitical limit to growth. Ezra Mishan (1977), for example, argues that in order to continue to drive the economy forward the western nations will need to use totalitarian measures to control resistance to develop projects, particularly those associated with nuclear energy, and to guard against misuse of nuclear materials. A similar thesis is advanced by Junck (1977). Both Mishan and Junck hence seek to attack the conceptual underpinning of the proponents of development and growth. The proponents have long claimed that growth is necessary to expand the economic opportunities which have in turn supported Western political freedoms and social peace (Ford, n.d.; Kahn in Oltman 1974; Kuznets 1966). The growth critics seek to turn this argument on its head: growth puts freedom at risk. We have here a case in which mobilized protest is legitimated by concepts which are in turn supported by or validated by such protest. Such mutual causality (Maruyama 1965) is typical of movements and one of the forces which produces sociocultural debate and change.

If we reflect upon the course of the "No-Growth Movement" we see this illustrated. In the 1970s people (including ex-president Ford in an address to the American Enterprise Institute) were speaking about this as a movement in and of itself. The big idea of "no-growth," or "limits to growth" itself grew from two directions, from the bottom-up actions of local groups seeking to find reasons to attack specific development projects, and from the top-down pronouncements of specialists and theoreticians making what became macro-ideology. For instance, on their own, people protesting nuclear and fossil fuel power plant construction in Minnesota in the late 1960s argued that energy consumption and hence energy production did not have to grow if Minnesotans' changed their mindset and began to conserve through changes in behavior and technical efficiency. But, when the experts at MIT pronounced the Limits to Growth (Meadows *et al.* 1972) principle and portrayed it with curves and graphs it became a very powerful and useful macro idea which people could then use either to reaffirm what they on their own had hoped to find, or to spark their new thinking.

Integrating Ideology

Indicated throughout our discussion so far is another use of ideology; that of contributing to the integration of the segments of the movement. We have summarized movement organization as segmentary, polycentric, integrated networks, or SPIN. It is ideology which is a big integrator. It provides the ideas which people share, and which bind them into a conceptual community. SPIN could also stand for a segmentary, polycentric, ideologically networked entity. Particular segments or local groups of course share their particular ideology, which helps them hang together, But it requires some large, master concepts to integrate the diverse segments into one system, under one umbrella. One of the characteristics of ideology is that it is often more ambiguous than tight and explicit. What, for example, is the meaning of "Participatory Democracy," or "Appropriate Technology?" Who would argue for inappropriate technology? Haven't we always had "participation" from the people? This ambiguity bothers many observers of movements, as well of course as those who are protested. At the very least it can be interpreted as producing debates which are called "Trans-Scientific" (Weinberg 1985) But it is this very ambiguity which helps different people and groups to accept the idea and to make of it their own thing. And it is this which helps in integration.

Opposition

Ideologically integrated, motivated, and legitimated; organized for predatory expansion through SPIN; recruiting and committing new participants — movement activists challenge not only projects and principles, but the proponents and perpetrators of them. They stand against an opposition, part real, part perceived and defined through the filter of ideology. Identifying, fighting, and defeating this opposition is not only a goal of the activists; it is also a factor which helps launch and then drive the movement system. No matter what the type of movement, there is a characteristic "we—they" ingroup-outgroup orientation on the part of most participants. A movement grows with the strength of its opposition much as a kite flies against the wind. Opposition, real or perceived, is necessary to promote a movement, to provide a common enemy against which it can unite its disparate segments to offer a basis for its commitment process.

We found that activists in social movements recognize this principle and express it in their basic doctrine. The charismatic, for example, characteristically warned new converts that they would face ridicule and worse for their beliefs and practices, such as "speaking with tongues." But they urged them to welcome this. After all, the converts can know that they are right. The Old Testament tells how such Baptism in the Spirit was part of the conversion of the first Christians (Acts 2, 10, 11, and 19; 1 Corinthians 12 and 14). And they know that the early Christians were called upon to be martyrs according to Christ's injunction: "And he that doth not take his cross and follow after me, is not worthy of me. He that findeth his life shall lose it; and he that loseth his life for my sake shall find it" (Matthew 10: 38—39). One indication that interest group protest is becoming a social movement is when it begins to turn the risks which should conventionally act to deter it, into its ritual of transforming commitment and a symbol of its unity.

As we think of this role of opposition now, it seems so obvious. Surely it must be generally understood by established orders as they seek to deal with insurgency against them. After all, most established orders share a national or religious history in which the very order they adhere to was born of "heroic" protest and movement — against a "tyrannical" opposition, overweening in its power. But time and again establishments do seem to fall into the trap of providing this necessary ingredient.

The police come out in Chicago during the Democratic Convention to control the anti-war protesters. The U.S. air force drops its bombs on suspected guerrilla bases in Viet Nam, the Russians wipe out Afghan villages, the Israelis, searching vainly for the center of its Palestinian movement opposition, invade Lebanon and take over Beirut. On a less catastrophic scale, the rural electric cooperative association builders of the d.c. power line in Minnesota hires a private investigator to harass the power line protesters with dirty tricks, and in the public hearings on the line the attorney for the cooperatives seeks to discredit the testimony of the protesters by pointing to their lack of technical training and implying their emotionalism. Or, the established Missouri Synod of the Lutheran Church seeks to stop the Charismatic Renewal challenge growing among some of its ministers and congregations by accusing these of theological and doctrinal error, and removing them from the Synod if they persist in such "error."

But as we think about these and other illustrations, we recognize that established orders face a dilemma more than a simple trap. If they seek to oppose the movement this provides the opposition which helps the movement grow, but if they do not try to stop it, their nonaction and the signals this gives out may also allow or promote movement growth. Could a Church be seen to condone alternative doctrine? If the attorney for the power cooperatives allowed damaging testimony from protesters to stand, would not this weigh against the position of the electric cooperatives in the public hearings and in legal cases? If military forces do not use their strongest weapons do they not put their men and their mission into jeopardy? If nations do not seek to retaliate against terrorist attack or if established orders do not show that they are able to maintain law and order as they define it, do they not seem to encourage attacks on their position? And do they also not risk having others in their country take some type of vigilante action?

In Viet Nam the U.S.A. sought to escalate slowly its counter-action, to provide some kind of measured response. This seemed simply to be met with measured escalation by the Viet Cong. But had the U.S.A. resorted to instant, massive action this would surely have been met with protest across the U.S.A. and the world. In 1977 in West Germany, government did successfully use the threat of massive action by anti-riot and border police to encourage protesters of a proposed nuclear waste management facility in Gorleben, Lower Saxony peacefully to terminate their occupation of a test site. But this may have worked because most of the groups in the SPIN organization of the protesters were pledged to nonviolent resistance.

In the d.c. power line protest in Minnesota government authorities knew that the use of state police to enforce the line construction order would likely raise the ire of the protesters and even make them dig in harder. But, they wondered, what other option did they have but to use force once they decided to build the line? The electric coops threatened to sue government if it did not protect their surveyors and workers. The coops have backed off, but it seems that they felt that their honor and credibility was at stake as well as their line. During their encounters with the protesters in hearings and courts they had themselves become as committed to fight as were the protesters. This kind of mutual escalation of opposition and commitment is a hard fact of movement dynamics and a powerful weapon in the hands of those who understand it.

In this power line case, even when some people seemed willing to come to terms to avoid further escalation and threatened violence, others were not. And once the electric coops advanced to survey and their protesters took a stand on Jenks/Slammer/Constitution Hill, there was no stopping further escalation. Word of this resistance spread rapidly by telephone and word of mouth across the movement grapevine and then beyond, through the media. Others who were in the network of resisters came out to join the stand on the Hill. Yet others became involved for the first time. Some started their own local groups, especially when surveyors sought to work in other counties, where people were supposedly acquiescent or at least quiescent. Once this happened, the best that could be hoped for, for a time, was that the police and the protesters could manage to conduct the confrontation with relative care and nonviolence.

The Minnesota case was complex enough to frustrate any attempt to manage conflict. At one point the line construction workers prepared to fight to defend themselves form the protesters and, it is alleged, to intimidate them. The governor odered the local sheriffs to swear in new deputies to enforce the construction order but also to stop any party from using force. When sheriffs said that they could not do this, since it meant fighting their own local people, the governor considered calling up the National Guard, and the Guard readied itself. But this seemed too drastic and symbolic a step to take, particularly when Guardsmen from this normally patriotic area said that they would also find it hard to take action against their kinfolk and neighbors in the protest. Finally, with great regret, the governor despatched state police — eventually several hundred — to do the job. It seems that the state police were trained and indoctrinated to act with the greatest caution. Though the analysis they used was probably different from that presented here, they certainly were aware their action could make things worse, could spark escalation, and that they would became not intermediaries but the symbolic and actual opposition. The protesters met them at times with hot coffee, donuts, and daisies (yes, plastic and paper daisies after the fashion of campus anti-war demonstrators) on the one hand, and at times with jeers — "Rudy's Redcoats" (one of the three governors to face this six year struggle was Rudy Perpich; the state police have maroon coats) — and at times with physical violence and even the spraying of chemical fertilizer. Police elsewhere, apparently across the world, have been trained to deal with protest by using procedures which are supposed to reduce committing, oppositional effect. Furthermore, officials sometimes advised by former activists now working as participation consultants (Bleiker 1978) have been seeking to improve procedures used at public meetings and bearings also to reduce the likelihood that these will exacerbate rather than counter protest. One former environmental activist now working as a consultant to help decision-making clients design more effective public participation procedures explains that while using these better methods involving the public may not necessarily produce agreements on a controversial project, not involving them would "guarantee failure of the project" (Bleiker 1978).

The cautiousness of this statement indicates that even a consultant on participation procedures knows that it is not process but protest movement which calls the tune.

Take-Off

As we have analyzed it, a social movement is a system produced by the action and interaction of its subsystems: organization, recruitment, commitment, ideology, and perception of opposition. The subsystems can each be dissected to examine their special characteristics. But in an actual event they are interwoven. It is through this interaction that they work or drive a movement forward. For example, commitment is generated through risk taking against an opposition which is defined by ideology as dangerous but not legitimate. It is shared ideology and opposition which integrate organization, even as differences in micro-ideology, and the actions of committed persons spark the birth of new groups and new leaders. Ideology and commitment drive people to go out and recruit others. But is is an act of commitment to face others and to try to recruit them. In the exchanges which take place in this process, people refine and elaborate ideology. It is SPIN organization which enables the movement and its ideas to survive and spread. In these and many other ways the subsystems of a movement are related in mutual causality. Each affects the state of the other through feedback. As movements grow, the mutual causality is of a type Maruyama (1965) terms deviation — amplifying, that is, the feedback is positive. As movements routinize (Wallace 1965) and stop growing, presumably the feedback becomes negative.

Movement growth — and decline — is also of course affected by the interaction of movement factors with established orders and the general sociocultural system within which it exists. Just as the parts of a movement interact with each other so does the movement as a whole interact in mutual causality with these larger systems. Established orders will characteristically seek to curb movement growth. People usually define such growth as a sign that something is wrong, very wrong, in the establishment, or among the movement participants, or both (Hine 1974). Established orders hence will intervene in ways designed to counter the deviant movement, and to prevent its further growth. But, there are times when this intervention has an opposite consequence. Deviation is increased, the movement grows. People know when this is happening. Those in the challenged established orders talk about being "damned if they do and damned if they don't" as they seek to respond to movement demands and actions. They talk about being caught up in "vicious circles," or "downward spirals." They seem to know that — for a time, at least — anything they do with the movement makes things worse, at least for them. But even participants in movements may feel that they also have lost control. The situtation is dynamic. To paraphrase W. W. Rostow (1961) the movements will have "taken off," from the conditions which gave it birth.

A most obvious type of counteraction is to apply force. It is often the proposed solution of first and last resort. Our analysis of movement factors shows us how the amount of force which established orders in the West can apply will likely amplify movement growth rather than control it. Established orders find it difficult to apply enough publicly sanctioned force, against the right targets, to protest. Instead, what they usually provide is enough risk to provoke commitment from resisters and sympathy for the resisters from the public. If they apply enough sanctions broadly and powerfully enough to stop the many segments and leaders in SPIN, they are likely also to hurt and then radicalize uninvolved bystanders and be accused of overkill. But if they are quite selective and cautious they will control only parts of the SPIN resistance and likely face a remainder which is more adapted and capable. If they escalate their application of negative sanctions slowly by stages, with due warning, they can claim to be acting responsibly. But then the movement can adapt steadily, and continue to grow. If they strike suddenly, and massively, to crush the movement before it can cope, again they are likely to lose the public relations battle as they are effectively accused of overreacting. Once a movement has "taken off" established orders find it difficult to win any kind of public relations battle, even if it is only a battle of words. Nestlé Corporation was hurt, not helped, when before a U.S. Senate hearing chaired by Edward Kennedy, one of its multinational executives from Brazil claimed that the campaign to curb abuse in infant formula marketing was a Marxist-inspired activity. Kennedy looked at the nuns from South America who were testifying to this alleged abuse and said in effect: What you mean is these sisters are communist agents.

In Western culture these days, forceful sanction, be it physical or verbal, does not provide a good solution to protest movement. Indeed, it is likely that most people wonder why it would be proposed as a solution instead of as an admission of failure to remove the causes of protest and to achieve negotiated settlement of dispute.

It is popular now to speak of solving disputes in ways in which everyone can win. This returns us to our proposition that disputes are most popularly conceptualized as problems in procedure and in risk assessment. This fits Western culture because it directs that solutions be sought in more, not less, participative democracy (Naisbet 1982; Nelkin and Pollak 1979) and in more, not less, science, technology and technique. Public hearings in the U.S.A. (called inquiries in the U.K.) have been steadily improved to provide just this; more carefully designed participatory procedure and scientific-technological assessment; more involvement from broader constituencies; and more accurate contributions by appropriate specialists. Where possible the search for solutions will include attempts to convert the claims of each side to that which can be roughly measured and stated in cost/benefit numbers so that the contesting parties can negotiate about trade-offs. One of the authors of the 1980 Waste Management Act in Minnesota said of the participatory procedures to be used in implementing it that these will "give people lots to negotiate about." Other contributors to the Act and the design of its implementation procedures explained that the public meetings and hearings to site hazardous waste management facilities would produce "focused debate on trade-offs" (Gerlach and Meiller 1986). Presumably through these negotiations and trade-offs people would agree upon some compromises in which all would win.

We can see the logic behind the development of such a design of participation informed by scientific-technological risk assessment. But if we examine its implementation, its actual interaction with movement factors, we see that it also is converted by these factors into being that which makes a movement grow, not stop. Just a few examples will show us this.

Simply stated, movements "use" the participatory events in public hearings as key episodes in organizing, recruiting, and committing their participants and then in developing and practicing their ideology and in defining opposition. For example, one of the first occasions people have to come together in emerging resistance groups is to plan for their appearance at public meetings and hearings. They will characteristically use these initial gatherings to remind themselves and others that it is necessary to "get the numbers out." This gives impetus to recruiting efforts, and helps them develop some initial motivating ideology. During preparation gatherings individuals will volunteer or "be volunteered" to prepare for the public events and to present testimony at the events. People will learn more about the capabilities and concerns of each other as they develop these statuses and roles. If some participants have been active in other kinds of political activism, perhaps in previous protest events, they are likely "to share" this and to help the new group build on this base of experience and to tap into the earlier information and action networks. This certainly happened in west central Minnesota as people prepared in 1981 to participate in public meetings on waste management facility siting by building on experiences and networks developed on the d.c. power line fight from 1974 to 1980.

Organization will be further elaborated as people do attend the public meetings and hearings and then as they review these events and react to their interpretation of the results. As established orders observe this growth of resistance they may wish to avoid participation, but general public opinion forces them to continue. Hence they agree that they simply did not do participation effectively enough. That is, they cannot admit that participation provides no solution, so they search for ways to improve it.

How can participation be improved?

A common complaint made of participation by protesters is that they were not involved early enough to shape the assumptions and basic plans underlying proposed projects (Bleiker 1978). Planners and decision makers respond with the complaint that protesters do not get involved to the extent that they share in solving the entire problem, or in accepting responsibility for failure to make or implement decisions, or in preparing themselves technically to understand the issue. Since these officials cannot dispense with participation they search for ways to involve people earlier and longer and more "constructively." Experience has led both decision authorities and grassroots activists to believe that people will only become involved if they see that their own future is seriously at stake through what is decided, and that they can affect these decisions. The designers of the 1980 Minnesota Waste Management Act demonstrated that they shared this belief when they said that people would have to be "hassled" to make them participate in making decisions about waste management. To provoke such involvement and also to demonstrate their concern for equity, the WMB declared that the whole of Minnesota was a search area for sites to build hazardous waste management facilities. In communities across the state various local officials and diverse local activists used this threat to mobilize neighbors to come out to the informational meetings held by the Waste Management Board. People came out in good number, but in most cases the ideology which was developed to motivate them to come and participate prepared them to protest even the format of the meetings, and to promise that they would fight any attempt to site such a facility in their community. It was an ideology which focused on this risk of "hazardous" waste, on the impossibilities of engineering a risk-proof management facility, environmental and social unsuitability of the community as facility location, and on the presumption that the whole siting effort was a conspiracy to make weaker, poorer, rural people carry the waste burden while powerful and wealthy communities and interests escaped.

The designers of this legislation and decision procedure anticipated that there would be protest of waste management decisions. They had experienced the resistance of Minnesotans to the d.c. power line and also to the siting of various solid waste facilities. Events around the country made them aware that solid and hazardous waste management issues are very controversial (Centaur Associates 1979). But they felt that they could use the controversy and conflict as a tool to alert the public to the waste management issue and to make decisions about waste management which would be broadly acceptable. They defined the problems of citizen resistance as one of errant procedure. They felt that they could solve it by shaping procedure as a part of this by communicating understanding of risk and benefit. They consulted with experts on participation to learn how to do it better — encouraged only by the promise that even though better participation cannot guarantee success, lack of effective participation guarantees failure (Bleiker 1978).

What happened?

By 1984, when the WMB was supposed to have recommended hazardous waste disposal sites to the legislature, the WMB instead was told by this legislature to go back to the proverbial drawing boards and determine if the state really needs such facilities. The legislature took this step because during the almost four years of elaborate participatory procedures and technical assessment grassroots protest against the WMB and its procedures and aims had indeed become a movement whose participants convincingly warned that they would use force to stop site construction. The elaborate procedure did not succeed in its stated objective of finding sites which were "technologically sound, ecologically and medically safe, economically viable and socially-politically acceptable." Above all, it failed to achieve acceptability. What it did help produce was a resistance movement which itself began to shape the debate away from trade-offs and to critique technology, economy, management, and understanding of the environment, indeed of the whole American way of life. This is what we would expect from a social movement.

The Waste Management Board continues to operate. It functions now to perform these technological and economic studies pertaining to waste management. It works with industry and local government to explore ways to control the generation of hazardous waste at its source. It seems determined to find ways to avoid need for a central hazardous waste disposal site.

The WMB newsletter, *Foresite*, describes these kinds of actions as but a logical progression of the original WMB mission. It can also say that modifications in its approach show that it followed its promise to be open to citizen input. But it has changed in ways it did not foresee, and the changes are a product of protest.

Conclusion

Established orders do not like to admit that protest movements made them change (though they may admit that these helped them to think about changing). Established orders want to say that they arrived at their decisions in ways which they controlled, using procedures appropriate to their organizations, and knowledge rationally obtained through their official means. If they admit to responding to pressures from outside their organization, it will seldom be protest movements which they will acknowledge as the source of this pressure. The nuclear energy industry, for example, says that it is the economy and government regulations which have made them change their timetables and programmes; not the antinuclear movement.

Perhaps this reluctance to admit to the success of movements is simply a good tactic to dissuade people from turning to social movements as a means to achieve their ends. (Much as governments do not want to admit that hostage-taking pays off.) But probably there are deeper roots to this reluctance. Social movements are characteristically regarded as forms of deviant behavior stemming from conditions of deprivation, disorganization, demoralization, devitalization (Hine 1974). This implies that the participants are themselves defective, looking for revitalization, reorganization, reintegration (Wallace 1956). From this perspective the best thing that can be said for participants in a movement is that they really are not troubled, but instead are using or making the movement in order to serve their very selfish ends. In the U.S.A. this means that ecology activists are regarded by their critics either as sincere deviants, pushed into butterfly protecting by their structural or economic marginality (Douglas and Wildavsky 1982), or as selfish suburban straights, using ecology or anything else to protect their backyards (Tucker 1982). The connection usually made between movements and change is that movements are a product of change which people cannot manage (Benjamin 1980).

But we have learned how movements do come to life, do grow through interactions with established orders designed to control them, and then to shape decisions and produce change. Their presence in a sociocultural system, a way of life, should be considered quite natural and ordinary, not something so unusual that specialists take it as a sign that the system is in extreme crisis. They can be more usefully considered as producers of change than as the products of change.

All of this means that movements are regarded as the result of factors external to them. It is thus understandable that technological disputes are also analyzed as products of external forces. I have offered an alternative perspective. I have shown how movements are important in their own right, how they come to life, grow through interactions with established orders designed to curb them, and then take over control of the disputes and shape outcomes. Disputes frequently take the form of dramatic, ritualistic and media-amplified encounters with established orders in official public meetings and hearings, in the courts, in legislatures, and through confrontation in the streets. Running through so much of this is risk as a theme, slogan, and impetus for commitment. But according to the approach presented in this essay, risk assessment is a product of movement structure and function, not cause.

There is much to be said for the proposition that public hearings are ritual events which function, as does ritual generally, to help people march to decisions with a minimum of disruptive anxiety (Wynne 1982). But there is more to it than this. Public hearings are but one of the settings within which movements work in interaction with established orders to debate ways of life and to explore avenues of change.

When we are ready to understand technological and resource management disputes as events involving the interaction of social movements with established orders, then we can go on to consider how such interaction can be facilitated as a kind of drama to produce sociocultural debate and change. Again, in this, risk will serve as theme and statement, but not as cause.

Acknowledgments

Research on which this chapter is based was funded over many years, primarily by generous grants from the Northwest Area Foundation, St. Paul, Minnesota. In the early 1980s small grants for the Graduate School and the Center for Urban and Regional Affairs (CURA) of the University of Minnesota helped fund abbreviated studies of the hazardous waste management facility siting process and a recent small grant from the U.K.

Fund of the Hubert H. Humphrey Institute of Public Affairs helped stimulate my exploration of modes of defining technology disputes.

Notes

1. Superficially, at least, technical controversy takes the same form across the Western industrial democracies. The constituents and their interactions appear similar. But it is likely that these controversies and the behavior of the constituencies will reflect deeper cultural differences. Among other things, I suspect that the culture of protest, of counter-culture, is itself cultural. Also, the techniques used by established orders to deal with protest vary across these Western societies, as do opinions held by managerial elites about lay critics of science and technology. Cross-cultural research is called for to investigate these deeper similarities and differences.

2. NIDR announced in 1984 that it would support research to resolve conflict in ways which would head off such experiment legal and administrative action.

3. T. Ranger referred to an early draft of a forthcoming book by Jean Comaroff, no date or publisher was announced. Ranger — a specialist on rebellions in Central Africa — seemed cautiously skeptical of Comaroff's interpretation.

References

Advisory Commission on Intergovernmental Relations. *Citizen Participation in the American Federal System*. Washington, D.C. 20575: U.S. Government, n.d.

Almond, Gabriel and Sidney Verba. *The Civic Culture*. Boston: Little Brown, 1965.

Barbour, Floys B. *The Black Power Revolt*. Boston: Proter-Sargent, 1968.

Bateson, Gregory. *Steps To An Ecology of Mind: The New Information Sciences Can Lead to A New Understanding of Man*. New York: Ballantine Books, 1972, 107—127.

Bell, Daniel. *The Coming Of Post Industrial Society*. New York: Basic Books, 1973.

Bell, Daniel. 'The Revolution of Rising Entitlements,' *Fortune*, April 1975, 98.

Benjamin, Roger. *The Limits of Politics: Collective Goods and Political Change in Post Industrial Society*. Chicago: University of Chicago Press, 1980.

Bentley, Arthur F. *The Process of Government*. San Antonio: Principia Press, 1949.

Bleiker, Hans. *Citizen Participation: Handbook for Public Officials and other Professionals Serving the Public*. Laramie, Wyoming: Institute for Participatory Planning, 1978.

Blumer, Herbert. 'Social Unrest and Collective Protest,' pp. 1—54 in Norman K. Denzin (ed.), *Studies in Symbolic Interaction*, Vol 1. Greenwich, Connecticut: JAI Press, Inc., 1978.

Carmichael, Stokely and Charles Hamilton. *Black Power: The Politics of Liberation in America*. New York: Random House Vintage Books, 1967.

Casper, Barry M. and Paul David Wellstone. *Powerline: The First Battle of America's Energy War*. Amherst: University of Massachusetts Press, 1981.

Centaur Associates. *Siting of Hazardous Waste Management Facilities and Public Opposition*. Washington, D.C.: U.S. Environmental Protection Agency (EPA) Publication SW—809, 1979.

Commoner, Barry, *The Closing Circle: Nature, Man and Technology*. New York: Alfred A. Knopf, 1971.

Committee on Facility Siting. Virginia Greenman, Chairman; *Citizens League Background Report Siting of Major Controversial Facilities*. Minneapolis: Citizens League Board of Directors, October 22, 1980.

Douglas, Mary and Aaron Wildavsky. *Risk and Culture: An Essay on the Technological and Environmental Dangers*. Berkeley: University of California Press, 1982.

DeBono, Edward. *New Think; the use of lateral thinking in the generation of new ideas*. New York: Basic Books, 1968.

Firth, R. *Symbols, Public and Private*. Ithaca: Cornell University Press, 1973.

Fischhoff, Baruch, Sarah Lichtenstein, Paul Slovic, Stephen L. Derby, and Ralph L. Keeney. *Acceptable Risk*. Cambridge: Cambridge University Press, 1981.

Ford, Gerald. A speech before the American Enterprise Institute, presented on Minnesota Public Radio, St. Paul, 1978.

Freeman, Jo (ed). *Social Movements of the Sixties and Seventies*. New York: Longman, 1973.

Gerlach, Luther P. and Virginia H. Hine. *People Power Change: Movements of Social Transformation*. Indianapolis: Bobbs-Merrill Company, Inc., 1970.

Gerlach, Luther P. and Virginia H. Hine. *Lifeway Leap: The Dynamics of Change in America*. Minneapolis: University of Minnesota Press, 1973.

Gerlach, Luther P. 'The Great Energy Standoff,' *Natural History*, January 1978, 87(1), 22—23.

Gerlach, Luther P. and Betty Radcliffe. 'Can Independence Survive Interdependence?' *Futurics*, Summer 1979, 3(3), 181—206.

Gerlach, Luther P. 'Energy Wars and Social Change,' in Susan Abbot and John van Willigan (eds.), *Predicting Sociocultural Change*. Southern Anthropological Society Proceedings # 13. Athens Georgia: University of Georgia Press, 1979, 76—93.

Gerlach, Luther P. 'The Flea and the Elephant: Protest, Response and Consequences in the Infant Formula Controversy', *Transaction/Society*, September/October 1980, 16(6).

Gerlach, Luther P. Program 8, 'Political Organization,' in 10 program series *Introduction to Social and Cultural Anthropology*. Interview with chair Robert Dunn on Decision Process. Minneapolis: University of Minnesota, 1982.

Gerlach, Luther P. and Betty Radcliffe. 'Methodology of the Anthropology Probe,' pp. 338—345 in *Faith and Ferment on Interdisciplinary Study of Christian Beliefs and Practices*. Joan D. Chittister, OSD, Martin E. Marty, and Rubert S. Bilheimer (eds.). Minneapolis: Augsburg/Liturgical Press, 1983.

Gerlach, Luther P. and Larry Meiller. 'Social and Political Process in Needs Assessment, the Hazardous Waste Management Case,' Chapter 4 in *Community Needs Assessment*, Gene Summers *et al* (eds.). Ames: Iowa State Press (in press).

Genevie, Louis E. *Collective Behavior and Social Movements*. Ithaca, Illinois: F. E. Peacock Publishers, Inc., 1978, xv—xxii.

Goodenough, Ward H. *Cooperation in Change*. New York: Sage, 1963.

Gross, Jonathan L. and Steve Rayner. *Measuring Culture: A Paradigm for the Analysis of Social Organization*. New York: Columbia University Press, 1985.

Gulliver, P. H. *Disputes and Negotiations. A Cross-Cultural Perspective*. New York: Academic Press, 1979.

Hadden, Susan G. (ed.). *Risk Analysis, Institutions and Public Policy*. A Policy Studies Organization Series Publication. Port Washington, N.Y.: Associated Press, 1984.

Hardin, Russell. 'Comments,' pp. 122—129 in Clifford S. Russell (ed.), *Collective Decision-Making: Applications from Public Choice Theory*. Resources for the Future. Baltimore: Johns Hopkins Press, 1979.

Hine, Virginia H. 'The Deprivation and Disorganization Theories of Social Movements,' pp. 646—661 in *Religious Movements in Contemporary America*, Irving I. Zaretsky and Mark P. Leone (eds.). Princeton, New Jersey: Princeton University Press, 1974.

Hoffer, Eric. *The True Believer*. New York: Harper and Row, 1965.

Huntington, Samuel P. 'Postindustrial Politics: How Benign will it be?' *Comparative Politics*, January 1974, 6, 163—91.

Huntington, Samuel P. 'The Democratic Distemper,' *Public Interest*, Fall 1979, 14, 9—38.

Junck, Robert. *The New Tyranny: How Nuclear Power Enslaves Us.* New York: Fred Jordan Books, Grosset and Dunlap, 1977. (Published in Germany under title: *Der Atomstaat.* Kendli Verlag GmbH, 1977, translated by Christopher Tramp).

Kahn, Herman. Interviewed and quoted in pp. 313—326, in Willem L. Oltmans, *On Growth: The Crisis of Exploding Population and Resources Depletion.* New York: Capricorn Books, G. P. Putnam's Sons, 1974.

Kahn, Roger. 'The Collapse of the SDS . . . Spectators Guide to Warring Factions,' *Esquire*, October 1969, 72, 140ff.

Kenniston, Kenneth. 'Youth, Change, and Violence,' *American Scholar*, 1968, 37(2), 37—45.

Kopytoff, Igor. 'Classifications of Religious Movements: Analytical and Synthetic.' Proceedings of the 1964 Annual Spring Meeting of the American Ethnological Society. Sattle: University of Washington Press, 1964.

Kunreuther, Howard, John Lathrop and Joanne Linnerooth. 'A Descriptive Model of Choice for Siting Facilities,' *Behavioral Science*, 1981, 27, 281—297.

Kuznets, Simon. *Modern Economic Growth Rate Structure and Spread.* New Haven: Yale University Press, 1966.

LaHaye, Tim. *The Battle for the Mind.* Flemming H. Revell Co., 1980.

Lewis, I. (ed.). *Symbols and Sentiments: Cross-Cultural Studies in Symbolism.* New York: Academic Press, Inc., 1977.

Lowi, Theodore. *The End of Liberalism.* New York: W. W. New York, 1969.

Lovins, Amory B. *Soft Energy Paths: Toward a Durable Peace.* New York: Harper Colophon Books, 1977.

Lowrance, William. 'The Nature of Risk,' in Schwing and Albers (eds.), *Societal Risk Assessment: How Safe is Safe Enough*? New York: Plenum Press, 1980.

Maruyama, Magorah. 'The Second Cybernetics: Deviation Amplifying Mutual Causal Processes,' *American Scientist*, 1965, 51(2), 164—179.

Mazur, Allan. *The Dynamics of Technical Controversy.* Washington, D.C.: Communications Press, Inc., 1981.

Meadows, Donella H., Dennis L, Meadows, Jorgen Randers and William W. Behrens, III. *The Limits to Growth.* New York: Universe books, 1972.

Mishan, Ezra. 'Extending the Growth Edbate,' in Kenneth Wilson (ed.), *Prospects for Growth: Changing Expectations for the Future.* New York: Preager, 1977, 283—292.

Mitchell, Robert Cameron. 'National Environmental Lobbies and the Apparent Illogic of Collective Action,' pp. 87—121 in Clifford S. Russell (ed.), *Collective Decision-Making: Applications from Public Choice Theory.* Baltimore: Published for Resources for the Future, by Johns Hopkins Press, 1979.

Moorhead, Alan. *The White Nile.* New York: Dell, 1960.

Morrison, Denton. 'Uphill and Downhill Battles and Contributions to Collective Actions,' pp. 130—136 by Clifford S. Russell (ed.) in *Collective Decision-Making.* Baltimore: Johns Hopkins Press, 1979.

Myers, Norman. *GAIA: An Atlas of Planet Management.* Garden City, New York: Anchor Press/Doubleday & Company, 1984.

Naisbet, John. *MEGATRENDS: Ten New Directions Transforming our Lives.* New York: Warner Books, 1982.

Nelkin, D. and M. Pollak. 'Public Participation in Technological Decisions: Reality or Grand Illusion,' *Technology Review*, September 1979, 55—64.

Oberschall, Anthony. *Social Conflict and Social Movements.* Englewood Cliffs, New Jersey: Prentice-Hall, 1973.

O'Hare, Michael. 'Not on My Block You Don't! Facility Siting and the Strategic Importance of Compensation,' *Public Policy*, 1977, 25, 407.

Olson, Mancur. *The Logic of Collective Action: Public Goods and the Theory of Groups.* Cambridge, Massachusetts: Harvard University Press, 1965.

Olson, Mancur. *The Logic of Collective Action: Public Goods and the Theory of Groups.* (rev. ed). New York: Schocken Books, 1971.

Ornstein, Norman J. and Shirley Elder. *Interest Groups, Lobbying and Policy-Making.* Washington, D.C.: Congressional Quarterly Press, 1978.

Rayner, Steve. 'Learning for the Blind Men and the Elephant, or Seeing Things Whole in Risk Management,' unpublished manuscript, n.d.

Rayner, Steve. 'Disagreeing about Risk: The Institutional Cultures of Risk Management and Planning for Future Generations,' Ch. 9 in *Risk Analysis, Institutions and Public Policy*, Susan G. Hadden (ed.). Port Washington, New York: Associated Faculty Press, 1984.

Rebuffoni, Dean. 'Physicist says Ground Currents Harm People,' Metro/State News. *Minneapolis Star and Tribune*, Minneapolis, Minnesota, Tuesday February 25, 1986, 5B.

Rostow, W. W. *The Stages of Economic Growth. A Non-Communist Manifesto.* New York: Cambridge University Press, 1960.

Sapir, J. D. and J. C. Crocker (eds.). *The Social Use of Metaphor: Essays in the Anthropology of Rhetoric.* Philadelphia: University of Pennsylvania Press, 1977.

Schaeffer, Francis A. *How should we then live? The Rise and Decline of Western Thought and Culture.* Old Tappan, New Jersey: Fleming H. Revell Company, 1976.

Schattschneider, E. E. *The Semisovereign People.* New York: Holt, Rinehard and Winston, 1960.

Self, Peter. *The Economists and the Policy Process: The Politics and Philosophy of Cost-Benefit Analysis.* London: Macmillan, 1975.

Slovic, Paul, Baruch Fischhoff, and Sara Lichtenstein, 'Facts and Fears: Understanding Perceived Risk,' in Richard C. Schwing and Walter A. Albers, Jr. (eds.), *Societal Risk Assessment: How Safe is Safe Enough?* New York: Plenum Press, 1980.

Stott, John, R. W. *Christian Mission in the Modern World.* Downers Grave, Illinois: Inter-Varsity Press, 1975.

Tillock, Harriet and Denton E. Morrison. 'Group Size and Contributions to Collective Action: An Examination of Mancur Olson's Theory Using Data from Zero Population Growth, Inc.,' in Louis Kreisberg (ed.), *Research on Social Movements, Conflicts and Change.* New York: JAI Press, 1979.

Theobald, Robert. *Habit and Habitat: A Call for Fundamental Changes to Solve the Environmental Crisis.* Englewood Cliffs, New Jersey: Prentice-Hall, 1972.

Todd, Thomas M. 'Presentation of Hazardous Waste Facility Development to Connecticut Legislature.' (Thomas M. Todd, Legislative Analyst, Research Department, Minnesota House of Representatives.)

Truman, David B. *The Governmental Process* (2nd edn.). New York: Alfred A. Knopf, 1951.

Tucker, William. *Progress and Privilege, America in the Age of Environmentalism.* Garden City, New York: Anchor Press/Doubleday, 1982.

Turner, Ralph H. and Lewis M. Killian. *Collective Behavior.* Englewood Cliffs, New Jersey: Prentice-Hall, 1957.

Turner, Victor. *The Forest of Symbols.* Ithaca, New York: Cornell University Press, 1967.

Wallace, Anthony F. C. 'Revitalization Movements,' *American Anthropologist*, 1956a, 58, 264—81.

Wallace, Anthony F. C. 'Mazeway Resynthesis: A Biocultural Theory of Religious Inspiration,' *New York Academy of Sciences Transactions*,1956b, 18, 626—38.

Wallace, Anthony F. C. *Religion: An Anthropological View.* New York: Random House, 1966.

Weinberg, Alvin M. 'Science and its Limits: The Regulator's Dilemma,' *Issues in Science and Technology*, Fall 1985, 2(1), 59—82.

Willems, Emilio. 'Protestantism and Culture Change in Brazil and Chile,' in D'Antonio and F. B. Pide (eds.), *Religion, Revolution and Reform.* New York: Praeger, 1964.

Willems, Emilio. 'Religious Mass Movements and Social Change in Brazil,' in E. Baklanoff (ed.), *New Perspectives on Brazil*. Nashville: Vanderbilt University Press, 1966.

Winner, Langdon. *Autonomous Technology, Technics-out-of-Control as a Theme in Political Thought*. Cambridge, Massachusetts: MIT Press, 1977.

Wynne, Brian. *Rationality and Ritual. The Windscale Inquiry and Nuclear Decisions in Britain*. Halfpenny Furze, Mill Lane, Chalfont St., St. Giles, Bucks., England: British Society for the History of Science-Monographs, 1982.

Zald, Mayer N. and J. D. McCarthy (eds.). *The Dynamics of Social Movements*. Cambridge, Massachusetts: Winthrop Publishers, 1979.

Chapter 6

The Environmentalist Movement and Grid/Group Analysis: A Modest Critique

Branden B. Johnson

What do environmentalists want? Why do people form environmentalist organizations? Why do people join such organizations? What factors explain the behavior of environmentalist organizations? To environmentalists, the answers to these questions are obvious. But to those in authority, the motivations and underlying goals and objectives of environmentalists are not so obvious.[1] For example, people with strongly held environmentalist values often do not join environmentalist organizations.

In this context, the publication in 1982 of *Risk and Culture*, by Mary Douglas and Aaron Wildavsky, startled the field of technological hazard management, which was still grappling with the validity of distinguishing between the objective reality of risks and irrational perceived risks. In a comprehensive attempt to demonstrate the social construction of risk identifications, Douglas and Wildavsky argued that environmentalist ideology is not a function of either objective reality or irrationality. Instead, it is the outcome of an attempt to maintain fragile organizational structures. Almost lost in the ensuing critical uproar, as the book's thesis was alternately reviled and lauded, was the larger context of grid/group analysis, upon which the thesis was based. This anthropological theory claims to explain the ideology of all social groups, not just social movement organizations, and of all issue areas, not just human-nature relations (Douglas 1982a). In this chapter the validity of grid/group analysis (GGA) as an explanation of environmentalism — and, by extension, of other risk constructs — will be examined. Details of GGA and the *Risk and Culture* thesis will be presented, and analyzed as to how well they deal with the issues of dissent expression and group behavior. Points from other GGA literature and the sociology of social movements will be used to suggest ways in which this thesis might be improved and tested.

Grid/Group Analysis and Environmentalism

The Basic Model

Douglas (1982a) began with the unexceptional proposition that individuals

B. B. Johnson and V. T. Covello (eds.), The Social and Cultural Construction of Risk, 147—175.

negotiate culture with their fellows, seeking agreement on rules for life. Out of this eternal flux, however, come a limited number of possible cosmologies, which contain "those beliefs and values which are derivable as justifications for action. . . ." These cosmologies are generated by the face-to-face interaction of individuals within particular social contexts (Douglas 1982a: 201). The accretion of individual decisions forms a "collective consciousness," which both makes and justifies the penalty-carrying rules which constrain future individual decisions (Douglas 1982a: 190). For example, assume that in a society there is a rule that low-ranking individuals may not eat in the presence of their betters. The phrasing of the rule will include its justification (e.g., "Eating before your betters is disrespectful; it implies you have some right to this food other than your superior's gracious permission"). Every time this rule is observed, it is reinforced for both the actor and the onlookers (including the next generation). Over time this reinforcement becomes automatic. The carrying out of the rule constantly buttresses and justifies a society which values strictly graded hierarchy and deference.

This linking of behavior and beliefs is not uncommon in anthropology. The concentration on cognitive rules, and the assertion that there are only a few ways to think, is a structuralist explanation of human behavior, stemming from British social anthropology and (ultimately) Durkheim. He saw social groups as having their existence apart from, and logically prior to, that of their individual members, who need the continuation of the group (Befu 1980).

Social structures

According to Douglas, there are a limited set of possible social contexts to generate such cosmologies; these social structures are constructed on two dimensions. *Grid* is a dimension of differentiation. At the high or strong end "an explicit set of institutionalized classifications keeps [individuals] apart and regulates their interactions, restricting their options" (Douglas 1982a: 192). For example, an Imperial Chinese offical whose rank defined his tasks, status, and even social etiquette would be in a strong grid situation. At the weak end of this dimension distinctions have broken down and only the rules governing culture-negotiation among individuals remain. An example of a weak-grid contract would be the free market agreement that people will negotiate over the exchange of money for goods and services (buying and selling), but otherwise any behavior is permitted.

Group defines the degree to which a boundary is drawn between members of a group and outsiders. Strong group implies exclusivity — stringent requirements for entrance into the group — weak group implies individual autonomy (freedom from group constraints in exchange for being on one's own).[2] When these two dimensions are combined, four potential, cosmology-generating social structures are produced: individualist (weak grid, weak group), atomized subordination (strong grid, weak group), hierarchist (strong grid, strong group), and sectarian (weak grid, strong group) Douglas (1982b: 4).[3]

In Douglas' model individualists are unconstrained by social rules (other than that of contract), except as they suffer from free competition with others and seek some public control of contracts or prices. A classic example is the *laissez-faire* entrepreneur. Rewards go to the innovating individual, who will therefore screen his or her acquaintances to retain only those who can help his or her personal enterprise. Success for the individualist is measured in terms of wealth or following. By contrast, Douglas argues, atomized subordinates — e.g., migrant workers or plantation slaves — are excluded from any existing groups but their allowed behaviors are fully defined by remote, impersonal others.

Hierarchists are highly organized and specialized within the group (e.g., bureaucrats). They measure success by the continuation of the group, at the expense of its members if necessary. As a result, hierarchists have a number of solutions to internal conflicts. Douglas distinguishes them from sectarians, who minimize differences between group members, so that the roles and rights of individuals are ambiguous and implicit. The internal conflicts of egalitarian religious or revolutionary groups, for example, can only be resolved by withdrawal of the privileges of membership. Given the harshness of the solution, disagreement will tend to be driven underground, leading to the formation of covert factions and the eventual fission of the group (Douglas 1982a: 205—208).

Douglas argues that these four social contexts will yield different cosmologies — about nature, time, human nature, and social behavior[4] — based on the different dynamics of cultural negotiation in each. An individualist involved in constant competition with others, for example, will not see the world in the same way as a hierarchist whose potential rewards and losses are both smaller and more certain. It is this grid/group analysis which formed the basis for the depiction of environmentalism in *Risk and Culture* (Douglas and Wildavsky 1982).

Environmentalism as Sect

The *Risk and Culture* discussion

Environmentalist groups are voluntary in nature, and therefore they cannot exercise sanctions to punish dissenters and defecters. Douglas and Wildavsky (1982) argued that the egalitarianism of these groups, and the paucity of solutions to internal conflict, are characteristic of sects. To maintain the cohesion of the sect, an ideology is formulated that posits an extreme, external danger that can only be fought successfully by the maintenance of group solidarity. The threat is seen as world-pervading and evil, promoted by organizations which are nonegalitarian, either explicitly (hierarchists) or implicitly (individualists).

In contrast to hierarchists' and individualists' views, sectarians see the future as being not only different from the present, but as *worse*. In most societies, the danger perceived by sectarians is to God, exemplified by the Anti-Christ or worldly clerics. In modern secular societies, Douglas and Wildavsky see this threat as less plausible: instead, the menace is polution and despoliation of a pristine Nature by Big Business and Big Government.

Douglas and Wildavsky also distinguish between two kinds of environmentalist sects. "Hierarchical sects" allow some internal differentiation of roles, do not seek to overthrow the existing social structure, and are willing to work within the system. Examples given are the Sierra Club and local intervenor groups against nuclear power plants. "Fraternal [archetypal] sects" maintain equality more strictly and are unwilling to compromise their purity by participating in less than sweeping change. Friends of the Earth and direct action antinuclear groups like the Clamshell Alliance are suggested instances (Douglas and Wildavsky 1982: 137, 182).

Critics' fallacies

This sectarian explanation for environmentalism has evoked many critical reviews, in part due to a feeling that the good guys were being attacked (cf. Douglas and Wildavsky 1982: 190). The objective existence of pollution is cited often by critics as the *obvious* reason for environmentalism (cf. McKean 1981: 266). Whatever their own biases,[5] Douglas and Wildavsky agree that pollution exists, but they maintain that Americans in nonsectarian social structures see this problem as less severe, urgent and pervasive than do sectarians. They are more concerned about such issues as maintenance of the social system (hierarchy) or personal resources (individualism). What counts as pollution is thus socially constructed, since scientific evidence is often highly uncertain and controversial (Douglas and Wildavsky 1982: 3).[6] This view is consistent with that of social movement theorists that the objective existence of a condition, such as deprivation, is neither sufficient nor necessary for the condition to be defined as a problem (e.g., McCarthy and Zald 1977: 1214—1215).[7]

Some criticism of the Douglas and Wildavsky thesis may be due to the difficulty of understanding a structural explanation for social movements. This has been often misunderstood as meaning that individuals' world views or cosmologies must be *determined* by their social structure (e.g., Ben-David 1981: 52; Barnes 1984: 196—197). Any individual's location in the grid/group map is potentially open to negotiation. Choices for the individual, if not the social group, cannot be taken as mechanistic and completely determined. However, GGA does not treat culture as completely negotiable either. The categories which emerge from such bargaining constrain what is seen as possible; what is fixed is not the individual's views, but the choice of alternative social structures.

Nor does the theory require that every member of a given group share precisely the same views. Cosmologies can have a normal distribution among a given social unit's members (Gross and Rayner 1985: 49).[8] GGA's focus is on *public* actions or *public* statements defending their actions to group members or to others (Douglas 1985: xxiii):[9]

> there will be moments of truth: when a misfortune strikes, the acceptable explanations will need to be plausible to people who have constructed their universe in a certain way . . . alignments that were loose and ambiguous will be tightened, and statements of value will be clarified.

Environmentalist Ethnography

Grid/group analysis is not intended to provide explanations for the existence of particular social structures. Although the general form of the four grid/group structures remains constant, the content will change with historical and cultural circumstances (Ostrander 1982: 25). Other social movement theories emphasize the origins and maintenance of movement organizations and downplay the development of their ideological content. This creates both opportunities and difficulties for comparison of GGA's analysis of environmentalism to those of other approaches. On the one hand, they may complement each other; on the other, they can provide directly competing hypotheses which the available data are not capable of falsifying. Both possibilities can be demonstrated with the literature on formation of social movement groups, declaration of membership in such groups, and the behavior of groups.

Forming Groups

Cultural tendencies

Douglas and Wildavsky cite several factors responsible for what they see as increasing sectarianism in the United States during the 1960s. They claim that Americans have always been attracted to "border" culture, as contrasted to the "center" cultures of hierarchism and individualism. This is exemplified by Americans' fascination with apocalyptic, millenarian religions. A low concern with weakness at society's center has led to a correspondingly high valuation of individuality, equality, and conspiracy theories of politics. Furthermore, Douglas and Wildavsky claim several trends since World War II provided economic resources, political opportunity, and an organizational model for environmentalist sectarians. These included the economic boom which made production of goods seem less important, and the expansion of higher education which produced more potential critics of capitalism. The erosion of political trust by such events as the Vietnam War and the example of the civil rights movement were also significant. A short-term factor in the growth of environmentalism is the existence of governmental support — through tax-deductible contributions — and the use of official coercive powers to leverage the power of sects.[10] Mass communication technologies have made mail-order memberships easier to solicit and maintain.

Much of this analysis — shorn of its GGA terminology — is supported by social movement theory generally. For example, collective action theory supports Douglas and Wildavsky's idea that when group action is political, the political system also has a general multiplier effect on group resources. If the group had to bribe industrialists not to pollute — or pay itself for all pollution control equipment — its aggregate resources would be unlikely to carry much weight, and contributors would be few in number (Hardin 1982: 118—122). On the specific issue of group formation, the social movement literature emphasizes explanations involving issue entrepreneurs and institutions.

Issue entrepreneurs

Resource mobilization theory (cf. chapter by Walsh in this volume) emphasizes the role of "issue entrepreneurs" in mobilizing potential supporters of a social movement who do not have the information or other resources necessary to organize themselves (Salisbury 1969). Ralph Nader is perhaps the quintessential example of an issue entrepreneur; David Brower of the Sierra Club and Friends of the Earth (McPhee 1971) is another. They use their organizational skills to establish one or more social movement organizations and then use advertising and other techniques to broaden their base of support. One study (Berry 1977) found that most environmentalist groups had been started by entrepreneurs. Their motivations have been complex, but appear to be more purposive and moral than self-aggrandizing (Frohlich *et al.* 1971; Johnson 1975; Tillock and Morrison 1979; Oliver 1983).

Institutional support

But organizational skill and willpower, whatever the motive, are insufficient to create an organization. Resource mobilization theorists argue that many political groups are instigated by existing institutions — government agencies, corporations, and foundations (e.g., Walker 1983, 1984). Groups that are devoted to causes (e.g., environmentalism) are the most dependent of all political groups on support from wealthy individuals, activist foundations, and other political patrons (Walker 1983; cf. Salisbury 1984). Curiously, groups dedicated to social change are often aided by powerful institutions of society which presumably benefit from the status quo. Institutional leaders perhaps see an opportunity to increase their net benefits without risking a loss of their institution's position due to revolutionary activity.

Institutions may help to foster social movements in an unintentional way as well. "The establishment of an agency authorized to deal with certain conditions generates dissatisfactions among populations about conditions that previously were unseen or routinely accommodated"; the apparent availability of "solutions" can generate hazard identifications (Spector and Kitsuse 1977: 84). For example, knowledge of civil rights legislation may lead to the reformulation of grievances as involving civil rights rather than merely isolated instances of personal prejudice. Institutional opposition to a movement can also legitimate it (cf. Walsh and Gerlach in this volume). But the *lack* of institutional, particularly government, action can also be a prime resource for dissident activism. For example, protest movements in Great Britain appear to flourish when citizens see political parties as failing to represent their interests (Messina 1985: 27; cf. Rhodebeck 1981: 257).

Douglas and Wildavsky argue that the professed goals of social movements should not be taken at face value; analysts should look for hidden agendas. Marxists also invoke this argument when they say that environmentalism protects capitalist or even pre-capitalist economic interests, and that this explains why institutions would support entrepreneurs in the founding of pseudo-dissenting social movements. For example, Lowe and Worboys (1978: 19) find "popular ecology" stemming from hierarchist wishes to see "men fulfilling their natural functions, socially and ecologically." The general argument is probably correct, but the importance of groups' ostensible agendas (the public statements of cosmology on which GGA focuses) in group formation should not be overlooked. Even the most cynical issue entrepreneur or institutional executive chooses an issue to exploit because he or she believes it to be one which will attract support. It will not attract support, however, unless the issue satisfies some previously unarticulated (or poorly articulated) hazard identification of potential public supporters. In short, it must (at least in part) be a hazard identification that predates the existence of the group itself. Yet Douglas and Wildavsky see the group's existence as a prerequisite for the identification of Nature as threatened by hierarchy.

Social movement theories thus fill the gap left by GGA's indifference to the question of group origins, while confirming some non-GGA explanations for increasing citizen group activity advanced by Douglas and Wildavsky. On the other hand, there is a conflict between the view that public statements of belief play no role in group formation — GGA is a theory which takes this approach — and the view that cosmology does have a significant role (see Gerlach in this volume).

Joining Groups

Grid/group analysis, by definition, does not explain why people become members of such groups after their formation (see Douglas 1984). However, besides Americans' alleged historical predisposition to sectarianism, Douglas and Wildavsky offer two membership explanations: selective incentives and communications innovations.

Selective incentives

Mancur Olson, Jr. (1965) analyzed the collective provision of goods the consumption of which by any person cannot be prevented. Assume that such a collective good — e.g., clean air or clean water — is desired by a large number of people, as these particular goods appear to be (e.g., Council on Environmental Quality 1980; Lake 1983). All can therefore expect to benefit from provision of this collective good, and presumably all have an incentive to organize to ensure its provision. But if they organize they must pay a share of the costs of such organization (dues, time, labor). It is entirely possible that the personal costs of participation will exceed its *per capita* benefits, making participation nonrational. Since consumption of collective goods by noncontributors cannot be prevented, the rational individual will see that the benefit can be gained at no cost. Unfortunately, if all individuals are rational, all will see the benefits of becoming free riders, no one will organize to obtain the collective good, and no one will obtain the resulting benefits.

This is a counter-intuitive result, since obviously people organize to obtain collective goods all the time: environmentalists for clean air and water, unionists for higher wages, business executives for tariffs and import quotas. Olson argues that collective action groups provide "selective incentives" to their members — items that cannot be obtained without being a member. For example, the Sierra Club provides its members with a glossy magazine, discounts on hiking and mountaineering expeditions, and discounts on books and gifts. Strictly interpreted, Olson's analysis leads to the conclusion that people join the Sierra Club and similar groups to obtain these selective incentives, and their dues provide the funds needed by group leaders to pursue their political ends.[11] Even ignorance of whether the benefits of group membership exceed the costs may not make membership nonrational if the member deems his or her costs to be trivially small (Olson 1982: 28) (plausible for high-income environmentalist group members).

Selective incentive critiques

This model powerfully underlines that movement members' avowed motives cannot be taken solely at face value. However, several empirical and conceptual criticisms can be made of its application to environmentalism by GGA theorists. First, there is evidence that selective incentives are rarely used by members (Mitchell 1979; Tillock and Morrison 1979; Johnson 1982), though this varies from group to group (Shaiko 1985). In fact, moral incentives appear to be more common motives for members of the more recently founded environmentalist groups (Hardin 1982: 117), which are also more sectarian according to Douglas and Wildavsky.[12]

Second, if selective incentives were sufficient for collective action, industry groups could buy public opinion on environmental issues (Tillock and Morrison 1979: 154). Although they certainly outspend environmentalists in lobbying and public relations, their success is partial at best. Third, the motive to be a free rider on others' contributions to obtain a collective good would seem to disappear in moral (strong group) social situations, of which Douglas and Wildavsky's sect is one. If free riders are relatively rare, so should be selective incentives.

Fourth, many voluntary-membership groups do not follow the Douglas–Wildavsky suggested history of environmentalist sects: (1) voluntary-membership group; (2) inherent instability; (3) need to posit threat from hierarchies; (4) justification for equality within the group. Douglas and Wildavsky (1982: 103) point to this possibility when they also posit hierarchism as a solution to the problems of voluntary organizations. However, this option is not explored further; since GGA is treated as a static model (but see Rayner 1982), no explanation is given for the differing outcomes. An alternative evolutionary path for sects might be (1) need to escape constraints of other structures (e.g., Rayner 1982: 271—272); (2) high value placed on egalitarian interaction; (3) sectarian cosmology as a justification of the group. In this case the conditions in their *previous* social situations that led people to form or join the group must be explained.

Defining group membership

A final problem with using Olson's thesis in conjunction with GGA to explain U.S. environmentalism concerns the anomalous nature of group membership. It is the small, exclusive distributional coalition which Olson's theory explains best (Olson, 1982: 67; Hardin 1982: 106) — exclusive because it has strict membership criteria, distributional because it seeks to limit the range of incomes and values among its members. An example is the trade association. Douglas and Wildavsky (1982: 117) themselves note that exclusive, not inclusive (few or no membership criteria), groups suffer the worst problems of voluntary organization. Since they are defined in part by strong group boundaries, sects must be exclusive. According to Douglas (1982a: 207—208), sects can only resolve conflict by expelling dissidents.

These tenets are at odds with empirical findings that citizen group members exit voluntarily if they become dissatisfied with the group's course. The difference between expulsion and voluntary departure is less stark in those cases where a sizeable group within the organization has lost a fight over policy and departs to form a new group. The formation of Friends of the Earth after David Brower lost a policy battle within the Sierra Club (McPhee 1971) might meet the Douglas–Wildavsky criteria for sectarian factionalism. Most departures from environmentalist groups, however, are individual and voluntary (e.g., Johnson 1975, 1982).

The Douglas–Wildavsky analysis is also at odds with the findings that environmentalist groups are highly inclusive. Membership criteria rarely extend beyond payment of dues, and sometimes not even that (Johnson 1975). GGA deals with this by suggesting that mere dues-payers are not true members of the group. An analogy can be made between most dues-paying ("mail order") members of environmentalist groups, and the resource-providing nonmembers of a radical (nonenvironmentalist) sect (Rayner 1982). In both cases these people provide needed resources but do not otherwise contribute to the group. Shaiko (1985) did a study of five environmentalist groups' members. He found low levels of self-reported "active" membership (2 percent for the liberal-radical Environmental Defense Fund, up to 10 percent for the moderate-liberal Sierra Club) and high levels of nonidentification with the group (25 percent for SC to 58 percent for EDF). Such nonidentification contradicts one criterion of group membership (Rhodebeck 1981: 248).

By contrast, political elites appear to be ideological more often, and more consistently, than their followers (Green and Guth 1984; cf. Stallings 1973 on environmental activists). The core group of Life of the Land, a liberal-radical environmentalist organization in Hawaii, had high levels of social as well as working contact with each other, and high self-identification with the organization, contrasting sharply with low levels on both items for general members (Johnson 1975). Not only is grid/group analysis of these core groups easier, but they obviously satisfy the face-to-face interaction criterion of membership used by GGA far better than do merely dues-paying members.

However, this distinction between elite and marginal supporters still provides no understanding of why the latter support environmentalism instead of some other cause available for purchase. Random choice can explain only a portion of these pledges. Most of the members in the five-group study cited earlier (Shaiko 1985) said they had formed their opinions on environmental issues *before* they joined the group (80 percent for the conservative National Wildlife Federation, 88–94 percent for the others). An average of 89 percent perceived their groups to be "representing their own views on environmental and conservation issues to the government." Furthermore, between 46 and 74 percent had individually sought to influence policy outcomes by contacting public officials. Elite coordination cannot explain the high level of individual activism (e.g., letter writing to officials). Only two of the groups (the moderate-liberal Wilderness Society and the liberal-radical Environmental Action) actively solicited and directed member action on specific issues. Furthermore, the environmentalist core group is not at all difficult to join as long as one is willing to contribute labor (as well as, or instead of, dues) (Johnson, 1975, 1982). Thus even a GGA focus on core members only fails the exclusivity criterion for proper application of Olson's model.

The other argument made by Douglas and Wildavsky for membership decisions — that communications innovations such as computerized mailing lists have facilitated groups' ability to contact and woo potential supporters — is far stronger. The availability of such technology has enhanced environmentalist leaders' ability to wield organizational power in policy actions.

Explaining Group Actions

Resource mobilization

Social movement leaders seek to transform adherents (believers in movement goals) into resource-providing constituents. But they must also neutralize or convert the general public and elite groups. Unfortunately for group leaders, tactics which achieve one of these purposes may reduce their chances of accomplishing the other (McCarthy and Zald 1977: 1217, 1221). Thus the ability of American women's antifeminist movements to attain their political goals was constrained by their need to conform to their rhetorical image of lady-like dignity and restraint. Ironically, the failure of antisuffrage partially removed this constraint and was a factor in the later success of the anti-ERA movement (Marshall 1985).

Furthermore, social movement organizations (SMOs) with relatively similar goals compete with each other for resources, as well as with other social movements. As the absolute resources available to social movements increase, as in post-war America, new movements and organizations develop to compete for them. People with discretionary income are likely to be more satisfied with their personal situations, so conscience adherents — supporters for moral reasons — become more important. They are more likely than self-interested adherents to demand change *and* to leave the SMO with their resources if they feel it is not proceeding toward their preferred goals. Both factors put pressure on SMO leaders to act 'correctly' in disposing of group resources, including choosing targets and tactics (see Cigler 1984: 6). Such strategic considerations are not necessarily incompatible with GGA: they could be part of the internal dynamics by which sectarian groups maintain themselves. But they are competing explanations for the public statements and actions on which GGA focuses. Its proponents have yet to address these.

Advertising

If one must deal with isolated constituents — as most American environmentalist groups do — then the flow of resources to the SMO is unstable without considerable advertising. Such advertising emphasizes the dire consequences of failure to attain the movement goal, but also outlines the extent to which the goal has or has not been accomplished and the importance of this particular SMO to goal attainment. To a varying extent "grievances and discontent may be defined, created, and manipulated by issue entrepreneurs and organizations" (McCarthy and Zald 1977: 1215, 1228; cf. Cole and Withey 1981). But issues are far from being completely controlled by movement leaders. The search for resources to improve contending parties' (i.e., SMOs and their targets) chances of success leads to an expansion and politicization of disputes (Spector and Kitsuse 1977: 146—147). Knowledge claims (e.g., expert opinions, laboratory studies, epidemiological data) are particularly potent tools in a society devoted to science as an allegedly value-free enterprise (see Downey 1986; Johnson 1986). Each party (both within and across social movement boundaries) seeks to convert onlookers into adherents by reformulating knowledge claims into salient value issues. These in turn affect resource mobilization and future value statements (Spector and Kitsuse 1977: 146—147; Petersen and Markle 1981). Again, there is a strategic element to environmentalist group statements which is ignored by the Douglas and Wildavsky analysis as either potential complement to or competitor with GGA.

Differentiation

Environmentalist groups tend to be inclusive and centralized for effective lobbying (Jenkins 1983: 542).[13] As a consequence of easy entry into, and exit from, the group, average environmental membership tenure is less than three years (Shaiko 1985). Resource competition should be less for inclusive than for exclusive groups, because a group need only keep someone interested in the same general issue area, and it will have a chance to receive the person's contribution (cf. Johnson 1975).

By contrast, exclusive groups (e.g., sects) cannot afford to share members. This competitiveness leads exclusive groups to attract members through product differentiation. Although product differentiation does occur among inclusive environmentalist SMOs as well, their strength is in tactical specialization. Tactics can range from the guerrilla warfare of Greenpeace through the demonstrations and street theater of Life of the Land, and from the lobbying and litigation by Audubon and the Sierra Club to electoral action by the Sierra Club and the Greens of West Germany.

Variety allows a given SMO to tap a base that might not be susceptible to the appeals of other groups. It also allows constituents committed to the movement to hedge their bets in the selection of their movement "portfolios" (Zald and McCarthy 1980; cf. Gerlach and Hine 1973). They can become affiliated with several groups pursuing similar goals in different ways; one study found an average of four memberships per person (Johnson 1975; cf. Shaiko 1985).

These choices of issues and tactics are shaped in part by group ideology. The vulnerability of the institutional target (cf. Walsh and Gerlach in this volume) also makes a difference (Walker 1977: 436) —

> activists in the consumer field interviewed in 1969 . . . said that their efforts were being devoted to safety and consumer protection at that particular time because it was a field where movement was taking place, one where large political payoffs were possible.

But issue choices are affected as well by the character of an SMO's potential support. For example, the Interfaith Center on Corporate Responsibility was created by the National Council of Churches in 1974 to advise church groups on the ethical implications of their investments. Although the staff had considerable latitude in issue selection, their need to obtain financial and moral support led them to eventually select infant formula marketing in the Third World. This was more attractive to ICCR constituents than other issues (including the environment), due to such factors as increasing debate in the U.S. over breast-feeding versus bottle-feeding, and over world food inequities. Ideology is only one factor in ICCR's identification of hazards, even if the National Council of Churches hierarchy appears to have spawned a group with sectarian cosmology: "ICCR's 'selection of issues appears to be based more on the intensity of moral offense than on an issue's widespread social impact' " (Ermann and Clements 1984).

The findings of resource mobilization theorists match much of the description of environmentalist groups in Douglas and Wildavsky: for example, use of extremist rhetoric and assertion of the centrality of the organization to attainment of movement goals. On the other hand, these strategic arguments may allow us to explain sectarian structures' *continuing* existence without recourse to GGA. Given that GGA's strength is its alleged ability to explain the maintenance (as opposed to the origin) of groups, this creates a problem. For example, a spirit of egalitarianism may be the outcome, rather than the cause, of a desire to provide insurance in an uncertain world (McCloskey 1976: 165). Suppression of internal dissent is seen by grid/group analysts as a manifestation of sectarian intolerance for disagreement. Yet it may allow alienated citizenry to send a single clear message without losing strength (Janeway 1981: 179—180), or any collective action group to achieve consensus and improve leadership efficiency (Olson, 1982: 25). And decentralized organizational structures need not only be the outcome of aversion to hierarchies, as GGA would have it. They can also be a function of opportunities to attract support from new populations or influence other social groups (Lowe and Goyder 1983: 23). The availability of alternate hypotheses requires that GGA's claims be investigated carefully, so that one set of hypotheses is rejected or the two are fruitfully combined.

Alternative Approaches to GGA and Environmentalism

Readers have applauded *Risk and Culture's* boldness or have been shocked by what they see as its ethnographic crudeness. Both responses do grid/group analysis an injustice, by over- or underestimating its potential contributions. Extending and improving the grid/group analysis approach to environmentalism may require that the Douglas–Wildavsky thesis that environmentalists are sectarian be abandoned. Instead, even within the GGA assumptions it can be argued that different kinds of environmentalism can stem from all four posited social structures (sectarianism, hierarchy, atomized subordination, individualism). If this premise is granted, there remain other obstacles to the proper testing of GGA, primarily concerned with issues of definition and measurement. These include possible nonenvironmentalist sectarians, intercultural differences, multiple group memberships in modern heterogeneous societies, and discrimination between grid/group categories.

Nonsectarian Environmentalists

The implications of grid/group analysis for social construction of risk are not fully treated in *Risk and Culture*, in which sectarian social structures are presumed to be the only generators of environmentalist cosmology. Even some who strongly support the concept of grid/group analysis question the description of environmentalism as solely sectarian. Wuthnow *et al.* (1984: 93—95) argue that "pollution [defined as anything "out of place"] concerns appear when social lines and boundaries are threatened." Because environmental pollution concerns are national, "they need to be examined in relation to national identity crises." Douglas and Wildavsky argue these concerns are the ideological positions of the interest groups that most vociferously promote them. Wuthnow *et al.* see this analysis as violating Douglas' commitment to macro-analysis of societal cosmologies.

The broader potential for application of GGA to environmentalism is based on Douglas' earlier discussion (1982a: 209—212) of the four basic cosmologies of nature. These are alleged to be projections from the respective social structures' organizational needs. This discussion allows for some interesting expansions on the Douglas–Wildavsky analysis of sect and hierarchy, and implies atomized subordinates and individualists as possible sources of different kinds of environmentalism.

Sect and hierarchy

Sectarians echo their alienation from wider social interaction and the need to unmask internal dissenters with visions of "lambs" and "wolves" in Nature as well as in society. "Vulnerable, lovable, natural victims" are distinguished from "menacing, predatory, ineducable nature" (Douglas 1982a). This early description is generally consonant with that in *Risk and Culture*. However, that book did not mention that sectarians might distinguish between good and bad portions of Nature in seeing a threat to Nature from hierarchy and individualism. This possibility poses as yet unexplored questions regarding environmentalist issue selection and the use of science-based knowledge claims.

The hierarchy presumes that its classifications (Douglas 1982a) reflect Natural Law, and that Nature's symmetries and regularities are on the side of the good society. Unnatural behavior in humans is also possible in animals. Douglas and Wildavsky did implicitly acknowledge the weakness of arguing only for a sectarian environmentalism by postulating the existence of hierarchical as well as fraternal sects (see earlier discussion). However, their distinction between the two emphasized only political and organizational differences, not differences in the perception of Nature. As yet no one has sought Natural Law or notions of unnatural animal behavior in the mental constructs of alleged hierarchical environmentalists.

Atomized subordination

By contrast with sectarians and hierarchists, Douglas (1982a) argues, atomized subordinates cannot be expected to have elaborate concepts of Nature, given their enforced withdrawal from social interaction. Passivity and a confused cosmology is their response. Controlled by others as they are, atomized subordinates may feel subject to involuntary risks, and see the wealthy and powerful as giving them "limited economic or political choice" (Nelkin 1982: 776; cf. Wynne 1980a: 286; Lowi 1983: 29). Risk-related beliefs and behavior — such as using confrontationist strategies — may both follow from that perception and offer a chance at escape from their predicament (Alinsky 1971; Simons and Mechling 1981: 429; Rayner 1984a: 163). For example, Rayner (1984b, and in this volume) suggests that concerns about radiation hazards among medical personnel may be high among both sectarian "free clinic" staffs and Physicians for Social Responsibility, and such atomized subordinates as maintenance staff, junior nurses, and cleaners and porters. Concerns of Love Canal residents and video terminal-using secretaries about cancer may be another expression of atomized-subordinate environmentalism (cf. Hall 1976).

Japanese citizens' movements (almost all concerned with environmental problems (McKean 1981: 5)) see the causes of pollution as large, powerful institutions, which seems a sectarian view (McKean 1981: 131—132). However, they do not accuse industrialization, as the Douglas–Wildavsky thesis would suggest. Instead, they place blame on particular firms' laziness, stinginess, or choice of inappropriate goals (e.g., economic growth), and perceive themselves rather than the environment as victims. This is due to the fact that these groups concentrate largely on local disputes (e.g., Minamata or itai-itai diseases), and see victims or potential victims of local pollution as the only legitimate members of citizens' movements (McKean 1981: 137—138). This view could reflect members' status as atomized subordinates (e.g., farmers and fishers) in an industrial Japan.

The greater the political and social distance between the social movement organization and its institutional targets, the more likely it is that the obstacles to, and opportunities for, staking political claims will be perceived differently by each party. Procedures that appear rational to the elite may appear "remote, arbitrary, dilatory and ineffective" to movement members; protest actions, on the other hand, may seem pathological behavior from the elite's viewpoint. This is not merely a matter of differing interests or resources — it concerns "the socially structured distribution of practical [political] knowledge" (Rootes 1983: 43—44). Apparently arbitrary actions against random technological targets may reflect an alienation from the dominant decision-making processes of society. For example, when public participation in project siting occurs late in the planning process, the "irrational and selfish" opposition which may result occurs at the last minute because only then does the meaning of the project for local residents become clear (Wynne 1980b: 193—194).

Completely atomized and alienated people would not have the economic or psychological resources to organize. Orr (1974) argues that they may in fact be against environmental protection (cf. Gross and Rayner 1985: 48). And perceived deprivation relative to other groups is not a guarantor of social movement formation and growth (see Rayner 1982: 250—251). However, these points do not invalidate the hypothesis that some expressions of environmentalism stem from atomized subordinate contexts. It is those with a few resources and some small freedom of behavior who are both most threatened by cultural change and most likely to overthrow the existing order in their attempts to prevent such change (e.g., Wolf 1969: 177, 292; cf. Janeway 1981: 113). People are also more sensitive to group memberships that carry subordinate status for them (Jackman and Jackman 1983: 218). This may make the incentive to exit atomized subordination by any means possible — including environmentalist activism — stronger. This may be particularly true if events indicate a breach of contract on the part of hierarchical powers.

Individualism

Douglas (1982a) describes the individualist as in constant struggle with others to gain rewards and avoid losses. As a consequence, he or she will tend to contrast the corruption, self-seeking and aggression of humanity with a good and simple Nature. This appears to provide a basis for an individualist environmentalism. However, sympathy for Nature's degradation conflicts with the individualist's simultaneous need to exploit Nature's resources for success in that social context. Some environmentalist stances by individualists may be due to opportunistic adoption of sectarian rhetoric (Rayner, 1984a: 163; cf. Lasswell 1976: 266—267; Cobb and Elder 1981: 404; Bloor, 1982: 205). Some individualist-based environmentalism could also be due to risk selection based on immediate self-interest (such as that of pollution control equipment manufacturers or hazardous waste disposal facility operators). Even in these cases the rhetoric helps to legitimate the called-for social changes (Gerlach and Hine 1973: 246; and Gerlach in this volume). But individualists' view of Nature as representing "all that is innocent and despoiled by civilization" (Douglas 1982a: 212) represents another source for environmentalist rhetoric and action. Its conflict with the individualist's need for personal resources need not eliminate this environmentalism if the individualist is skillful at sharing this resource cost with competitors.

These speculations about nonsectarian origins of environmentalism must await confirmation or rejection from a fuller analysis of such groups than was undertaken by Douglas and Wildavsky (cf. comment by Rayner in this book). An exploration of these hypotheses that different grid/group social structures yield different environmentalist cosmologies should benefit both grid/group analysis and social movement theory.

Nonenvironmentalist Sectarians

Is it true that all modern sectarians in secular societies (excluding such exclusive minority groups as the Amish) are environmentalists? That is the implication of *Risk and Culture*. Yet some of the attributes of sectarian ideology — such as concern for a future that looks to be worse than the present, and extremist rhetoric about the need for action against powerful hierarchical conspiracies — seem to be more widely distributed than environmentalism. "We ... often tend to have divine and diabolical models for allies and adversaries, especially in polarized politics of domestic or international strife" (Cook 1984: 14). It is not clear why sectarians must be concerned only with God and Nature, and hierarchists and individualists concerned only with Economy and War. Douglas and Wildavsky (1982: 123) explain that God and Nature are arbiters external to the large-scale organizations of society's mainstream. This proposition needs some empirical justification, since the same God and Nature are often used as justifications for the mainstream's actions. Furthermore, if sectarians are afraid of subversion by corporate interests, environmental protection is not the only, or even the most obvious, modern issue with which to attack such interests.

There are several possible examples of nonenvironmentalist sectarians. "[S]tagnating small business interests" join the John Birch Society or Posse Comitatus to defend themselves against the threat posed by Big Business, Big Labor, and Big Government (Cook 1984: 24, fn 2). Such movements may result from a need to restore or maintain a favored social position (Lipset and Raab 1970). Certain proponents of "national security" portray defense needs as necessarily limitless to defend against a pervasive foe, while their domestic opponents portray our own military strength as the global threat (Yergin 1977: 196; Fallows 1981: 180; cf. Holdren 1983: 36). Ranchers in economic straits identify the cause of their problems as unchecked predation on their livestock rather than changes in their markets or in industry structure (Schueler 1980). Gerlach in this volume refers to the Satanism of "secular humanism" (a threat to God, rather than to Nature) perceived by the Conservative Christian Action movement. Upon closer examination some or all of these may prove to be examples of non-sectarian groups. However, if they are really slipping hierarchists, failing individualists, or atomized but organizing subordinates, how does one tell the difference between these and sectarians (see Rayner 1982: 271—272)?

Douglas' original cosmological derivatives were "conditional" and "provisional" (Douglas 1982a: 209). She admits (Douglas 1982b: 116) that a grid/group analysis of Enlightenment ideas of nature and history (Kelly 1982) found results incompatible with the original grid/group formulation. Her classifications have been criticized as "strongly colored by . . . subjective choices" (Hofstede 1980: 47; cf. Downey 1986). To date, however, no one has ventured to alter those original propositions linking particular social structures with particular world views. Given the possibilities of nonsectarian environmentalism and nonenvironmentalist sectarianism, it is understandable why many will doubt the validity of *Risk and Culture's* discussion of modern sectarianism until confirmatory research is done.

Cultural Differences

Subcultures share enough traits to be recognizable to foreigners as members of the same society (Hofstede 1980: 26). Members of an individualist society who are members of a sect within it may thus still appear strongly individualist to outsiders. Americans who are in service and nonprofit sectors of the economy and who express antibusiness opinions are prime candidates for sectarianism (Douglas and Wildavsky 1982). They also appear to be among the most adventurous in seeking out and enjoying exotic cuisines, a trait alleged to be characteristic of individualists (Douglas 1982a: 215). Assuming this observation is correct, either the postulated links between structure and cosmology must be further revised, or American sectarians may be more individualist than sectarians in other cultures. (Similarly, the Japanese environmentalists discussed earlier as possible atomized subordinates may actually be the singularly Japanese version of sectarians.)

Multiple Group Memberships

Perhaps the most difficult problem posed by grid/group analysis is that of defining group membership in a heterogeneous society where any given individual may have multiple memberships. Strong group (extreme hierarchism or sectarianism) is not difficult to identify. It joins people "in common residence, shared work, shared resources and recreation, and by exerting control over marriage and kinship" (Douglas 1982a: 202). Such a strong group rating is characteristic of exclusive tribal or peasant societies, or some religious sects in modern industrial societies. But if the individual "spends the morning in one [group], the evening in another, appears on Sundays in a third, gets his livelihood in a fourth, his group score is not going to be high" (Douglas 1982a: 202). This fragmentation seems typical of industrial societies, suggesting their members will tend more toward individualism and/or atomized subordination.

Such fragmentation poses no problem for grid/group analysis as long as it is assumed that in each social situation the individual plays essentially the same role: individualist, sectarian, hierarchist, or atomized subordinate. The stress of cognitive dissonance alone might lead the individual to seek continuity of grid and group across these various life realms (Thompson 1983: 15; Hill 1984: 17). In fact, as noted at the beginning of this chapter, each of these structural archetypes is supposed to yield a universal cosmology. That cosmology is alleged to be consistent across attitudes toward nature, foreign places, gardening, cookery, medicine, old age and youth, sickness and health, and justice, among others (Douglas 1982a).

But what if such continuity cannot be maintained? What if different social structures are sought in the home, for example, to offset perceived inadequacies of the structure in which one works? This has been given as the meaning of homeownership for lower-middle class workers (Cox 1983: 114; cf. Fowlkes and Miller in this volume). Then one would have difficulty specifying which structure will dominate cosmological stances during times of stress. It is also unclear whether one structure in one part of an individual's life would affect the cosmology expressed in another part.

Several authors have in fact suggested that multiple identities of individuals, and even groups, exist in industrial societies. Individuals have myriad roles, relationships, activities, and even manufactured personal identities from which to choose (Kilbourne and Richardson 1984: 239; cf. Bender 1978: 136—137; Hampton 1982: 78). Interacting groups and individuals may operate in different roles from one relationship to another (McCombs 1981: 126, 128; Smith 1982: 231). Environmentalist values may cluster around different aspects of personal experience, such as work and home (Cotgrove 1982: 29), or become salient for risk selections occurring at different times (Stallings 1973: 476; Mazur 1981: 62; Fischhoff *et al.* 1981: 164; cf. Campbell 1964). People may "identify with those seeking radical change in one aspect of our social structure while at the same time . . . resisting change in some other area" (Gerlach and Hine 1973: 219).

It is not known whether people do in fact switch grid/group cosmologies as they move from one part of their lives to another. Earlier, social movement theories (e.g., resource mobilization) were shown in some circumstances to complement grid/group explanations of environmentalism. This may prove true here as well: for example, economic circumstances or strategic considerations may force behavior which falls outside a person's or group's central (grid/group) situation. Research is needed to test these propositions.

Grid/Group Category Discrimination

Environmentalist categories

Reality is inevitably more complex than our models of it can convey. GGA is no exception, but it is difficult to test the validity of a hypothesis if there is no apparent agreement on how to measure the significant variables. The issue of how (or even whether) to distinguish between hierarchical and fraternal environmentalist sects (Douglas and Wildavsky 1982) exemplifies this GGA difficulty.[14]

The public statements of environmentalists are supposed to justify their social structure and reveal the distinctive cosmologies of the various grid/group social locations. Metlay and Hoberg (1984) found that Sierra Club members (the Douglas and Wildavsky exemplar of hierarchical sects) and members of Environmental Action (an example of a fraternal sect) are almost indistinguishable in their support for radical rhetoric. Examples include their assessments of the threat to the environment, alienation, willingness to negotiate, and the perceived evil of central societal institutions. Rhetoric alone does not appear to adequately distinguish between the sectarian categories Douglas and Wildavsky present. Moderate environmentalists have been called upon to follow both the political (hierarchist?) and spiritual (sectarian?) paths to social change (Kennard 1984). Hierarchical sects use the apparently sectarian rhetoric of "wilderness purists" (Graber 1976). Single-issue organizations cannot afford to make major compromises for fear of losing members and resources (Cook 1984: 4). Yet multi-issue groups like the Sierra Club and Friends of the Earth appear to be more radical than the relatively single-issue Wilderness Society or Ducks Unlimited.

On the other hand, the transformation of fraternal into hierarchical sects does not necessarily make them more conservative. It may merely foster professional staffs with continuing reformist impulses (Pratt 1972). The fraternal sect Friends of the Earth has been accepted as a legitimate expert intervenor in nuclear power and some other regulatory proceedings in the United Kingdom (Jasanoff 1983). It has shifted its objectives to cope with the last decade's economic pessimism in Great Britain, from being against economic growth to promoting house insulation for energy conservation and job creation (Lowe and Goyder 1983: 181—182). These are pragmatic steps one would not expect from a fraternal sect; they suggest that either FOE is shifting to a more hierarchical stance, or the Douglas—Wildavsky analysis is overstated.[15]

The picture is further complicated by the possible existence of "deep ecologists" (sectarians-without-a-group), who avoid confrontations with either ideological reformist environmentalists or the Establishment. Instead, they "prefer to act as exemplary models and to teach through acting" (Devall 1980). (They are thus kin to the "hermits" who are at zero grid and zero group, but provide new ideas and images for society (Douglas 1982a: 204—205, 234—235).) They also interact in loose networks rather than in untrustworthy hierarchies or "charismatic-fascistic" groups (Devall 1980).[16] If sectarian views can be held by groupless individuals, can those same views be the outcome simply of a group's need to maintain itself?[17]

It is not clear why Douglas and Wildavsky classify the Sierra Club as a hierarchical sect rather than as a sectarian hierarchy. The Sierra Club is commonly seen as a member of the American environmentalist mainstream, if its most radical member. Certainly the far more conservative members of that mainstream do not appear to be sectarian in organization (cf. Agar 1983: 103). Yet Douglas and Wildavsky do not provide a basis for clearly distinguishing hierarchy from sect, making it difficult to tell whether they are using these terms in the same way as their critics.

General categories

These problems of classification appear to be generic to grid/group analysis, at least in its current formative stage. Douglas (1982a: 205) sees intermediate mixtures of the four archetypal social structures as a possibility (cf. Douglas 1982b: 3, 6; Kelly 1982: 121; Ostrander 1982: 15;

Rudwick 1982: 238). Hampton (1982: 79) goes so far as to fear no clear differentiation of groups may be possible. But Thompson (1982b: 50—52; cf. Douglas 1982b: 4) finds such intermediate positions highly unstable, with little prospect for their members sharing world views and thus developing strategies for action.[18]

Gross and Rayner (1985) have proposed a quantification of grid and group variables which transforms the four cell nominal classification into a continuous two-dimensional range. This may allow future studies to distinguish grid/group types more clearly and resolve the classification difficulty. Certainly the widely varying sophistication with which GGA has been applied to date[19] is due in part to lack of consensus on which data and analytic methods are appropriate. Both proponents and critics of GGA must be more rigorous in both areas. There is no agreement even on whether individuals, or only groups, can be mapped onto the grid/group dimensions (see Douglas 1982b, and Gross and Rayner 1985). Questionnaires are poor instruments to identify either the structural position of a group (Gross and Rayner 1985) or its cosmology. Expensive and time-consuming observational methods must thus be used to test GGA postulates, which could be an obstacle to swift resolution of the issues raised in this chapter.

Closing Remarks

Risk and Culture does not succeed in describing modern American environmentalism in all its complexity. In some instances other theories concerned specifically with explaining social movements can supplement the usually static analysis of GGA with dynamic analysis. These theories include the approaches used by Walsh and Gerlach in their respective chapters. Olson's selective incentives theory is also helpful in this regard.

In other cases these hypotheses are direct competitors, and either GGA falls short or the data are inadequate to allow rejection of any hypotheses. The "modesty" of this critique refers primarily to this weakness of the data. Survey data can be useful in explaining environmentalism. But much of the information used in this chapter to criticize certain aspects of GGA derived from the same questionnaire approach which grid/group analysts see as not properly testing the theory. This problem is enhanced by both sides' tendency to seize upon apparently confirmatory data on group cosmologies, while failing to grapple with the more difficult problem of linking cosmologies with particular social structures.

Grid/group analysis is a provocative and potentially powerful theory. Its critics have failed to disprove its contention that cosmology stems from the varying cultural negotiations of individuals within different social structures. But grid/group analysts have generalized from too narrow and sketchy a data base. Sectarianism does not explain American environmentalism by itself, but there is not enough knowledge to reject sectarianism as a partial explanation. At best the current theory will yield correlations of systems of thought with systems of social organization. The causal mechanisms of this correlation must be explained more clearly. Grid/group analysis cannot be restricted to an elucidation of social rules, as in an older anthropology (Knorr–Cetina 1982; Whitten and Whitten 1972: 247).

This chapter has attempted to suggest ways in which the data base can be expanded, and the full richness of the grid/group model left untapped in *Risk and Culture* can be drawn upon. It has also suggested that GGA might be modified to take into account the contributions of social movement theories. In the spirit of this volume, more, and more rigorous research must be undertaken to resolve the controversy.

Notes

1. Many research approaches pursue explanations of the unusual and deviant (in fact they *define* something as being deviant), to the exclusion of analysis of widely-held norms and institutional behavior. These studies risk serving, or appearing to serve, the interests of the powerful who are the target of the social movement (Edelman 1977).

2. Erikson (1966) links group boundaries directly to hazard identifications. A collectivity will engage in ritual persecution — e.g., scapegoating, witchhunting — and thus actually manufacture a certain amount of deviance. This helps to redraw group boundaries or identities that appear threatened.

3. These names for the cells of the model have been criticized as indicating a bias in favor of individualism and hierarchism. Other titles that have been suggested include "manipulative individualist" (Thompson 1982b) and "independent/charismatic individualism" (Rayner 1984b) (low grid, low group); "survival individualist" and "stratified individualism" (high grid, low group); "manipulative collectivist" and "complex groups" (high grid, high group); and "survival collectivist" and "egalitarian groups" (low grid, high group).

4. Schein (1985) appears to have independently formulated these cosmological categories as basic assumptions of organizational culture.

5. Douglas and Wildavsky (1982: 198) confess a preference for hierarchism and individualism. Rayner (personal communication, May 1985) suggests grid/group analysis is an individualist theory.

6. Holdren (1983: 36) points out that the risk agenda of society as a whole may be more consensual than Douglas and Wildavsky allow (cf. Downey 1986).

7. Rhodebeck (1981) suggests that the deprivation thesis is more valid if the focus is shifted from personal grievances to feelings that one's reference group is deprived. Political action is more likely as a consequence if the political legitimacy of the current regime or its success in alleviating group deprivation is questioned, or if government intervenes in a group activity.

8. Ben-David (1981: 52) argues that there is no evidence for permanent links between one's social location and one's cosmology.

9. What counts as "public" is not clear in the GGA literature. One definition is what is said or done whenever three or more group members are gathered (Rayner, personal communication, May 1985).

10. More recently formed environmentalist groups are said to be more sectarian (Douglas and Wildavsky 1982: 130). Walker (1983: 403) found that the formation of new interest groups generally was a *consequence*, not a cause, of the new social regulation legislation of the 1960s and early 1970s. Grid/group analysis and resource mobilization theory would see this as evidence of the Douglas and Wildavsky notion of sects leveraging government resources. Another interpretation would suggest that these are in some respects sectarian environmental statutes. If sectarian groups were a consequence of these laws, what non-sectarian groups were responsible for creating them?

11. A looser interpretation of Olson's model is that the decision to join is due to the *combined* perceived net benefits of selective and purposive (i.e., ideological) incentives. But this version still leaves selective incentives as the dominant explanation of membership. Purposive incentives by themselves are insufficient (and unnecessary) to guarantee that enough people will join the group to ensure provision of the collective good.

12. For alternative models of group membership incentives, see Wilson (1973), Margolis (1982), and Conway (1984), among others.

13. Social movement organizations dedicated to personal change (as opposed to institutional change) tend to be decentralized and exclusive (Jenkins 1983: 542). This does not contradict Gerlach's chapter in this volume, which emphasizes the strength of decentralized social *movements* devoted to social change. Where organizations within such movements are decentralized, they tend to be less effective.

14. Thompson (1982b: 45) defines members of the lower levels of a hierarchy as sectarians, since they lack prescriptions on their behavior but cannot manipulate anyone else. These are contrasted to atomized subordinates, who also lack power but whose behavior is severely constrained.

15. However, Shrum (1984) suggests that antinuclear waste groups which include scientists on their staff have fairly high status and visibility in the policy debate. This is because "it takes fewer resources to oppose a technology than to create one." In other words, the FOE case may be an example of the strength of the resource mobilization perspective more than it is an example of the weakness of GGA.

16. However, Mitchell (1980) finds deep ecologists in conventional environmentalist organizations, where they are *less* likely to support income redistribution and *more* committed to the group than radical (sectarian?) environmentalists.

17. There are at least two types of hybrid groups. One type combines inconsistent values. For example, unions may be simultaneously concerned with preservation of the corporation (and thus jobs) and maintaining an adversary relationship with the bosses. The other type primarily mediates between contending social groups (e.g., the U.S. Congress), and thus emphasizes procedure over content.

18. Thompson's (1982b) three-dimensional grid/group model postulates *five* structures, rather than four. He allows for smooth, gradual transitions between high- and low-grid situations (e.g., between hierarchy and sect, or between atomized subordination and individualism). But movements from high to low group and vice versa are deemed sudden and "catastrophic." His use of cusps at the boundary between high group and low group allows "overlapping [cosmological] alternatives" to be "available as justifying resources to individuals in moral predicaments" (Douglas 1982b: 12; cf. Thompson 1982a: 305]. This idea is reminiscent of Edelman's (1977: 7, 10) discussion of contradictions in political rhetoric and belief, though Edelman considers this a universal, rather than structure-specific, condition. Although Thompson's model is superior to Douglas' earlier one in its allowance for intermediate positions, he does not provide evidence *for* gradual grid–grid movements and *against* gradual group–group movements. Nor is a definition of "instability" provided so that empirical tests could be made of a possible polarization of the environmentalist movement over time into pure hierarchist and sectarian groups. Both Downs (1972) and Mitchell (1980) suggest that this is unlikely for some time to come.

19. Sometimes only one data point is used per structure by grid/group analysts. Structuralist theories such as grid/group analysis are often more analytically than empirically driven (Befu 1980: 211).

References

Agar, Michael. Review of *Risk and Culture* by Douglas and Wildavsky. *Anthropological Quarterly*, April 1983, 56(2), 102—104.

Alinsky, Saul D. *Rules for Radicals: A Pragmatic Primer for Realistic Radicals*. New York: Vintage Books, 1971.

Barnes, Barry. 'The Conventional Component in Knowledge and Cognition,' in Nico Stehr and Volker Meja (eds.), *Society and Knowledge: Contemporary Perspectives in the Sociology of Knowledge*. New Brunswick, New Jersey: Transaction Books, 1984, 185—208.

Befu, Harumi. 'Structural and Motivational Approaches to Social Exchange,' in Gergen *et al.*, 1980, 197—214.

Ben-David, Joseph. 'Sociology of Scientific Knowledge,' in James F. Short, Jr. (ed.) *The State of Sociology: Problems and Prospects*. Beverly Hills, California: Sage, 1981, 40—59.

Bender, Thomas, *Community and Social Change in America*. Baltimore: Johns Hopkins University Press, 1978.

Berry, Jeffrey. *Lobbying for the People: The Political Behavior of Public Interest Groups*. Princeton, New Jersey: Princeton University Press, 1977.

Bloor, David. 'Polyhedra and the Abominations of Leviticus: Cognitive Styles in Mathematics,' in Douglas, 1982b, 191—218.

Campbell, Donald T. 'Social Attitudes and Other Acquired Dispositions,' in Sigmund Koch (ed.), *Psychology: a Study of Science*, Vol. 6. New York: McGraw-Hill, 1964, 159—162.

Cigler, Allan J. 'From Protest Group to Interest Group: Coping with Internal Factionalism and External Threats.' American Political Science Association, Washington, D.C., August 30-September 2, 1984.

Cobb, Roger W. and Charles D. Elder. 'Communication and Public Policy,' in Nimmo and Sanders, 1981, 391—416.

Cole, Gerald A. and Stephen B. Withey. 'Perspectives on Risk Perceptions,' *Risk Analysis*, June 1981, 1(2), 143—163.

Conway, M. Margaret. 'Social Values and Mobilization to Political Participation.' American Political Science Association, Washington, D.C., August 30-September 2, 1984.

Cook, Terrence E. 'Stands, Stances and Posturings: Levels of Sincerity in Political Positions.' American Political Science Association, Washington, D.C. August 30-September 2, 1984.

Cotgrove, Stephen. *Catastrophe or Cornucopia; The Environment, Politics and the Future.* New York: John Wiley and Sons, 1982.

Council on Environmental Quality. *Public Opinion on Environmental Issues.* Washington, D.C.: U.S. General Printing Office, 1980.

Cox, Kevin R. 'Residential Mobility, Neighborhood Activism and Neighborhood Problems,' *Political Geography Quarterly*, April 1983, 2(2), 99—117.

Devall, Bill. 'The Deep Ecology Movement,' *Natural Resources Journal*, April 1980, 20(2), 299—322.

Douglas, Mary. 'Cultural Bias,' Royal Anthropological Institute, Occasional Paper 35, 1978. Reprinted in Mary Douglas, *In the Active Voice*. London: Routledge and Kegan Paul, 1982a, 183—254.

Douglas, Mary. 'Introduction,' in Gross and Rayner, 1985, xvii-xxvii.

Douglas, Mary. Personal communication, May 26, 1984.

Douglas, Mary (ed.). *Essays in the Sociology of Perception*. London: Routledge and Kegan Paul, 1982b.

Douglas, Mary and Aaron Wildavsky. *Risk and Culture: An Essay on the Selection of Technological and Environmental Dangers*. Berkeley: University of California Press, 1982.

Downey, Gary L. "Risk in Conflict: The American Conflict Over Nuclear Power," *Cultural Anthropology*. November 1986, 1(4), 388—412.

Downs, Anthony. 'Up and Down With Ecology — The "Issue-Attention Cycle",' *Public Interest*, Summer 1972, 28, 38—50.

Edelman, Murray. *Political Language: Words That Succeed and Policies That Fail*. New York: Academic Press, 1977.

Erikson, Kai T. *Wayward Puritans*. New York: Wiley, 1966.

Ermann, M. David and William H. Clements II. 'The Interfaith Center on Corporate Responsibility and Its Campaign Against Marketing Infant Formula in the Third World,' *Social Problems*, December 1984, 32(2), 185—196.

Fallows, James. *National Defense*. New York: Random House, 1981.

Fischhoff, Baruch, Paul Slovic, and Sarah Lichtenstein. 'Lay Foibles and Expert Fables in Judgments About Risk,' in Timothy O'Riordan and R. Kerry Turner (eds.), *Progress in Resource Management and Environmental Planning*, Vol. 3. New York: John Wiley and Sons, 1981, 161—202.

Frohlich, Norman, Joe A. Oppenheimer, and Oran Young. *Political Leadership and Collective Goods*. Princeton, New Jersey: Princeton University Press, 1971.

Gergen, Kenneth J., Martin S. Greenberg, and Richard H. Willis (eds.), *Social Exchange: Advances in Theory and Research*. New York: Plenum, 1980.

Gerlach, Luther P. and Virginia H. Hine. *Lifeway Leap*; *The Dynamics of Change in America*. Minneapolis: University of Minnesota Press, 1973.

Graber, Linda H. *Wilderness as Sacred Space* (Monograph 8). Washington, D.C.: Association of American Geographers, 1976.

Green, John C. and James L. Guth. 'Who Is Right and Who Is Left? Varieties of Ideology Among Political Contributors.' American Political Science Association, Washington, D.C., August 30-September 2, 1984.

Gross, Jonathan L. and Steve Rayner. *Measuring Culture: A Paradigm for the Analysis of Social Organization*. New York: Columbia University Press, 1985.

Hall, I. M. *Community Action versus Pollution: A Study of a Residents' Group in a Welsh Urban Area* (Social Science Monograph No. 2). Cardiff: University of Wales, 1976.

Hampton, James. 'Giving the Grid/Group Dimensions an Operational Definition,' in Douglas, 1982b, 64—82.

Hardin, Russell. *Collective Action.* Baltimore: Johns Hopkins University Press, 1982.
Hill, Stuart. 'The Problem of Time for Policy Analysis.' American Political Science Association, Washington, D.C., August 30-September 2, 1984.
Hofstede, Geert. *Culture's Consequences: International Differences in Work-Related Values.* Beverly Hills, California: Sage, 1980.
Holdren, John P. 'The Risk Assessors' (review of *Risk/Benefit Analysis* and *Risk and Culture*), *Bulletin of the Atomic Scientists*, June/July 1983, 39(6), 33—38.
Jackman, Mary R. and Robert W. Jackman. *Class Awareness in the United States.* Berkeley: University of California Press, 1983.
Janeway, Elizabeth. *Powers of the Weak.* New York: Morrow, 1981.
Jasanoff, Sheila. 'Legitimating Private Sector Risk Assessment: A U.S.-European Comparison.' Society for Risk Analysis, New York, August 2, 1983.
Jenkins, J. Craig. 'Resource Mobilization Theory and the Study of Social Movements,' *Annual Review of Sociology*, 1983, 9, 527—553.
Johnson, Branden B. *Hazard Identification: Toward a Theory of the Social Construction of Risk.* Manuscript, 1986.
Johnson, Branden B. 'Life of the Land: An Environmental Action Group in Hawaii' (B.A. thesis). Honolulu: University of Hawaii, 1975.
Johnson, Branden B. 'Sierra Club Members in Western Pennsylvania.' Manuscript, 1982.
Kelly, George A. '"Les Gens de Lettres": An Interpretation,' in Douglas, 1982b, 120—131.
Kennard, Byron. 'Mixing Religion and Politics,' *Audubon*, March 1984, 86(2), 14—19.
Kilbourne, Brock and James T. Richardson. 'Psychotherapy and New Religions in a Pluralistic Society,' *American Psychologist*, 1984, 39(3), 237—251.
Knorr-Cetina, Karin D. 'The Constructivist Programme in the Sociology of Science: Retreats or Advances?,' *Social Studies of Science*, May 1982, 12(2) 320—324.
Kriesberg, Louis (ed.). *Research in Social Movements, Conflicts and Change*, Vols. 1—5. Greenwich, Connecticut: JAI Press, 1978—1983.
Lake, Laura M. 'The Environmental Mandate: Activists and the Electorate,' *Political Science Quarterly*, Summer 1983, 98(2), 215—233.
Lasswell, Harold D. 'The Continuing Revision of Conceptual and Operational Maps,' in Lasswell, Harold D., Daniel Lerner, and John D. Montgomery (eds.), *Values and Development: Appraising Asian Experience.* Cambridge, Mass.: MIT, 1976, 261—283.
Lipset, Seymour Martin and Earl Raab. *The Politics of Unreason: Right-Wing Extremism in America, 1790—1970.* New York: Harper and Row, 1970.
Lowe, Philip and Jane Goyder. *Environmental Groups in Politics* (Resource Management Series 6). London: Allen and Unwin, 1983.
Lowe, Philip and Michael Worboys. 'Ecology and the End of Ideology,' *Antipode*, July 1978, 10(2), 12—21.
Lowi, Theodore J. 'A Jaundiced Public Eye,' review of *The Confidence Gap* by Seymour M. Lipset and William Schneider, in *New York Times Book Review*, April 10, 1983, 7ff.
McCarthy, John D. and Mayer N. Zald. 'Resource Mobilization and Social Movements: A Partial Theory,' *American Journal of Sociology*, May 1977, 82(6), 1212—1241.
McCloskey, Donald N. 'English Open Fields As Behavior Towards Risk,' in Paul Uselding (ed.), *Research in Economic History*, Vol. 1. Greenwich, Connecticut: JAI Press, 1976, 124—170.
McCombs, Maxwell E. 'The Agenda-Setting Approach,' in Nimmo and Sanders, 1981, 121—140.
McKean, Margaret A. *Environmental Protest and Citizen Politics in Japan.* Berkeley: University of California Press, 1981.
McPhee, John. *Encounters with the Archdruid.* New York: Farrar, Straus and Giroux, 1971.
Margolis, Howard. *Selfishness, Altruism, and Rationality.* New York: Cambridge University Press, 1982.

Marshall, Susan E. 'Ladies Against Women: Mobilization Dilemmas of Antifeminist Movements,' *Social Problems*, April 1985, 32(4), 348—362.

Mazur, Allan. *The Dynamics of Technical Controversy*. Washington, D.C.: Communications Press, 1981.

Messina, Anthony M. 'The Political Life-Cycle of British Protest Movements.' American Political Science Association, New Orleans, August 29-September 1, 1985.

Metlay, Daniel and George Hoberg, Jr. 'Interest Groups and Activists: Their Motivations, Incentives, and Objectives in Risk Controversies.' American Association for the Advancement of Science, New York, May 26, 1984.

Mitchell, Robert C. 'How "Soft," "Deep," or "Left"? Present Constituencies in the Environmental Movement for Certain World Views,' *Natural Resources Journal*, April 1980, 20(2), 345—358.

Mitchell, Robert C. 'National Environmental Lobbies and the Apparent Illogic of Collective Action,' in Clifford S. Russell (ed.), *Collective Decision Making: Applications from Public Choice Theory*. Baltimore: Johns Hopkins University Press, 1979.

Nelkin, Dorothy. 'Blunders in the Business of Risk' (review of *Risk and Culture*), *Nature*, August 19, 1982, 298, 775—776.

Nimmo, Dan D. and Keith R. Sanders (eds.), *Handbook of Political Communication*. Beverly Hills: Sage, 1981.

Oliver, Pamela. 'The Mobilization of Paid and Volunteer Activists in the Neighborhood Movement,' in L. Kriesberg (ed.), 1983, Vol. 5, 133—170.

Olson, Mancur. *The Logic of Collective Action: Public Goods and the Theory of Groups*. Cambridge, Massachusetts: Harvard University Press, 1965.

Olson, Mancur. *The Rise and Decline of Nations: Economic Growth, Stagflation, and Social Rigidities*. New Haven: Yale University Press, 1982.

Orr, Robert H. 'The Additive and Interactive Effects of Powerlessness and Anomie in Predicting Opposition to Pollution Control,' *Rural Sociology*, Winter 1974, 39(4), 471—486.

Ostrander, David. 'One- and Two-Dimensional Models of the Distribution of Beliefs,' in Douglas, 1982b, 14—30.

Petersen, James C. and Gerald E. Markle. 'Expansion of Conflict in Cancer Controversies,' in L. Kriesberg (ed.), 1981, Vol. 4, 151—169.

Pratt, Henry J. 'Bureaucracy and Interest Group Behavior: A Study of Three National Organizations.' American Political Science Association, Washington, D.C., September 1972.

Rayner, Steve. 'Disagreeing About Risk: The Institutional Cultures of Risk Management and Planning for Future Generations,' in Susan G. Hadden (ed.), *Risk Analysis, Institutions, and Public Policy*. Port Washington, New York: Associated Faculty Press, 1984a, 150—178.

Rayner, Steve. 'The Perception of Time and Space in Egalitarian Sects: A Millenarian Cosmology,' in Douglas, 1982b, 247—274.

Rayner, Steve. 'Radiation Hazards in Hospital: A Cultural Analysis of Occupational Risk Perception,' *Royal Anthropological Institute News*, February 1984b, 60, 10—12.

Rhodebeck, Laurie A. 'Group Deprivation: An Alternative Model for Explaining Collective Political Action,'*Micropolitics*, 1981, 1(3), 239—267.

Rootes, C. A. 'On the Social Structural Sources of Political Conflict: An Approach from the Sociology of Knowledge,' in L. Kriesberg (ed.), 1983, Vol. 5, 33—54.

Rudwick, Martin. 'Cognitive Styles in Geology,' in Douglas, 1982b, 219—241.

Salisbury, Robert H. ' An Exchange Theory of Interest Groups,' *Midwest Journal of Political Science*, February 1969, 13, 1—32.

Salisbury, Robert H. 'Interest Representation: The Dominance of Institutions,' *American Political Science Review*, March 1984, 78(1), 64—76.

Schein, Edgar H. *Organizational Culture and Leadership*. San Francisco: Jossey-Bass, 1985.

Schueler, Donald G. *Incident at Eagle Ranch: Man and Predator in the American West.* San Francisco: Sierra Club Books, 1980.

Shaiko, Ronald G. 'Grass Roots Lobbying: Environmentalists as Constituents.' American Political Science Association, New Orleans, August 29-September 1, 1985.

Shrum, Wesley. 'The Organizational Context of Public Interest Science,' *Technology in Society*, 1984, 6(4), 299—312.

Simons, Herbert W. and Elizabeth W. Mechling. 'The Rhetoric of Political Movements,' in Nimmo and Sanders, 1981, 417—444.

Smith, Ken K. 'Social Comparison Processes and Dynamic Conservatism in Intergroup Relations,' in L. L. Cummings and Barry M. Staw (eds.), *Research in Organizational Behavior*, Vol. 5. Greenwich, Connecticut: JAI Press, 1983, 199—233.

Spector, Malcolm and John I. Kitsuse. *Constructing Social Problems.* Menlo Park, California: Cummings, 1977.

Stallings, Robert A. 'Patterns of Belief in Social Movements: Clarifications from an Analysis of Environmental Groups,' *The Sociological Quarterly*, Autumn 1973, 14, 465—480.

Thompson, Michael. 'To Hell with the Turkeys! A Diatribe Directed at the Pernicious Trepidity of the Current Intellectual Debate on Risk.' Center for Philosophy and Public Policy, Working Paper RC-5. College Park, Maryland: University of Maryland, Mary 1983.

Thompson, Michael. 'The Problem of the Centre: An Autonomous Cosmology,' 1982a; in Douglas, 1982b, 302—327.

Thompson, Michael. 'A Three-Dimensional Model,' 1982b; in Douglas, 1982b, 31—63.

Tillock, Harriet and Denton E. Morrison. 'Group Size and Contributions to Collective Action: An Examination of Olson's Theory Using Data from Zero Population Growth, Inc.,' in L. Kriesberg (ed.), 1979, Vol. 2. 131—158.

Walker, Jack L. 'The Origins and Maintenance of Interest Groups in America,' *American Political Science Review*, June 1983, 77(2), 390—406.

Walker, Jack L. 'Setting the Agenda in the U.S. Senate: A Theory of Problem Selection,' *British Journal of Political Science*, October 1977, 7(4), 423—445.

Walker, Jack L. 'Three Modes of Political Mobilization.' American Political Science Association, New York, August 30-September 2, 1984.

Whitten, N. E., Jr. and D. S. Whitten, 'Social Strategies and Social Relationships,' in B. J. Siegel (ed.), *Annual Review of Anthropology.* Palo Alto: Annual Reviews, 1972.

Wilson, James Q. *Political Organizations.* New York: Basic Books, 1973.

Wolf, Eric R. *Peasant Wars of the Twentieth Century.* New York: Harper and Row, 1969.

Wuthnow, Robert, James D. Hunter, Albert Bergesen, and Edith Kurzweil. *Cultural Analysis: The Work of Peter L. Berger, Mary Douglas, Michel Foucault, and Jurgen Habermas.* Boston: Routledge and Kegan Paul, 1984.

Wynne, Brian. 'Discussion Paper on J. Conrad: Society and Risk Assessment — An Attempt at Interpretation,' in Jobst Conrad (ed.), *Society, Technology and Risk Assessment.* London: Academic Press, 1980a, 281—286.

Wynne, Brian. 'Technology, Risk and Participation: On the Social Treatment of Uncertainty,' in Jobst Conrad (ed.), *Society, Technology and Risk Assessment.* London: Academic Press, 1980b, 173—208.

Yergin, Daniel. *Shattered Peace; The Origins of the Cold War and the National Security State.* Boston: Houghton Mifflin, 1977.

Zald, Mayer N. and John D. McCarthy. 'Social Movement Industries: Competition and Cooperation Among Movement Organizations,' in L. Kriesberg (ed.), 1980, Vol. 3, 1—20.

PART IV

Part IV

Agenda-Setting, Group Conflict, and the Social Construction of Risk

The focus of this section of the book is on the influence of social and cultural factors on the setting of risk agendas. The two chapters specifically examine the role played by group conflict in modifying and revising risk agendas.

The first chapter, by Harold Sharlin, focuses on the role of the media in transmitting risk information and in setting public risk agendas. In recent years, these topics have been the subject of intense scrutiny and debate. Considerable evidence exists that the media engage in selective and biased reporting that emphasizes drama, wrongdoing, and conflict (Combs and Slovic 1979; Freimuth *et al.* 1984; Kasperson and Gray 1982; Nimmo and Combs 1983). Evidence that such biased coverage affects public risk agendas is, however, contradictory. Some studies suggest that public perceptions of risk and levels of public concern are substantially affected by the amount and content of media coverage (e.g., Combs and Slovic 1979; Mazur 1981; Hawkes *et al.* 1984). Other studies have failed to find such an association (e.g., Fagnani 1977; Tichenor *et al.* 1973: 77). A partial explanation for this discrepancy is that public knowledge, attitudes, and behavior are significantly influenced by media coverage only when the media are the exclusive source of information on the risk (Dunwoody 1984: 7). For other risks, the data are inconclusive. The general literature on media agenda-setting (e.g., MacKuen and Combs 1981; Erbring *et al.* 1980; Shaw and McCombs 1977; Howitt 1982; Gandy 1982) is also inconclusive on this matter.

Sharlin's chapter specifically focuses on media reporting of the risks associated with the pesticide ethyl dibromide (EDB). In 1984, EDB was the subject of considerable media attention. At the center of the controversy was the mismatch between the information goals and objectives of the Environmental Protection Agency and those of the news media. Sharlin points out that the Environmental Protection Agency tended to emphasize in its press releases and other public communications the macro-risks of EDB — that is, the aggregate risk for the nation as a whole. These aggregate statistics were, however, of little interest to the media or individual citizens, who were more concerned about risks to individuals and about answers to such questions as: "Can I eat the bread? Should I

B. B. Johnson and V. T. Covello (eds.), The Social and Cultural Construction of Risk, 179—181.

throw away the muffin mix?" EPA's failure to provide micro-level risk information apparently exacerbated public fears and concerns. These fears and concerns were a major factor in the decision by EPA to ban the use of EDB. Sharlin's study illustrates, in a graphic and dramatic way, the ability of the media to influence the risk agenda of a federal agency — especially for risks that are unfamiliar and that are characterized by large scientific uncertainties. The study also illustrates the substantial extent to which the media are dependent on information sources in establishing their own risk agendas.

The second chapter in this section, by Janet Bronstein, focuses on the use of symbols in defining risks and in creating risk agendas. Bronstein's analysis of the controversy surrounding pneumoconiosis (black lung) and byssinosis (brown lung) illustrates an important theme in the literature on the social construction of risk — that attempts to define a substance or activity as a health or environmental problem frequently pose an actual and symbolic challenge to the legitimacy and authority of existing institutions. In the black lung and brown lung cases, Bronstein chronicles the struggle between workers and authorities over whether the risks of black and brown lung were real, significant, and widespread. Authorities — who originally denied the existence of a significant risk — later attempted to regain credibility by creating new definitions of risk. Bronstein shows how this process created fundamental changes in the attitudes and behavior of both workers and managers.

References

Combs, Barbara and Paul Slovic. 'Newspaper Coverage of Causes of Death,' *Journalism Quarterly*, Winter 1979, 56(4), 837—843.

Dunwoody, Sharon. 'Communicating Risk Information: What Role Do the Media Play?' Paper presented at the Annual Meeting of the American Association for the Advancement of Science, New York, May 25, 1984.

Erbring, Lutz. Edie N. Goldenberg, and Arthur H. Miller. 'Front-Page News and Real-World Cues: A New Look at Agenda-Setting by the Media,' *American Journal of Political Science*, February 1980, 24(1), 16—49.

Fagnani, Francis, *Le Debat Nucleaire en France*. Paris: Cordes, 1977.

Freimuth, Vicki S., Rachel H. Greenberg, Jean DeWitt, and Rose Mary Romano. 'Covering Cancer: Newspapers and the Public Interest,' *Journal of Communication*, Winter 1984, 34(1), 62—73.

Gandy, Oscar H., Jr. *Beyond Agenda Setting: Information Subsidies and Public Policy*. Norwood, New Jersey: Ablex, 1982.

Hawkes, Glenn R., Marc Pilisuk, Martha C. Stiles, and Curt Acredolo. 'Assessing Risk: A Public Analysis of the Medfly Eradication Program,' *Public Opinion Quarterly*, Summer 1984. 48(2), 443—451.

Howitt, Dennis. *Mass Media and Social Problems*. International Series in Experimental Social Psychology, Vol. 2. Oxford: Pergamon, 1982.

Kasperson, Roger E. and Arnold Gray. 'Societal Response to Three Mile Island and the Kemeny Commission Report,' in Howard C. Kunreuther and Eryl V. Ley (eds.), *The Risk Analysis Controversy: an Institutional Perspective*. Berlin: Springer-Verlag, 1982, 61—77.

MacKuen, Michael B. and Steven L. Coombs. *More Than News: Media Power in Public Affairs*. Beverly Hills, California: Sage, 1981.

Mazur, Allan. 'Media Coverage and Public Opinion on Scientific Controversies,' *Journal of Communication Research*, 1981, 31, 106—115.

Nimmo, Dan D. and James Combs. *Mediated Political Realities*. New York: Longman, 1983.

Shaw, Donald L. and Maxwell E. McCombs. *The Emergence of American Political Issues: The Agenda-Setting Function of the Press*. St. Paul, Minnesota: West, 1977.

Tichenor, Phillip J., Jane M. Rodenkirchen, Clarice N. Olien, and George A. Donohue. 'Community Issues, Conflict, and Public Affairs Knowledge,' in Peter Clarke (ed.), *New Models for Mass Communications Research*. Beverly Hills: Sage, 1973, 45—79.

Chapter 7

Macro-Risks, Micro-Risks, and the Media: The EDB Case

Harold Issadore Sharlin

Introduction

Although special in many ways, the EDB (ethyl dibromide) case illustrates the problems that regulatory agencies have when they must take regulatory action and assure the public that the risks in question have been dealt with adequately. It also illustrates issues that the press faces. Above all, it illustrates the barriers to communication presented by the different perspectives of regulatory agencies and individuals as well as the type of information each are most interested in.

An agency does risk analysis according to strict scientific methods that strive for exact answers. Communication with the public is a matter of explaining the meaning of the scientific results but these results do not answer the questions that individuals are asking. As a consequence the newspapers and television are able to set the news agenda by responding to the individual's concerns.

The specific question for this study was: What did EPA try to tell the public about the risk issues in the EDB case and what information did the public actually receive about the risks? The study covered the period from September 1983 to April 1984, when public anxiety reached a peak and rapidly subsided (Sharlin 1985).

The study was based on a variety of sources. The Agency's approach to the problem of conveying risk information was determined by examining EPA press releases and technical documents plus interviews and statements from EPA officials. A content analysis of fifty newspapers from thirty-two states, television programs as well as national press and weekly magazines was conducted in order to determine what message was sent.

Since reporters represent a significant segment of public opinion, the viewpoint expressed in this representative sample of newspapers reflects, I believe, more accurately the public perception than the spur of the moment responses to contrived public opinion poll questions.

In the short space of six months, during 1983—1984, the American public was faced with fast breaking stories about the threat to the public health from a widely used, effective pesticide — EDB. News stories reported that EDB was a potent carcinogen and that residue of EDB was

B. B. Johnson and V. T. Covello (eds.), The Social and Cultural Construction of Risk, 183—197.

being found in packaged food on supermarket shelves. A tidal wave of public anxiety swept the country. The public wanted specific easily understood information about the risk and assurance that timely action was being taken to eliminate the threat.

A basic assumption of the study was that the public perception of risk cannot be said to be right or wrong since perception is a matter of attitude. Public perception has come to mean, in the technical literature, public *misperception*(Otway and Kerry 1982). The rise in public anxiety over EDB was caused, according to EPA, by the public not receiving the EPA message correctly. But EPA did not address the question in individuals' minds, so television and the local press held the public attention and set the news agenda.

Two terms, micro-risk and macro-risk, will be used to identify what information EPA was conveying and how the public interpreted that information. Macro-risk represents EPA's technical assessment of the threat that EDB posed to the public health. Micro-risk, on the other hand, is the individual citizen's perspective and is the means of answering the unexpressed question: What does that mean to me, personally?

The EPA Message

On September 30, 1983, EPA announced that it had ordered an immediate suspension of EDB as a soil fumigant for agricultural crops and, at the same time, announced a cancellation proceeding against all other pesticide uses of EDB.

The reason for the suspension, the Agency explained, was that EDB had contaminated groundwater in several states and had been shown to be a carcinogen and mutagen in test animals. Therefore, EDB would be regulated as though it were a carcinogen. The EPA news release went on to explain that a suspension was the most restrictive measure that the Agency could take against the use of the pesticide under the law and that the action would immediately halt the sale and distribution of EDB as a soil fumigant, which represented 90 percent of the pesticide use of EDB. The EPA action removed the largest part of EDB in a timely manner. The Agency's actions supported the reassuring words of the press releases. The deeds and words were intended to convey the idea that EPA had identified an unacceptable risk and was moving to eliminate that risk.

The second part of the announced regulatory action, the cancellation proceeding, was intended to end all other major uses of EDB as a fumigant for stored grain, for spot fumigation, and for use in flour milling machinery. The principal risk to the public, EPA said in its decision document, occurred in the EDB residue, which remained in the fumigated grain that became part of the food chain. Because of EPA's concern that

EDB contaminated the food supply, the Agency was going to collect data to obtain a better understanding of the risk to the public. The EPA action was also intended to phase out the uses of EDB as a fumigant for citrus and tropical fruits. This phase out was to be completed by September 1984.

The September 30, 1983, EPA news release explained that under the Federal Insecticide, Fungicide, and Rodenticide Act (FIFRA), it was EPA's responsibility to insure that pesticides do not cause unreasonable adverse effects on the public or the environment. Whenever EPA determines that continued use of a pesticide poses an *imminent hazard* to the public health, the Agency is authorized to take emergency action.

The EPA order for emergency suspension of EDB was to take effect immediately, that is on September 30, and the cancellation action was to be effective in thirty days unless there were objections by registrants or users of EDB. If there were objections the cancellation proceedings might take as long as two years to complete, according to the news release.

A number of EPA technical program and information staff said that the Agency did not have sufficient information to take emergency action against EDB as a grain fumigant. There was serious concern in September by the Agency and by the public about contamination of groundwater and, therefore, the immediate and swift action was justified to eliminate the dietary risk from fumigating grains. According to one Agency official, EPA took responsible action by informing the public of areas of uncertainty in the scientific understanding of the effects of EDB.

The sense that the Agency tried to convey through action and through public information was that there was cause for concern about EDB but not for alarm; the Agency was moving quickly and in an orderly fashion to bring macro-risks under control. The situation was serious, but EPA tried to convey a sense that the risk was chronic or long range in nature and that there was no cause for alarm. But what of the micro-risk problem? As one EPA professional said, the information conveyed by the Agency did not answer the public's question: What does EDB exposure mean to me, the average American?

For the moment, EPA turned its attention to the immediate macro-risk problem of EDB's invasion into the groundwater. But for the reporters from the press the suggestion that EDB was in the food supply, that is, a micro-risk, affecting the very bread people eat, gave the reporters the local angle that they needed for their newspapers. As one EPA public information person described it, there was an absolute frenzy, after the September 30 announcement, of reporters trying to obtain copies of the press releases. They lined the halls of EPA scrambling for copies of the technical Position Document #4 of which 400 to 500 copies were distributed to reporters. What message would these reporters transmit to the public?

Macro-Risks, Micro-Risks, and the Media

1. *Television*

The major impact that the television medium had on the public micro-risk perception was in the creation of dramatic images. During the course of regular seven o'clock news programs and in news specials, several images were conveyed repeatedly about the risk from EDB. In one 'home movie' sequence shown on all the major networks, two workers who had come in direct contact with liquid EDB were being washed down with a hose in a futile attempt to detoxify them. The next scene showed an ambulance in the distance rushing the workers to the hospital where, the viewing audience was told, they died soon after.

Another repeated sequence shown was a close-up of a grain elevator worker who had been exposed to EDB over a thirty year period and had succumbed to a terrible, debilitating nervous disorder. The television images, the siren-sounding ambulance and the tragic picture of a man who could not control his erratic movements, transmitted the latest news about EDB. *Individuals* were harmed. All these images conveyed a micro-risk. EDB was toxic and, therefore, an immediate danger to individuals, while the risk from EDB in the food supply that concerned EPA was a chronic or long-term risk. The news stories missed the point that EPA tried to make and the television mistranslated the macro-risk assessment. The television images often lacked subtlety. One network displayed a background of a skull and cross-bones every time a reporter discussed the latest news about EDB.

2. *Newspapers*

The Twentieth Century Fund report, *Science in the Streets*, states that news is not education but information; ". . . it provides acquaintance with, not knowledge about dramatic events." The Report continues by saying that economic, social, and political influences converge on journalists to shape the definition of what is news (Twentieth Century Fund 1984). How was the information released by EPA interpreted by the press who reflect such influence and who inform but do not educate?

If the hypothesis of this study holds true that there is a macro- as well as a micro-risk perspective, how would that affect newspaper treatment of the EDB risk issue? Did the national press, such as the *New York Times*, the *Wall Street Journal*, and the *Washington Post*, treat questions about EDB differently from the local press? If so, what was the difference?

National press

The national perspective can be seen in viewing EPA and the Occupational

Safety and Health Administration (OSHA) as part of the national government so that their actions were seen as related to each other and as part of national policy. The *Times* frequently linked OSHA and EPA in the same article. Although OSHA's actions and EPA's actions do overlap, they affect different publics; in the local press the two agencies are usually dealt with separately. As an example of the *Times*' national status, there is its query to the Dow Chemical Company about their plans for EDB. The paper printed the statement from Dow that they were going to cease production of EDB for agricultural applications. Whereas local papers called chemical companies for technical information, the *Times* was interested in company policy.

Also, the *Times* was consistently more accurate and detailed than local newspapers on technical questions. Few local newspapers noted the difference between an emergency suspension, a cancellation, and the more general term, "ban", in their coverage. The *Times* had an expectedly high degree of accuracy on regulatory and scientific matters, since such major newspapers have specialized staffs of science writers as well as better access to expert information than do small local papers (Twentieth Century Fund 1984).

In September and October 1983, the *Times* coverage was of aggregates. The paper reported, among other things, that approximately 100,000 residents in Florida were affected by EDB groundwater contamination and that fifty-three companies manufactured EDB. In one article, the *Times* referred to the "emergency suspension" of EDB as a soil fumigant and noted that this was only the second time EPA had taken the emergency suspension action.

The perspective of the *Times* as a national newspaper was very much the same as that of EPA. That perspective as expressed in the newspaper's articles are at the macro-risk level. When 100,000 Florida residents are affected that is news for the *Times*, but the threat to the individual in terms of micro-risk is rarely news to a national newspaper.

The *Wall Street Journal* also had a national perspective and combined coverage of EPA's actions with those of OSHA. The technical competence of the paper, particularly in this regulatory area, was somewhat lower than that of the *New York Times*, as is evident in the *Times*' specific treatment of laboratory tests and the *Journal*'s general reference to "recent scientific evidence."

One *Washington Post* story in September 1983 dealt knowledgeably with the technical aspects of EDB risk from the macro-risk point of view. At the end of the article the micro-risk issue was raised when someone at the National Institute for Occupational Safety and Health (NIOSH) was asked by the reporter what the safe level of EDB in food and water was. The NIOSH scientist answered, "I'm a consumer, too. I eat all those things. And I don't know how to answer that question."

Time magazine, in an October 1983 feature article about EDB, flippantly referred to EDB as "the chemical of the month," saying that EDB, along with other "once obscure substances such as dioxin and PCB," was suddenly catapulted into the public spotlight. The article dealt more with the politics of regulatory action in Washington and minimized the macro-risk aspect while belittling the micro-risk view.

Local press

For most newspapers outside of the national press, reporters must translate EPA information into something newsworthy and find some handle that is local in interest. The reporters on most papers surveyed in this study used local sources of information, such as the resources of nearby universities.

The local newspapers analyzed in this study were the *Honolulu Star-Bulletin*, the *Honolulu Advertiser*, the *Arkansas Gazette*, the *Dallas Morning News*, the *Miami Herald*, the *St. Petersburg Times*, the *Tallahassee Democrat*, and the *San Jose Mercury*.

The *Honolulu Star-Bulletin* and the *Honolulu Advertiser* carried a number of articles during the early EDB regulatory action. These newspapers served a dual constituency of tropical fruit growers and that of consumers. The major emphasis in the papers during the September and October 1983 period was critical of EPA's actions from the micro-risk view and showed an anxious concern that the regulatory action might adversely affect Hawaii's fruit growers.

Hawaii was one of the four states, the others being Georgia, Florida, and California, in which EDB was discovered in the groundwater. The *Star-Bulletin* and the *Advertiser* gave some space in mid-September 1983 to the so-called seven year EPA delay in dealing with EDB. But, as time went on, the newspapers minimized the EDB risk and, in the first part of October, the report was that the amount of EDB in Hawaiian wells was only 0.09 parts per billion (ppb) compared to Florida where the EDB level in some wells was 800 ppb.

Just prior to the 30 September EPA announcement, the *Star-Bulletin* and the *Advertiser* reflected the growing anxiety in Hawaii about the possible effect of an EDB suspension. Banning EDB, the newspapers reported, was "a life-death matter" for tropical fruit growers. Officials in the pineapple industry said that there was no substitute for EDB and that possibly 85 percent of the crop shipped to Japan and the United States mainland would be stopped. Governor Ariyoshi and the pineapple growers association were reported to be arguing their case against the ban at EPA headquarters.

The Papaya Administration Commission conferred with the Animal and Plant Health Inspection Service (APHIS-USDA). APHIS was quoted as saying that there was little EDB residue in papayas; USDA was reported as favoring continued use of EDB for fruit flies; and the governor was quoted as saying that the threat from EDB, minimal as it was, might actually be coming from EDB treated gasoline used by the Department of Defense on the Islands.

The EDB threat to the public health in Hawaii was minimized in the local press and any severe restrictions on the usage of the pesticide was considered unnecessary as well as disastrous to the economy of Hawaii. The *Star Bulletin* and the *Advertiser* reported views as expressed by tropical fruit growers and the governor against a background of centralized decision making in Washington. The Hawaiian papers' reporting typified the local view taken by most newspapers in the country.

There was a sharp contrast between local newspapers and national television news on EDB in groundwater. National televised news about Hawaii showed viewers the beautiful, lush Hawaiian forests, and sparkling brooks with an announcer's voice-over saying ominously that EDB was poisoning this paradise. The local view expressed in newspapers was quite different. The argument was that, indeed, Hawaii is different from the rest of the United States and, if there was going to be a ban on EDB, it should not apply to Hawaii.

This slant on news began with selection of experts as background sources for articles. From EPA, the two Honolulu newspapers obtained information directly by telephone. But for the most part, the newspapers used sources such as the president of Maui Pineapple and representatives from the Pineapple Grower's Association, the Papaya Administration Commission, and the Hawaiian Department of Agriculture pesticide chief.

The (Little Rock) *Arkansas Gazette* and (Phoenix) *Arizona Republic* took up the EDB story from the point of view of the consumers. The *Gazette* came out with its story on 1 September, before the EPA soil fumigant announcement, because the paper was interested in the detection of EDB, "a potent cancer causing agent" in subsurface water samples. The *Arizona Republic* saw a story in the September 30 EPA suspension of soil fumigation by EDB and the cancellation of EDB as a fruit and grain fumigant. The paper quoted the 1977 National Cancer Institute report on the "Unprecedented high risk of cancer" from EDB. What made this report a story for the *Republic* was that EPA had found EDB in groundwater of farming states.

Both the *Arizona Republic* and *Arkansas Gazette* used local sources of information to obtain the local perspective and to be independent of centralized Washington-based information by seeking a skeptical analysis of EPA data by local experts. The *Republic*, for example, used the University of Arizona Council on Environmental Studies, the state Agricultural Economist, and the Arizona Department of Health Services Ambient Quality Bureau. The *Gazette* used the Arkansas Pollution Control Ecology Department, the state Health Department, and an enterprising reporter contacted the president of the Great Lakes Chemical Company, one of the manufacturers of EDB.

The *Dallas Morning News* and the *Miami Herald* wrote the stories from the dual perspective of consumers and citrus farmers. The *Morning News* reported that public pressure was mounting to do something about EDB in its October 1, 1983, story, but tempered their report by describing EDB as a "suspected carcinogen" that was used on Texas citrus fruit. The paper's sources of information included both an environmentalist group and the Texas Citrus Mutual Organization.

The *Miami Herald* began its coverage of EDB with two stories on October 1 and 3, 1983. The consumer perspective consisted of a complaint about the slow manner in which the EDB issue was being handled. That view was also tempered. Although EDB causes cancer in animals, there were no known human cancer cases that would demonstrate a health hazard in Florida. Farmers were not worried. They neither believed that a chemical that had been used for two decades was harmful nor that there was serious danger from EDB "if it is used carefully." The *Herald* used as its sources the Florida Department of Agriculture, industry sources, such as Monsanto Chemical Company, and interviews with farmers.

EDB had become news in September 1983 and, as evidence of its newsworthy status, other Florida papers picked up the story in November and December with their own perspectives. The *St. Petersburg Times* reported in November that EDB had become political in Florida. The state legislature had hearings on how the issue was being handled and at least one state representative attacked the Florida Agricultural Commission for being, as he said, too lax in the handling of the EDB problem.

In November the problem was seen as contamination of some Florida wells where, the *St. Petersburg Times* said, "dangerous concentrations" were found. The *Miami Herald* reported that EDB had spread to "private wells beneath populated subdivisions into kitchen sinks and now to the carefully landscaped, tree-lined country clubs." The contamination of wells near golf courses was the result of heavy use of EDB on those golf courses.

But the state's efforts to do something about the contaminated wells was hampered by public defiance and disbelief. The *Miami Herald* reported that home owners near the golf course neither trusted the government's warning nor believed that EDB levels represented a serious danger. A special Florida task force on EDB received no volunteered information to aid them in their investigation. The public was not only uninterested in EDB, in spite of newspaper reports, but were resisting government efforts to deal with the problem by refusing to respond to a public appeal for eyewitness accounts.

The NBC television program "First Camera" tried to raise a clamor about EDB. In one program a reporter asked a Florida housewife what she thought about the government not warning her about EDB in the drinking water. The housewife looked more puzzled than angry and replied, "it upsets me."

The Discovery of EDB Residue in Food Grains

The situation changed drastically in December. The Florida State Agriculture Commission, probably prodded by criticism from the legislature, began to test grain products on food store shelves on the supposition that, since EDB was used as a grain fumigant, there might be some residues in these products. The Commission found EDB in corn meal, grits, and hush puppies. The *Tallahassee Democrat* asked the question: "Should people be alarmed?" Florida Health and Rehabilitation officials, the *Democrat* reported, thought there was a potential risk but did not want to frighten the public. The extent of the threat was unclear, but the Health and Rehabilitation Service official said, "We could be sitting on a time bomb." He was right.

No longer was EDB contamination limited to a few isolated wells in Florida. It pervaded foodstuffs on grocery shelves. The news story had been elevated to the position of being a serious public concern. People were now asking: What does this news mean to me? Am I in danger? The *St. Petersburg Times* quoted a state agriculture inspector that people should not be alarmed if they had eaten the tainted products and they need not go to a doctor. On the other hand, he said, "I wouldn't make a diet of it now, don't misunderstand." The inspector said that he would return boxes from the contaminated batches to the store and that it was not their intent to scare people to death because the level of EDB was not dangerous. But even so, "We don't want it to be there."

Public confusion rose and a growing anxiety agitated the public in spite of the general reassurances of no danger. The very direct micro-risk questions, "Am I in danger?" and "What am I to do?" could not be answered in a straightforward manner. There was not enough data to give authoritative answers, but without any official direction the result was bound to be public confusion and a rising anxiety which was heightened by the sight on television of local stores removing boxes of grain products thought to be contaminated with EDB.

The information from Florida about EDB residue in food products made from grains called for a change in EPA's risk management strategy. But how was the Agency to respond to such headlines as the one in the *Tallahassee Democrat*: "How Dangerous is EDB? No One Knows For Sure." In the *San Jose* (California) *Mercury*, the headline read: "EDB: It Causes Cancer, It's in our Food, But at What Peril?" Grain food products were nationally distributed items and soon after the Florida discovery, officials in twelve states, including Arizona, (one of the local media study sites) were contacting government officials in Florida. In response to this new development, EPA took steps in January to shift strategy, convey to the public added reassurances, and regain control of the news agenda.

Risk Management — Grains

The discovery of EDB residue in food grains came as no surprise to EPA, which had noted EDB residues in grain, flour, and finished baked goods in Position Document #4 (page 58) issued on September 27, 1983. In 1956, before improved instrumentation, only parts per million (ppm) of EDB could be detected and no EDB residues were found in grain products at that level of instrumentation. On the basis of that information, EDB was given an exemption from residue regulation. Now in the 1980s, residues were detected by instruments that could measure in parts per billion (ppb). The questions remaining were the following: How large were the residues in grains, both raw and cooked? What was the safe level of residue?

The sense of those involved in making the EDB decision at EPA in January 1984 was that "the data available at the time the decision must be made" was inadequate but the public concern was forcing the issue. The dilemma faces all regulatory agencies at one time or another: "How can public officals make policy decisions on the basis of complex, poorly understood, and controversial technical data?" The next question is directly pertinent to this study: "And how can they [the public officals] communicate such data to those directly affected by policy decisions?" (Twentieth Century Fund 1984).

The EPA Message

On January 13, 1984, EPA announced through a news release that William D. Ruckelshaus, Administrator of EPA, was sending a letter to the governors of the fifty states requesting data on food products that contained EDB residue. The letter said, in part, "We have to first assess the risk involved before we can act." Mr. Ruckelshaus was considering recommending a national standard for permissible EDB residue, and he was considering whether further regulatory action was appropriate. There was an urgency to this request for data. Ruckelshaus was asking the governors to reply within the week that is, by January 20.

In addition to the information in the EPA January news releases, the Agency conveyed a different type of message through interviews that Mr. Ruckelshaus gave. The theme of his message was candor. He was quoted in the *Dallas Morning News* (January 8, 1984), "The truth is we don't know. We're operating in an area of enormous scientific uncertainty. We are operating with substances that the public is terribly afraid of. If they want absolute information, we can't give it to them." He added that EPA was "trying to proceed as sensibly and rationally as possible. I don't want to unduly alarm the public nor do I want them not to know about it." For a period of almost a month, that is, until the EPA decision was announced in February, newspapers around the country were interpreting this message in their own way.

Transmitting the Message

1. *Television*

In late December 1983, the major networks were showing the same film clips of food store employees removing boxes of grain products that were supposedly contaminated with EDB. Here was a visual and vivid portrayal. Products that individual consumers could identify with, located in stores that looked all too familiar, were being grabbed off the shelves and hurried off the scene.

The carrying off of cake mixes and other food grain products became a familiar scene on television news programs, as Massachusetts and California followed Florida's lead. There was no need for commentary. After the shock effect of seeing shelves cleared, the television reporters interviewed supermarket customers who usually exhibited one of three attitudes: (1) very serious concern about the threat to them of EDB; (2) a confidence that the situation was being handled by the authorities; or (3) an indifference to the whole situation expressed by a shrug of the shoulders and a repeat of the bromide, You have to die of some thing some time.

The other side of the story presented on television during December and January was EPA's point of view. But for EPA to maintain its aspect of candor, no guidance could be offered to those who were asking, what ought I to do. Mr. Ruckelshaus appeared on television at the end of December 1983 saying, "If we thought it [EDB] was a hazard, we would remove it." That same message was delivered on television at the beginning of January and then, near the middle of January, Mr. Ruckelshaus, in an interview, assured the public that EPA was gathering data and would set tolerance levels for EDB residues according to "our best estimate." The Administrator had to admit that the Agency's decision was being made more difficult because of scientific debate caused by scientific uncertainty. "We have," he said, "a lot of speculation and few facts."

2. *Newspapers: the National Press*

The *New York Times*, during December 1983 and January 1984, maintained its national perspective in reporting about EDB. First of all, the paper's focus was on EPA in Washington and much of the information reported in the news articles was either a direct quote by a spokesman for EPA or by EPA technical staff. What was considered newsworthy by the *Times* was the drastic action taken by the state of Florida. The paper went so far as to publish a list of the food products banned by Florida.

One very long article on January 31, 1984, the result of assiduous investigative reporting, gave a good overview of the whole EDB situation up to that point. Industry, environmentalists, lobbyists, and EPA staff were quoted liberally. An editorial, under the headline, "EDB: A NEEDLESS CANCER SCARE," on January 21, 1984, accused the government of ten years of foot dragging. The *Times* noted that Mr. Ruckelshaus faced a tough decision in setting tolerance levels for EDB residues. The editorial asked, "How did the Government get into so tight a spot? The law regulating pesticides is impossibly cumbersome."

The *Wall Street Journal* expresses stronger opinions on most subjects than the *Times*, and the *Journal* concentrates more on issues that affect business. So one would expect to find an article in its pages, such as the one that appeared February 3, 1984, under the headline, "Consumer Fears on Ethylene Dibromide Hurt Sales of Cereal, Bread and Other Foods." The *Journal* did a survey of supermarket chains around the country and reported that customers *were* concerned about EDB and that sales *were* affected. Although not drastically.

An article in the *Journal* on January 13, 1984, "Prefer Worms to EDB in Your Cereal?" Produced a flurry of letters from annoyed *Journal* readers. The author of the article was a former Dow Chemical Company employee. The author began by asking when was the last time that the reader opened a package of cereal product and found it wormy? He then described the effectiveness of EDB as a pesticide and its benefits. On the risk side, the author questioned the validity of translating the results of laboratory tests on mice to an effect on human beings, and raised again a long-standing scientific dispute. (The bulk of scientific opinion seemed to favor the validity of bioassays.) A recent report by a group of scientists concluded, "Although data from studies of rats and mice may not always be predictive of adverse health effects in humans, the scientific validity of this approach is widely accepted" (Committee on the Institutional Means for Assessment of Risks to Public Health 1983). The *Journal* article concluded, "Certainly it is inappropriate to allow highly problematic and minuscule risks to completely overshadow proven benefits."

The *Wall Street Journal* article at first appeared to be a micro-risk assessment, but after a brief reference to worms in your cereal, the author shifted ground to macro-risk questions, such as causing one in 100,000 persons to contract cancer and the problems of feeding a growing world population. One letter writer, a physician and father, giving his own micro-risk assessment, replied to the question, "Prefer Worms to EDB in Your

Cereal?" "You bet I do," he retorted. "I would take worms any day. Worms are harmless which is more than can be said for EDB." Of the four letters that replied to the *Journal* article, three found the personal, the micro-risk unacceptable.

Newspapers: the Local Press

In Florida, one of the states with legislative authority for setting residue standards on foodstuffs, the residue level for EDB was established as one part per billion (ppb) and for drinking water 0.1 ppb. The (Jacksonville) *Florida Times-Union* quoted the chairman of the Department of Entomology and Nemotology at the University of Florida Institute of Food and Agricultural Sciences as saying that one ppb was "ultraconservative." On the other hand, the paper noted that the state health officer thought that, "It is a bad chemical, one of the worst." No absolute statement could be made about the risk from EDB or, as the *Tallahassee Democrat* wrote: "Risk assessment is a complicated matter." Scientists examining similar situations can reach "wildly varying conclusions."

The *Tallahassee Democrat* also quoted a biostatistician at EPA who said that the government's risk assessments were not realistic. "There's too much unknown. The scientific knowledge is not there to do it." The risk assessment and the animal doses for the laboratory tests may have been invalid, but there was no way to prove that contention, the biostatistician said. Another official at EPA was quoted in the *Democrat* as say that EDB had to be eradicated because the slightest exposure caused cancer. This official said, "Any level of exposure entailed some risk."

The more sources questioned (the *Democrat* had a wide variety of them), the more difficult it was to state the micro-risk perception. The *Democrat* consulted with a government epidemiologist and entymologist at the University of Florida, four officials at EPA, representatives of the grain milling industry, and a representative of Dow Chemical (an EDB manufacturer). It also reviewed a 1978 test report in the University of Florida library. No authority was willing to give definitive answers and a conflicting statement could be found for any assertion. Public confusion and consternation was on the increase all during the month of January 1984, agitated by images on the television screen that included interviews with store customers whose faces exhibited puzzlement and frustration.

The *Raleigh News and Observer* noted that North Carolina could not ban any food products as Florida had done because of North Carolina law. What the state needed, the *News and Observer* claimed, was for EPA to set standards for the EDB residue, and this standard could be enforced in the state.

The *News and Oberver* took a less anxious position than the *Tallahassee Democrat*. The National Toxicology Program in Research Triangle in

North Carolina was quoted in the paper as saying "There is sufficient evidence for carcinogenicity (cancer causing properties) of EDB in experimental animals," but "the evidence of carcinogenicity in humans is inadequate." In addition, the North Carolina Food and Drug Protection Division found that EDB residue was either zero or below one ppb after the food product was baked, they assured the newspapers readers that, if their technical staff thought EDB was a really serious threat, the Division would develop some way of taking the contaminated food stuffs off the grocery shelves.

Conclusion

In February 1984, EPA issued an emergency suspension against use of EDB in grain milling or storage and, at the same time, established tolerance levels for residue EDB in foodstuffs. This last action answered the question: Can I eat the cake mix? The answer was yes, if the residue falls within the recommended levels. EPA started the process for removing all EDB from the food chain by a combination of actions. Public anxiety over EDB subsided and EDB disappeared from the newspepers and was heard of no more on television.

What should EPA have done? As far as this analysis goes, EPA performed in a manner that is standard for all federal regulatory agencies, that is, to use the best scientific advice in risk assessment and in risk management, while trying to reassure the public by explaining the risk-management process as clearly as possible. Then why the public panic? Did the media create the hysteria?

No, Mr. Ruckelshaus insisted at a press conference. The media did not create the turmoil, although (he might have added) the local newspapers and television did much to abet the anxiety. What started all the trouble? The answer to that troublesome question is unsettling: scientific investigation began the rise of public anxiety over EDB in the food chain. There is nothing in the scientific method that can be used to reassure the public once fears have been aroused over the threat of cancer in food containing parts per billion of EDB.

Scientists performing risk assessment must by the nature of the scientific method deal with uncertainty, and they are unable to give an exact answer to the question of micro-risk assessment: Can I eat the bread? One of the limits of science is that a negative cannot be proved. In scientific terms, one cannot say that there is no risk in eating bread with EDB residue. That limit to science explains why all during the increasing public concern EPA did not issue a simple statement about the safety of the food on grocery shelves. The public affairs office was carefully translating the precise statements made by the scientists, and the precise statements were absolutely correct but not reassuring.

With no reply available to the micro-risk question from EPA, the responsible media felt free to set the news agenda. The local newspapers sought local authorities, and these experts, not constrained by regulatory responsibility, spread the alarm. If scientists cannot prove a negative, one can add that spreading an alarm cannot injure the public health.

What solved the situation? Why did the media grow quiet so quickly? EPA decided to answer the micro-risk question not with scientific exactitude but by means of candor that relied on sound political judgment. The decision to establish acceptable levels of EDB in food was neither exact nor justifiable by any accepted scientific method. The media willingly allowed EPA to set the agenda as soon as EPA ended the crisis by making what was, in the best sense of the expression, a political decision.

EPA, like so many other regulatory agencies, has a dual role. The first is to make decisions about allowing and prohibiting and to set standards, in order to protect the health of the public. In that regulatory process, the public's views and comments must be taken into account. The second role is public information. Since the success of regulation depends on public participation in that process, a part of the agency's responsibility is to see that public participation is informed.

The regulatory role is that of macro-risk assessment and management. The public information role is that of micro-risk assessment. The distinction between the two is extremely important and until an agency learns to manage both roles equally well it will be confronting crisis after crisis.

References

National Research Council. Committee on the Institutional Means for Assessment of Risks to Public Health, Commission on Health Sciences, *Risk Assessment in the Federal Government: Managing the Process*. Washington, D.C.: National Academy Press, 1983.

Otway, Harry and Thomas, Kerry. 'Reflections on Risk Perception and Policy,' *Risk Analysis*, 1982, 2, 69—82.

Sharlin, Harold Issadore. 'EDB: A Case Study in the Communication of Health Risk,' Unpublished report submitted to Derry Allen, Associate Director, Office of Policy Analysis, U.S. Environmental Protection Agency, January 9, 1985.

Twentieth Century Fund. Report of the Twentieth Century Fund Task Force on Communication of Scientific Risk, *Science in the Streets*. New York: Priority Press, 1984.

Chapter 8

The Political Symbolism of Occupational Health Risks

Janet M. Bronstein

Introduction

In the space of a decade between the late 1960s and 1970s, coal mines and cotton mills in the U.S. were publicly acknowledged to be dangerous places to work because of the dusts in the air. Respiratory diseases endemic in the two workforces became legally compensable, while major capital investments were made to improve ventilation and make the workplaces safe. This sudden perception of danger in the workplace was supported by the labor movement and activist social reformers, but opposed by major business interests. What accounts for the success of both social constructions of risk in winning official legitimation and in altering the flow of resources in these industries?

This chapter explores a three stage process in the social construction of risk in the coal and cotton textile industries. The first stage was the genesis and spread of the belief that mills and mines are not simply bad, but dangerous places to work. The second stage was the official acknowledgment that these places are dangerous, while the third stage involved the implementation of solutions to resolve the problem and neutralize the public perception of workplace danger. These three general stages provide a useful framework for understanding the dynamics of public perception and action on risk and danger.

Theoretical Framework

"Danger" is a meaning which we selectively attach to elements of the environment. The theoretical framework for this chapter is based on an understanding of the social function of symbolic systems: symbolic meanings such as "danger" influence action directly by changing the definition of situations, and indirectly by affecting the social structures which constrain decisions. This discussion is drawn primarily from interpretive cultural anthropology and symbolic interactionist sociology.

Social Action and Society

People make decisions and take action based on understandings derived

B. B. Johnson and V. T. Covello (eds.), The Social and Cultural Construction of Risk, 199—226.

from culture (that is, the shared vocabulary of symbols) and within constraints imposed by other groups. By continually using symbols to accomplish our goals and by acting in reference to constraints, we continually recreate and reinforce our culture and our social structure (Ortner 1984: 152—158: Hall 1972: 37—42). As a whole, societies comprise groups and individuals acting through symbols and within constraints for two basic motives: to further our material interests and to maintain our sense of the meaningfulness of the world. These two sorts of motives are usually discussed in terms of *interest theory* and *strain theory* (Ortner 1984: 151: Geertz 1973: 201—204).

The decisions which further our interests and the activities which reduce cognitive strain do not take place in isolation; they are joint actions undertaken with others. Much energy is put into aligning our tasks with those of others and in matching their and our interpretations of the situation at hand (Hall 1972: 40—41). These joint actions involve give and take and various sorts of bargaining; they can be broadly termed negotiation (Strauss 1978: 2—11).

We are always negotiating with others in ambiguous situations where the implications of our own decisions are unclear. However, the negotiations themselves are patterned by differences in the resources available to the different parties. We and the people we interact with have different skills, are part of different social networks with different degrees of influence, and have differing amounts of power over others. To the extent that the negotiations which comprise our social lives are frequently repeated and are patterned by power differences, we perceive a sense of social stability and predictability, which can be termed a negotiated order (Strauss 1978: 2—11). In Hall's (1972: 45) words:

> The model of society that derives from the negotiated order is one characterized by a complex network of competing groups and individuals acting to control, maintain, or improve their social conditions as defined by their *self* interests. The realization of these interests, material and ideal, are the outcomes of negotiated situations, encounters, and relationships.

Although the negotiated order seems stable, it is always in flux, always emerging into a new form as a consequence of the multiple contingent ongoing negotiations on issues involving basic conflicts of interest (Maines 1977: 244).

The Role of Symbolic Systems

Negotiations comprise verbal and nonverbal modes of communication. In negotiations we attempt to convince others to define situations in a way that is beneficial for us. We establish mutual expectations and formulate our own role identities. The more deeply the verbal and nonverbal language of negotiation taps into a symbolic system which we and our negotiating partners share, the more effective the interaction will be in satisfying (or seeming to satisfy) both of our needs. Edelman (1971: 5) writes of the shared systems of meaning evoked in such interaction:

> It is through their power to merge diverse perceptions and beliefs into a new and unified perspective that symbols affect what men want, what they do, and the identity they create for themselves.

Negotiations best fill their desired ends of protecting our material interests and reducing cognitive strain when we share vocabularies of symbols (that is, our culture) with our negotiating partners. Yet the differences in our backgrounds, our experiences in life, and the range of interactions which we maintain, differences which are especially marked in complex societies, guarantee that in some interactions we will not be able to invoke symbols which bring us closer to mutual goals. Instead, the same objects will signify different concepts, evoking different value-related emotions in ourselves and others. Alan Batteau (1978—1979: 19) refers to negotiations which occur between groups with differing meaning systems as "working misunderstandings":

> Examples of such situations would include nearly all class relationships, such as those involved in labor relations in the mining industry. While at one level both the miner and the operator share certain ideas (such as common lore about the mines), at a more significant level what divides them is their *different* ways of thinking: for many operators, the bottom line on the balance sheet is the basic goal: anything that enhances profits is good, while anything, such as safety measures, that costs money without yielding a greater return is, at best, a nuisance. The miner, on the other hand, is interested in his life, his family, his safety, and his paycheck. That the miner is not religously devoted to mining more coal causes some operators to see him as lazy. The miner, on the other hand, often sees the operator as callous and exploitive. Thus, when the two face each other across the picket line, the crucial fact for determining their behavior towards each other is not their common understandings but their common *misunderstandings*.

Batteau points out that while these misunderstandings are inevitable, they can also be cultivated either by the dominant partner as a way to maintain control over an interaction (as welfare workers will define all autonomous actions of clients as cheating), or by both partners as a way of expressing solidarity with potential allies and thus altering the power differential of the interaction. It is also important to note that symbolic systems are uniquely mutable. New objects or acts can substitute for traditional ones in symbolizing a stable value or concept; old symbols can be attached to new concepts to increase their acceptability (see Cohen 1974). Batteau (1978—1979: 23) provides an example of both changes in symbolic systems and the purposeful cultivation of misunderstanding from the situation of the coal miners, whose meaning systems shifted as they tried different model of interacting with coal operators.

In this process of adjustment, continuously provoked by negotiation and testing, relationships, alternatives and also values are often redefined. In their collective relationship to the mine bosses, coal miners redefined relationships (de-emphasizing personal loyalty to the employer and weakening family ties), alternatives (adding peer group solidarity and confrontation to the range of survivial strategies available to them) and values (for the first time, for some, placing greater value on economic position than on attachment to place.)

Whether negotiations proceed on the basis of shared understandings or shared misunderstandings over meanings and values, the skillful use of symbols affects the outcome, helping to determine which parties' goals are best satisfied.

The Meaning of Danger

"Danger" is a concept which means vulnerability to harm. When it is attached as a meaning or symbolic loading to an object or act, that object then evokes fear and avoidance behavior. In her classic study of the opposition of the concepts of purity and danger, Mary Douglas (1966: 96) defines danger as the perceived disorder which spoils the patterns we create together to make sense of the world. Marginal states of passage (e.g., between childhood and adulthood) are dangerous because their subjects are neither one thing nor another. People with the power to harm but without predictable social roles (e.g., witches or criminals) are dangerous. A study conducted in a multi-ethnic low income housing project found that danger was perceived to lie in anybody or any place which was unfamiliar. On the other hand, teenage street-fighters who were known by name and residence were not considered personally dangerous (Merry 1981: 124—165).

In a different domain, Kenneth Hewitt (1983: 20) makes the point that natural events are labeled as disasters as part of an ideology which maintains that technology can control nature. In less technocratic societies, calamities are part of the human calculus. As our definitions of purity and order become more restricted, however, more elements in life are seen as dangerous, and thus as disruptive but potentially controllable: "Natural hazards, like diseases, poverty, even death, become simply the unfinished business of our endeavors."

This increase in confidence about what can be controlled helps to explain the paradox of modern society described by Douglas and Wildavsky (1982: 13): as our lives become safer from a statistical point of view, we seem to grow more and more concerned about danger. In complex societies some aspects of experience become increasingly known and familiar, and thus controllable and less dangerous. Other aspects of life undergo change, become less familiar, less predictable, and potentially more dangerous. It is possible that more people are harmed in non-mechanized than in mechanized workplaces, but mechanization will seem more dangerous because it is unnatural, uncontrollable, and unfamiliar. Changes in the perception of danger are not uniform, because the fragmentary nature of complex societies means that different groups of people are familiar with different experiences, and have differential amounts of power to control the circumstances of their lives.

As the symbolic loading of danger is removed from some objects and attached to others, the roles which these symbols play in social interactions change as well. Symbols of order which once functioned to unite people, like the smoke-belching factory which once represented prosperity, now divide those who perceive the smoke as dangerous from those who do not accept the redefinition. At the same time, new shared understandings of danger now unite groups who never interacted effectively before. In this chapter we shall explore how the new understanding of coal mines and cotton mills as dangerous united workers and social activists, while dividing them and the public from the coal and textile industries. Use of the new symbols altered the pattern of social interaction, eventually changing the distribution of resources in these economic sectors.

Stage One: The Definition of Workplace Danger

How did the dust in coal mines and textile mills come to be defined as dangerous? To answer this question we need to identify the groups who first used workplace danger as a communication device in their regular negotiations, and examine why the use of this symbol was effective in helping them accomplish their goals. The redefinition of mines and mills as dangerous was an outcome of interactions between activist social reformers, politicians, and employed and retired workers. By adopting the vocabulary of workplace danger, both groups established newly legitimate social roles and the moral credibility to demand change in the economic and social orders.

Danger in the Coal Mines

The setting

The physical danger of mine accidents has been part of the lore of the mines for centuries. Periodically this danger has been recognized by government authorities, who pass laws which mandate safety precautions, but rarely provide for effective enforcement. Curran (1984: 13—15) has suggested that these mine regulations have had the primarily symbolic values of demonstrating the government's intention to protect miners from danger. The laws are written and enforced in such a way that serious disruptions of the industry are avoided.

Lung disease has also been recognized for centuries as being prevalent among coal miners (cf. Rosen 1943). However, with the growing dominance of the germ theory in medicine in the late nineteenth century, miners' lung disease was discredited as a medical category. Leading occupational medical specialists in Britain came to believe that coal dust was inert, and simply collected in miners' lungs without damaging them. Only miners exposed to silica crystals while mining were thought to be at risk for disease. However, in the 1930s physicians on government-sponsored commissions re-examined the evidence on coal miners' disease in light of the numbers of coal miners rejected for compensation benefits because their severe lung problems could not be diagnosed as silicosis. Coal workers' pneumoconiosis (CWP) was established as a unique compensable disease in the British Factory Act of 1942 (Meiklejohn 1952: 215).

The belief that coal dust is inert and that only silica causes disease held sway in the United States through the 1950s, despite ample discussion of CWP in the British medical literature. Smith (1981: 344—348) asserts that the belief that lung disease was normal for miners, or that respiratory symptoms were a manifestation of anxiety, fit the social context in which medicine was practiced among miners. Once the prepaid medical group practices financed by the union-controlled Welfare and Retirement fund replaced company doctors as the source of health care of U.S. miners, reports of high rates of lung disease associated with coal dust became more common. The Fund employed a physician to educate the medical profession about coal workers' dust diseases (Smith 1981: 348; cf. Kerr 1968). Yet it was not the absence or presence *per se* of medical information about lung disease in coal miners which finally brought the danger of the coal-mine atmosphere to public attention in the late 1960s. Rather it was the use to which these medical facts could be put by those fighting for social change in the Appalachian coalfield region which led to the new public understanding of danger in the mines.

Between 1950 and 1970, coal mines in Appalachia underwent massive mechanization; their workforce was reduced by 70 percent. Unemployment devastated the coal industry-dependent communities and there was large-scale migration to northern industrial cities. Although the mine workers' union maintained its militant rhetoric of defending miners' rights, health and pension benefits were reduced, and president John L. Lewis cooperated with the economic transformation of the industry. Lewis maintained tight control over the leadership of the union. Over these two decades miners and their families experienced poverty, unemployment and the accompanying sense of social marginality, the dispersal of their kinship networks across the country, and the loss of faith in important social institutions like the union (Smith 1981: 350—352; Marschall 1978). They were victims, vulnerable to unfamiliar forces outside their control.

Once proverty in Appalachia came to national attention in the early 1960s, federal, church, and university related programs were implemented to improve the education, health, and income of the poorest residents. Community Action Agencies became sources of local jobs and distributors of government resources. The VISTA program brought young volunteers from outside the region to live in communities and organize people to work for self-improvement. These War on Poverty programs generated two sorts of local opposition. There was generalized resentment that decisions were made without consulting the residents of communities and that outsiders were brought in as role models, and there was specific resentment on the part of elected officials and established leaders against rival sources of jobs and resources which they did not control (Ball 1969: 152—156; Plaut 1976: 298—299; Good 1967: 166).

Initial explanations of why poverty existed in Appalachia and why government poverty programs failed focused on the supposedly unique cultural traits of Appalachians: fatalism, personalism, and present-orientation. This culture of poverty rhetoric (Fisher 1976: 146) was useful for the religious, government, and medical professionals who were frustrated in their attempts to help the Appalachian poor behave more like the upwardly mobile middle class. However, those VISTA volunteers and others who were critical of middle-class values saw these explanations as attempts to blame the victims for their economic conditions; they sought instead to blame the absentee coal corporations and state and federal governments for promoting the root causes of poverty in the mountains. Likewise, the symbolic system inherent in the culture of poverty language did not help unemployed miners and their families understand their social conditions, nor negotiate for better treatment by the union, the coal companies, and the government. By the late 1960s, the time was ripe for new sorts of explanations and new symbols to use in acting on the social and economic conditions in the Appalachian coalfields.

The Black Lung Movement

Lung disease from coal-mine dust and from cigarette smoke was endemic among older miners in the 1960s. In 1968, three physicians who were concerned about the ongoing lack of recognition of miners' lung disease in the U.S. formed the Committee of Physicians for Miners' Health and Safety (Fox and Stone 1980: 46). With the help of a few VISTA volunteers, they made contact with some West Virginia coal miners who were discontented with their union's complacency toward the problems of unemployed miners. To the union activists, the black lung story fit with their other experiences of frustration with the union and the coal industry. A deadly explosion in a West Virginia mine in November 1968 added fuel to the sudden public and worker interest in mine safety. In her oral history of the black lung movement, Smith (1984: 21) recounts:

Throughout the winter of 1968—1969, black lung rallies in the coal camps of southern

West Virginia grew in size and exuberance. Every weekend, miners, miners' wives, and widows met in churches, schools, local union halls, and — when there wasn't a place large enough to accommodate them — out of doors in the snow. The main attraction was usually "a real dog and pony show" put on by the physicians, who would bring along a set of real lungs from a victim of black lung. Clara Cody remembers the impression the doctors' performances made:

> "We would open with a prayer. [Dr. I.E.] Buff would take them lungs and act . . . well, he went a bit too far sometimes. It's like this, you see, very few people ever saw a set of lungs. He told the men about the bad lungs, and he'd yell, 'Feel' em! Feel 'em!' "

The rallies galvanized the spirit of unity and promoted the belief that people were *entitled* to redress for the injustices they had endured, as Mildred Mullins recalls:

> "I probably was fighting mad. 'Cause, you know what got me, you'd go to these rallies, there'd be these old men there, much older than my husband, couldn't get no breath at all. You'd think they'd go through the ceiling trying to get air. Before they'd get done telling you how dirty they'd been done, they'd be crying. Didn't get no pension, didn't have no hospital card, didn't have nothing to live on. Now, that gets next to you. It still makes me mad when I think about it."

The miners began to campaign for black lung to be added to the West Virginia workers' compensation statute as a compensable disease. The activists were aware that the union leadership would oppose such unauthorized political activity, but as the black lung cause grew more popular, union president Tony Boyle began to lose credibility by opposing it. During the state legislative session in February 1969, a series of wildcat strikes brought 10,000 West Virginia miners off their jobs. The massive strike convinced state, federal, and union leaders that coal-mine dust disease was an issue of major concern to rank-and-file miners.

Catalyzed by the strike, the insurgent movement grew within the union. In May 1969, Jock Yablonski mounted the first serious election challenge to the UMW leadership. In December 1969 he and his family were murdered. Over the next three years, the rank-and-file Miners for Democracy worked in tandem with the community based Black Lung Association to challenge the UMW hierarchy, to monitor black lung legislation, and to advocate the cause of retired, impoverished miners. Both organizations were led by retired miners and designed to appeal to and serve miners and their families. Young activists from the War on Poverty who remained in Appalachia found roles as staff, strategists, and trainers for these organizations. Smith (1984: 24) quotes community organizer Gibbs Kinderman:

> The black lung, the strike in 1969, was a real spontaneous thing, as far as I know. It never developed any organizational structure to keep continuing. There was no communication structure; the organization was practically gone. So we decided in DRA (Designs for Rural Action, an organization formed of former War on Poverty volunteers) that our main priority was going to be the kind of thing that we knew how to do. That we could get people together, get them enough back together on black lung. But it was more using black lung to get at the union. The black lung thing everybody could agree on. We didn't have to directly attack Boyle; you could show by contrast that the union wasn't doing anything.

The rhetoric of danger

With the birth and growth of the black lung movement, physicians, miners, and activists came to believe that prolonged exposure to coal-mine dust inevitably causes lung disease. Lung disease in coal miners was no longer considered a natural or inevitable occurrence; rather it was an avoidable condition which arose because the atmosphere in the mines was dangerous. As the theoretical section of this chapter suggested, once beliefs about what is "pure" or naturally ordered became more restricted, the scope of what was dangerous and what could potentially be ordered and cured expanded. Lung disease was abnormal; dust in the mines is dangerous. Mines can be cleaned up; mine operators and the government have the responsibility to clean them up.

The symbolic system surrounding black lung converted feelings of victimization into feelings of anger against mine operators, the government, and the unions. As indicated, the feelings of being victimized and vulnerable were widespread among the Appalachian poor and among the miners facing the transformation of the coal industry. Thus the appeal and the utility of this new vocabulary of mine-atmosphere danger, symbolized by a pair of blackened lungs, reached far beyond that segment of miners with chronic lung disease. Smith (1981: 351) writes:

> Black lung disease in a sense became a metaphor for the exploitative social relations that had always characterized the coalfields, but worsened during two decades of high unemployment, social dislocation, and rank-and-file weakness *vis-à-vis* the coal industry. The goal of black lung compensation represented, in part, a demand for retribution from the industry for the devastating human effects of its economic transformation.

The language of black lung united working and retired miners and their families, who might otherwise have been competing for scarce jobs and pension benefits. It gave them highly respectable moral grounds on which to oppose their employers, although their anger and feelings of victimization may have stemmed more from economic grievances. Activists and miners could talk to each other about political action using the language of black lung, circumventing the communication gaps which had plagued other Appalachian social programs. Doctors could rouse the interest of the public and of lung disease victims themselves about a serious health problem which had been ignored; certainly it was in their interest to define disease as an abnormal state. Finally, black lung symbolized in a meaningful, but not directly threatening, manner the serious complaints which rank-and-file miners and activists had about the leadership of the United Mine Workers Union. The symbolic loading of danger placed on the atmosphere of coal mines was multifunctional, and thus was rapidly adopted in the United States at this time.

Danger in Textile Mills[1]

The setting

The high prevalence of lung disease among textile workers, like the disease among miners, was known to physicians in nineteenth-century Britain. The term byssinosis (from *byssus*, or fine white fiber) was coined in 1872 to describe a progressive disease, much like chronic bronchitis in its final stages, which affected mill workers. Byssinosis was declared a compensable disease in the British Factory Act of 1942. During the 1920s, improved mill ventilation in Britain was thought to be resolving the problem, so lawmakers did not anticipate having to pay benefits to very many disabled textile workers. However, sophisticated epidemiological studies in the late 1940s and 1950s documented continued widespread obstructive lung disease among workers exposed to raw cotton dust. By 1960, one thousand British workers were receiving compensation for byssinosis (Schilling 1970: 8—12; Harris *et al.* 1971: 199—200).

Since the cotton being processed in Britain was grown in America, researchers on both sides of the Atlantic were puzzled by the apparent absence of byssinosis in U.S. textile mills. A few investigations of U.S. textile workers in the 1930s had found no elevated rates of chronic lung disease. In 1964 a Dutch researcher, Arend Bouhuys, received a U.S. Public Health Service grant to conduct an epidemiological study among Georgia textile workers, using methodology identical to the British studies. No textile manufacturer would allow him access to textile workers, and Bouhuys studied instead the rate of lung disease among inmates working at the cotton mill of the Atlanta Federal Penitentiary. In 1967 and 1968, physicians with the North Carolina Board of Health surveyed workers in two North Carolina mills. Although the data from these studies were not immediately published, investigators and public health officials became aware that the rates of byssinosis were as high among workers in the southern U.S. textile states as in Britain. About 30 percent of the cotton textile workforce was thought to have significant, occupationally related lung disease.

Meanwhile, in the economic and political arena, the plight of the southern textile workers was already a cause for outrage and a rallying point for labor, civil rights, and social activists. The textile industry dominated the political and economic life of small southern towns. Textile employees were the lowest paid manufacturing workers, and had minimal pension benefits. The textile industry systematically refused to hire blacks, and employers were known to use every conceivable strategy to oppose unionization of their workforce. Supreme Court cases, National Labor Relations Board rulings, and Congressional hearings on the union organizing issue received considerable publicity over the decade of the 1960s. In

1969 Cannon Mills was charged with civil rights violations for racial discrimination in housing. Still the situation in the southern textile region differed from that of the Appalachian coalfields in the same decade. Impoverished textile workers were neither a major nor a powerful segment of the population, and their living conditions were gradually improving, not worsening like those of miners and their families. The textile industry was not controlled by a few absentee corporations, but was dominated by locally based, often family owned mills. The extremely limited degree of unionization meant that few textile workers had a sense of belonging to a distinct class of people, a consciousness which was already highly developed among coal miners.

The Brown Lung Movement

In 1967 the Johnson Administration began to work with labor groups on an Occupational Safety and Health (OSH) bill which would give the federal government the power to regulate working conditions in all industries. Hearings on a proposed bill in 1968 featured testimony on danger in coal mines and other industries, but made no mention of any danger in textile mills. This bill failed to reach the floor in that session of Congress, but was introduced again in the spring of 1969. By then, separate legislation had been introduced in Congress to regulate conditions in coal mines, in response to the widely publicized Black Lung Movement. Lobbyists for the general OSH bill needed new evidence to support the need for federal regulation of health conditions in other industries.

In May 1969, the Textile Workers Union of America passed its first resolution calling for some government and industry action of byssinosis; the move was in response to the first publications and Public Health Service announcements of the ongoing studies in textile mills. Congressman James O'Hara of Michigan, sponsor of the OSH bill, promptly introduced the union's motion into the Congressional record, commenting that it indicated the serious occupational health problems lurking in U.S. industries. His action provoked a comment from a conservative textile trade journal that O'Hara was doing the bidding of the labor union hierarchy, and that there was no such disease as byssinosis.

In August 1969, the new Nixon Administration introduced a second OSH bill into Congress. The Republican version included fewer of the strict provisions favored by most Democrats and the labor movement. Consumer advocate Ralph Nader responded with a widely publicized letter to Secretary of HEW Robert Finch, in which he described the occupational danger of dust in textile mills to support the need for a stricter bill. Using the widely known and broadly supported black lung movement as a reference point, Nader invented a new, popular sounding name for byssinosis; he called it "brown lung." In his letter to Secretary Finch (Nader 1969: 23117—23118) he linked the entire occupational health issue to existing beliefs about the poor treatment which textile workers received from the textile industry. He blamed brown lung on the profit motive of big business, and justified the intervention of the federal government on the grounds that, as in the civil rights situation, the southern textile states were failing to protect the rights of the poor. Just as rank-and-file miners could respond to physicians' talks about black lung because the story fit their experiences with the coal industry, the media, Congressmen, labor groups, and other liberal supporters could respond to Nader's discussion of brown lung because it fit the beliefs and experiences they had with the plight of textile workers.

Throughout 1969 and 1970, members, officials, and staff of the textile workers' union testified on brown lung at all of the Congressional hearings on the OSH bill. Senator Harrison Williams held an entire day of hearings in Greenville, South Carolina, on the issue. In the summer of 1970, students at Nader's Center for Responsive Law made a documentary on Kannapolis, North Carolina, the Cannon Mills company town, which included a segment on lung disease among textile workers. The documentary aired on television the week before Congress voted on the bill, and was followed by a press conference in which Nader endorsed the labor-backed version. The Occupational Safety and Health Act was passed and signed in December 1970. A few months later, the new Occupational Safety and Health Administration announced that cotton dust was one of the top five priority substances for regulation (see Ashford 1976: 86; Page and O'Brien 1973: 180).

The rhetoric of the Brown Lung Movement was initially directed at the national level political groups whose support was needed to pass the OSH bill. Few textile workers heard the term brown lung until five years later. In 1974, a small group of young community organizers came to North and South Carolina with the funds to begin a broadly defined citizens' self-help movement. Although for several years it had been legally possible in these two states to claim workers' compensation benefits for byssinosis, very few textile workers had done so; neither their employers nor their physicians had informed them that cotton dust exposure cound be a factor in chronic lung disease. The organizers concentrated on locating textile workers with chronic lung disease, informing them that their health problems stemmed from years of work in the mills. They encouraged the workers to file compensation claims for themselves and to publicly condemn and press for change in the industry's policies. As these activists explained in a request for funds to finance a region-wide Brown Lung Association:

> The forging of the Brown Lung Association not only provides a tool for pressuring local, state and federal agencies to address the problems of Brown Lung, but serves as a model for what people can do in the Carolinas. Despite the success of civil rights activities a decade ago, the tool of social activism is still hardly known here. In states with the lowest level of social services in the nation, the Brown Lung Project is creating a vehicle for

profound social protest, and a force of tremendous social change. The Brown Lung Associations are challenging the omnipotence of the textile manufacturers head on, in a way that has never happened before. If they win, the southern textile scene, and the lives of the 700,000 people who live in it, will never be the same.[2]

As in the black lung situation then, a major motive of the activists who were promoting the belief that cotton dust is dangerous was to encourage people to challenge established institutions and to work for social change.

The small group of disabled textile workers initially contacted by the community organizers became the membership base of local chapters of the Brown Lung Association throughout the Carolinas, and eventually in Georgia and Alabama. The activists became staff members, and funding was secured from federal and state agencies and private foundations. Chapters sponsored brown lung screening clinics, where retired textile workers received pulmonary function tests. If those who attended the clinics had signs of obstructive lung disease and a history of exposure to cotton dust, they were given the names of plaintiff lawyers who specialized in brown lung workers' compensation claims. While before 1975 only sixty workers in North Carolina had requested compensation for byssinosis, by the end of 1982 almost 2,400 cases had been filed.

The symbolic language of brown lung and the danger of dust in the cotton mill served to express feelings of exploitation and victimization which already existed among the older generation of textile workers. One retired worker who was pursuing a claim explained in an interview in 1981:

Regardless of what Dr. ——'s diagnosis is, I believe that I really have brown lung. If you have ever been in there — we had to brush or fan the machines off, and I think that just gradually, it was so gradually that I didn't notice the build up until just practically all my lungs became [a knot].

When you give a plant the best years of your life, which I did from the 1930s until I retired in the early '70s, and they don't feel any compassion or anything, it's just your pay check stops. Of course they are having fringe benefits now, I was just born too soon. I just missed out on everything. Except I've got this brown lung, that's all I've got to show for all these years I worked. Or I feel like I've got it, I really do, I'm sincere. I ready do feel I've got it — that my lung condition is caused by all those years in that dust.[3]

The Brown Lung Movement gave retired workers a chance to express their bitterness against their employers by publicly declaring that they were brown lung victims. The activists who staffed the chapters supported the expressions of anger by the victims. The workers filed compensation claims and joined public demonstrations, both against mills which contested claims and against the state compensation agencies which were slow to process requests and strict in the enforcement of eligibility criteria. The national and regional media showed great interest in the stories of brown lung victims, and the textile industry was soon on the defensive over the issue of dangerous dust in the mills.

Still, very few employed textile workers joined the Brown Lung Movement. Many expressed doubt that the disease really existed, and many also expressed concern that they would lose their jobs if their employers thought that they were about to file compensation claims or to support the victims. Although many textile workers were affected by lung disease by the time they were in their mid-fifties, the average age of byssinosis compensation claimants in 1980 was sixty-four (Stephenson 1981: 133); at that age people could afford to leave their jobs since they would soon be eligible for Social Security pensions.

The rhetoric of danger

Brown lung was shaped as a symbol of danger in the textile mills with explicit reference to the Black Lung Movement of the coal miners. As in the case of black lung, the language of brown lung captured the plight of impoverished southern industrial workers, transforming their feelings of victmization into anger against the textile industry. While the mutual understandings of black lung were first shaped in a dialogue between activists, doctors, and miners, brown lung was developed in the context of the labor-liberal lobbying efforts for the Occupational Safety and Health Act. Brown lung was a logical extension of the widespread belief that textile workers were exploited and that government intervention was necessary to improve their lives. The symbol was thus an excellent vehicle for convincing policy makers and the public that the federal government had the responsibility to regulate private workplaces.

A few years later, brown lung came into active use in the textile region as a way to express previously unexpressed resentment of textile workers against their employers. While the black lung language was being used as a vehicle to challenge the credibility of the Mine Workers Union, brown lung served as an organizing principle for workers who, for the most part, had no direct experience with unions and who had been taught to oppose them. The brown lung language provided a basis for social activists to talk to textile workers and to encourage them to take political and social action to improve their lives. Brown lung provided a recognized social role for workers who were critical of the mills, and a rhetoric for grievances which was more socially acceptable in the southern textile region than any direct discussion of labor issues. Still, employed textile workers risked losing their jobs by angering their employers if they used the brown lung symbol, and local physicians had little to gain from promoting public education about the disease. The belief that dust in textile mills is dangerous spread slowly through the nation over the 1970s; the symbolic language of brown lung was less useful than that of black lung in the social interactions of groups in the southern textile region.

Stage Two: Legitimation of the Claims of Danger

Black lung and brown lung diseases symbolized a new belief that the dusts in coal mines and cotton mills were dangerous. In order to define the dusts as dangerous, actors altered their notions about order and disorder. Endemic lung diseases in miners and textile workers were redefined as abnormal. Dusts in mines and mills were not part of the natural order but represented disorder, and therefore workplaces had the potential to be purified or cleaned.

Black lung and brown lung captured the sense of exploitation and victimization of miners and textile workers. The symbolic system provided a language in which those who wanted to organize these workers to press for social change could communicate with them. Clearly the symbols were useful in bargaining for more resources, compensation benefits for example, and in helping some people explain their life experiences in a meaningful way. That is, the language was used in negotiations conducted to promote material interests and to reduce cognitive strain. Still, given that the rhetoric of workplace danger was employed to demand political and social change, why were social authorities so willing to recognize the danger and take remedial action? Why were these claims of danger and demands for change accepted as legitimate? How does the redefinition of danger become a threat to those in power?

Both power and authority involve control over the actions of others, although authority is exercised with the knowledgable consent of those being controlled (Hall 1972: 47). The manipulation of symbols can be a key technique of social control; if the public accepts the definition of a situation put forth by those in power, it will generally consent to the actions which the powerful wish to take. The language of danger is very useful in maintaining social control in this manner. In *Purity and Danger* Douglas (1966: 5) writes:

> [T]he ideal order of society is guarded by dangers which threaten transgressors. These danger-beliefs are as much threats which one man uses to coerce another as dangers which he himself fears to incur by his own lapse from righteousness. They are a strong language of mutual exhortation. At this level the laws of nature are dragged in to sanction the moral code: this kind of disease is caused by adultery, that by incest; this meteorological disaster is the effect of political disloyalty, that the effect of impiety. The whole universe is harnessed by men's attempts to force one another into good citizenship.

Douglas goes on to note that the person with the responsibility to mediate between purity and danger in a society inevitably has power. This power arises from the ability to protect others (at least symbolically) from the dangers that pervade the universe. She hypothesizes that in societies with amorphous structures of social control, this power is involuntary, uncontrollable, and vested in witches who arbitrarily cause disorder. In societies where positions of authority are clearly articulated, it is the authorities who assume the role of maintaining the natural/social order (Douglas, 1966: 99).

It follows that when definitions of danger shift, people in power lose claim to this mediating position. They can no longer evoke the standardized threats of danger to promote their positions as keepers of safety, because the public no longer agrees with them about what is safe. When people in a technocratic society begin to suspect that technology is not protecting them from danger but is causing danger, the authorities wielding the technology and the scientists supporting them lose credibility (Hewitt 1983: 18). Similarly, if industries are publicly seen to be creating danger, not economic well-being, with their manufacturing processes, they lose the public support which enabled them to dominate negotiations with their labor force and their host communities. To the extent that government assumes the ultimate responsibility for preventing danger, it too loses support if people suspect that the (newly defined) dangers were knowingly condoned.

This next section of our discussion of danger in mines and mills examines how people in positions of authority reacted to the use of the rhetoric of danger by workers and social reformers. Efforts to deny the assertions that the dusts in coal mines and cotton mills were dangerous quickly gave way to efforts to co-opt and take charge of the issues, as industries and government agencies acted to maintain control of their positions as protectors of the public welfare.

Legitimizing Danger in Coal Mines

The West Virginia state legislature responded to the black lung strike of February 1969 by adding Coal Workers Pneumoconiosis to the list of scheduled diseases whose victims were eligible for workers' compensation benefits. The striking miners declared a victory, although leaders of the movement knew that very few disabled miners would qualify for benefits under the law (Smith 1984: 23). In March Congressional representatives from the region introduced federal legislation which required coal mine operators to reduce dust levels in mines. The law included a program, initially funded from the federal treasury, to provide compensation benefits for black lung victims.

Although the Coal Mine Health and Safety Act was a precedent setting and costly piece of legislation, it passed through Congress in only nine months, and was signed by President Nixon in December 1969. The rapid passage of the bill was partly due to the key legislative positions held by some coalfield representatives; for example, Carl Perkins of Kentucky chaired the House Education and Labor Committee, while Dan Flood of Pennsylvania chaired the Appropriations Committee. The black lung issue gave these Congressmen something highly visible and obviously beneficial to support on behalf of their constituents (McGillicuddy 1978: 1128). At the same time, Congress as a whole and the executive branch agencies were under pressure to disavow their earlier lack of action on mine safety and health problems. One legislative summary presented to a 1975 National Academy of Sciences Committee reviewing the black lung issue gives an indication of the pressures faced by federal policy makers in 1969. The speaker noted that the explosion in West Virginia that killed 78 miners was the original focus of the legislation, but that attention turned to the problem of coal dust as the heartrending stories of black lung victims were presented. He added (Mittleman 1976: 95—96):

> At the time this legislation was pending, there were strikes and some union unrest which appeared only to lend urgency to the problem. The overriding evidence was that a gross injustice had been perpetrated on coal miners for many years, which Congress was very desirous of rectifying, so that at the time of passage of the law there was very little opposition to the program that was eventually proposed and great momentum for its passage. There was some concern about the eventual costs of the program. It was reported on the House floor that this could cost as much as $40 million per year and some estimates made in the Senate ranged as high as $100 to $120 million per year. It is to be noted that much higher estimates than these were given in conference, but these tended to be ignored in the face of the other contingencies.

A second speaker told the National Academy Committee that Congress took the lead on black lung legislation because the new Nixon Administration was not yet well organized enough to delineate a position. Liberal Congressional Staff people along with some individuals in the Bureau of Mines crafted the law and congress supported it as a way of demonstrating disapproval of previous government inaction (O'Leary 1976: 100—101).

Clearly some elements within the federal government, including coalfield representatives and liberals seeking the support of labor groups, adopted the rhetoric of coal-mine dust danger to advance their material interests, just as the miners, activists and doctors had done. But it is also clear that the government as a whole felt that the public perceived that authorities had failed in a responsibility which had already been assumed (at least symbolically) to protect miners from danger (see Curran 1984:15). The coal industry, dominated now by major oil and power companies, did not oppose the demand to reduce coal-mine dust levels, although they resisted the very strict limits which were proposed. The industry was also willing to allow the government to pay compensation benefits to unemployed miners with disabling lung disease.

After the passage of the law, the focus of black lung activity shifted to a dialogue between black lung benefits claimants and the branches of the federal government responsible for administering the compensation program. The key issue was how lung disease related to coal-mine dust exposure could be distinguished from lung disease to which miners were as susceptible as anyone else. Under pressure from activist physicians and disabled miners organized by the Black Lung Association, Congress intervened in 1972 to force the Social Security Administration to adopt liberal eligibility criteria. Under these interim regulations, any miner with fifteen years of work exposure and signs of obstructive lung disease was eligible for compensation. At the same time Congress extended the deadline by which the industry had to take over compensation payments (Smith 1984: 24—26; Fox and Stone 1980). Progressively stricter standards for permissible levels of coal dust in the mines were also in force by this time.

Legitimizing Danger in Cotton Mills

Both the government and the textile industry responded promptly to Ralph Nader's framing of the byssinosis issue as parallel to the black lung cause of the coal miners. Secretary Finch of HEW urged Congress to support some form of the Occupational Safety and Health bill as the only comprehensive solution to danger in workplaces. The statement in Finch's reply to Nader (Finch 1969: 2938) —

> It is clear that byssinosis is a serious occupational disease that has been too long ignored in the United States. For many years it was thought that the disease was not a significant problem in this country. That comfortable illusion no longer prevails.

— was widely quoted by brown lung advocates as proof that the problem was real. For its part the textile industry responded by allocating money for industry-organized research on the disease, and began making public statements about the issue (Hanes 1970: 60). As the American Textile Manufacturers' Institute explained in a statement to Senator Williams' committee hearing on the OSHAct in April 1970 (U.S. Senate, Subcommittee on Labor, 1969: 1004—1005):

> As we understand it, your Subcommittee is directly and specifically interested today in a symptom complex which is called byssinosis. We heartily agree that this is an area which needs additional light thrown upon it and we want you to know what our industry is doing about it . . .
>
> We were not surprised to find that in general industry representatives at these [regional] meetings had not heard of byssinosis, nor in most instances had they been aware of any syndrome or symptom complex such as was described in these [medical journal] articles.
>
> Nevertheless, the Board of Directors of the Institute was sufficiently concerned about the matter to ask its Safety and Health Committee to investigate whether or not there were respiratory problems which were peculiar to textile workers and if there were, to search out the causes for such problems and to recommend remedial measures.
>
> The Committee immediately made contact with the U.S. Public Health Service, with manufacturers of air cleaning equipment, with individual textile companies and their medical staffs, and with various research agencies in an effort to determine what was actually known about byssinosis and to separate cold, hard scientific fact from speculation. It became immediately apparent to the Committee that a great deal of research would have to be done before many of the unanswered questions could be answered.
>
> When this became apparent, the Committee sought to determine the organization best qualified to conduct an intensive research program on this subject, and decided upon

the Industrial Hygiene Foundation, Pittsburgh, Pennsylvania, an organization with a tremendous amount of experience in the field of respiratory ailments, and this group was immediately engaged to make the ATMI study.

It is significant that both the government and the industry statements on the brown lung issue stressed that the dangers of cotton dust had only just been discovered, and that these authorities were taking immediate action to study and to solve the problem. It was the accusation that the authorities had known but not done anything about the danger which was most damaging to their positions as mediators of safety.

With the passage of the OSHAct in December 1970, brown lung achieved national recognition as a problem of textile workers, and cotton dust was labeled a hazardous substance by a federal government agency. Labor groups mounted a campaign to have byssinosis declared an occupational disease in the workers' compensation statutes of major textile states. In the spring of 1971, the North Carolina legislature amended its compensation statute to include "any disease characteristic of and peculiar to a particular trade, occupation or employment." A news report of the bill's approval (*Raleigh News and Observer* 1971: 5) indicates how state authorities attempted to stake out a position which could win the confidence of the public and of workers by acknowledging the byssinosis issue, without offending the textile industry or appearing to endorse organized labor.

[State Senator] Staton said he believes the language of the bill is more acceptable to industry than adding byssinosis by name, which has been a goal of the state AFL-CIO.

Howard Bunn, chairman of the N.C. Industrial Commission, said industry officials have studied the bill and found it acceptable.

"We feel without any question this will include byssinosis but will not include other diseases that might be contracted outside employment," Bunn said.

He noted that researchers have determined the disease is caused by a dust or substance in cotton, and the disease appears most frequently in the preparation rooms of textile mills. The disease is exclusive to the textile industry.

Bunn said the Commission could not determine how much the addition of byssinosis would cost the state. He said there may be "a fair number of claims the first year or two" but research and industry efforts are expected to sharply reduce the incidence of the disease in the near future.

Sen. Marshall Raush, D-Gaston, said the measure was a "good bill." He said the textile industry is trying to eliminate the disease by installing various cleaning devices in the mills, but he said help should be given to workers who do develop Brown Lung.

As indicated earlier, the danger of dust in textile mills proved to be a difficult issue for the state and the textile industry, despite these accommodating positions. The social activists in the Brown Lung Association and the textile union argued successfully in law suits and compensation hearings that textile companies showed negligence by not informing their employees of the results of health surveys conducted in the early 1970s.

The North Carolina state compensation agency was paralyzed by the flood of claims unleashed by the brown lung screening clinics in the mid-70s. The delays in claims processing became a political embarrassment for North Carolina Governor James Hunt, who convened a task force on the issue in 1980. Over this time the image of the industry began to shift from that of family enterprises which provided jobs and assured local economic prosperity to that of distant corporations, more concerned about profits than public welfare.

By the early 1980s, however, the textile industry was moving to reclaim its position as benefactor of textile workers and mill communities. It launched a media campaign to publicize how much money had been spent on ventilation equipment and compensation payments. Many companies began sponsoring their own brown lung clinics for retirees so that employers, not the Brown Lung Association, could take the role of informing workers of their chances for byssinosis compensation. For example, Cannon Mills informed its employees in a May 1979 letter (*Stanley News and Press* 1979):

> You are probably aware of the respiratory condition associated with the textile industry called byssinosis. Byssinosis, which is sometimes incorrectly referred to as Brown Lung, has been found by us to occur in a relatively small number of people who have worked in areas involving the processing of cotton from the bale through weaving.
>
> Our company has an extensive voluntary screening program which has been doing an exceptional job of identifying employees in these processing areas who are developing any chronic obstructive lung disorders, whether work related or due to other causes such as smoking, emphysema and asthma. Where there is a possibility that a lung disorder is work related, the Company provides an opportunity for further diagnosis and treatment as needed, at no expense to the employee.

By legitimizing brown lung, the textile industry was able to regain control over the symbols of order and disorder, purity and danger, as these categories affected workers and the general public. Not only did this help to re-establish the industry's authority over workers, it also allowed them to set the criteria for who was and who was not a brown lung victim. Legitimization of the issue and co-optation of the rhetoric of danger by the authorities form a fairly predictable second stage in the social construction of risk.

Stage Three: Implementation of a Solution

The final stage in the social construction of risk is the institutionalization of the new definitions of danger through the implementation of solutions to the problem, which are created and managed by social authorities. Once solutions are implemented, a new negotiated order arises among workers, employers, doctors, activists and the government. The symbols of danger are back under the control of the dominant partners in these negotiations.

The new definitions of danger are no longer a threat to the credibility of the authorities because they have reassumed the role of keeping people safe.

Still, such changes in the terms used in the regular negotiations among groups and individuals may well have changed the distribution of resources and altered the power differentials which structured the old order. It is useful to think clearly about how this structural change takes place. The power and resource differences which pattern negotiations are themselves a consequence of negotiations made in the past or made by those at different levels of power in society. Policy makers bargain over the federal budget, for example, and their trade-offs determine the resource levels of welfare recipients all over the country a few years later (cf. Estes and Edmonds 1981: 85). Negotiations over danger have a similar ripple effect; eventually they change both the actors' access to resources and the structural constraints within which they must operate.

This discussion of the third stage in the social construction of risk in coal mines and cotton mills focuses on the actions taken by authorities to control the danger, and the impact of these risk management routines on the resources of actors in the coal mine and cotton textile industries.

Solution to the Coal-Mine Dust Danger

The Coal-Mine Health and Safety Act of 1969, the law passed by Congress in response to the perceived danger crisis in the mines, contained provisions mandating reductions in the level of coal-mine dust and providing compensation for miners with lung disease. With both programs, the federal government placed itself visibly in support of the health and safety of miners, and altered the resources necessary and available to miners and mine operators.

The black lung benefits program resulted in a massive transfer of income to Appalachian miners and their families. The rate of acceptance of claims varied dramatically over the years, as eligibility criteria were altered in response to miners' lobbying and changes in the political climate. The Department of Labor reported in 1980 that 450,000 miners, widows, and dependents were receiving benefits at a level equal to half the compensation rate of a totally disabled first step federal employee; in 1980 this was $4,180 annually for a claimant and one dependent. Claimants were eligible for Social Security disability, union pensions, and state workers' compensation benefits as well. Since 1977 these benefits have been paid from a federally administered trust fund financed by a severance tax on coal tonnage (U.S. Department of Labor 1980: 88—90). Black lung benefits are considerably higher and easier to secure than compensation for other occupational diseases. Through the program, retiring miners enter into unique new negotiations with their employers, doctors, lawyers, and judges. If their claim is successful miners can relieve some of the financial pressures of retirement.

The visibility of the government's role in reducing coal dust levels in mines is indicated by the fourfold growth in the budget of the Mine Enforcement and Safety Administration (MESA) between 1970 and 1977. The ratio of mines per MESA inspector declined from 11 to 3.5 over the same period (Curran 1984: 18). Yet Curran notes that, despite the increased visibility of the government's role, the number of mine inspections actually declined over this period, and the rate of recovery of the moderate fines for safety violations was only 44 percent of the sanctions issued. Curran concludes (1984: 21—25) that the function of the enforcement agency was largely symbolic, and that it remains more economical for companies to pay or appeal fines than to strictly adhere to the dust requirements.

However, the creation of MESA did force coal mine operators to engage in a whole new set of bargaining interactions, including submitting to inspections, completing extensive paperwork, and hiring lawyers to challenge fines. For the first time federal safety regulations were extended to "nongassy" as well as "gassy" mines. The nongassy mines were predominantly small, locally owned mines which employed nonunion labor. Although they had fairly good safety records before the 1969 law went into effect, they simply could not meet the costs of ventilating equipment, nor the legal fees required to comply with the law. Soon after the law went into effect, there was a 22 percent drop in the number of operating coal mines, mostly in the Appalachian region, and an 18 percent increase in Western surface mines, which were mostly owned by large corporations (Curran 1984: 17—18). A change in the resources required to operate coal mines altered the balance of power among the coal operators, and restructured the U.S. coal industry to the benefit of large coal operators and the United Mine Workers of America.

Solution to the Cotton Dust Danger

Although cotton dust became a priority substance for regulation in 1970, it was not until 1978 that the Occupational Safety and Health Administration published its final mandatory cotton dust standard. The delays were due in part to objections by the cotton and textile industries that the proposed reductions were too strict, would raise consumer prices and would bankrupt all but the largest textile manufacturers. The cotton dust regulation was also delayed by other legal and political challenges which OSHA faced as it attempted to carry out its mandate to regulate working conditions, with weak support from other government agencies and strong opposition from the business community (see McCaffrey 1982: 49—67; Calavita 1983: 439). However, the 1978 OSHA cotton dust standard was fairly strict: it required that all textile workplaces reduce levels of exposure to cotton dust below 200 micrograms per cubic meter (750 micrograms in weaving areas), on the grounds that the substance caused lung disease.

After its release, the cotton dust standard was challenged in court. The Supreme Court's ruling that OSHA was justified in demanding major reductions in cotton dust levels in textile plants came in June 1981, six months after the start of the first Reagan Administration. The new Administration had already shifted the official government position on the issue, from a determination to eliminate the danger of dust in the mills to a disinclination to mistrust the goodwill of the textile industry or to disrupt the business climate. Accordingly, in March 1981, and again in February 1982, OSHA announced its intention to review and probably relax the cotton dust standard.

However, by the time evidence was requested to support an official revision of the government's position on the dangers of cotton dust, thirteen years had passed since brown lung disease had been introduced as a symbol of workplace risk. Many textile companies had adopted a moderated form of the rhetoric of brown lung in order to reassure workers and the public that they still assumed the role of protecting them from danger. They had invested considerable money in new ventilating equipment for textile plants. There was some concern that if the regulation was eliminated, companies which had chosen not to act against the cotton dust danger would have an unfair competitive advantage over those which had. A North Carolina paper reported the reaction of executives of two major textile companies to the announcement in 1982 that the cotton dust standard would be reviewed (Alston 1982):

> [W.O.] Leonard [of Cone Mills] said he favors a cotton dust standard. But he said the debate is over what standard is needed to protect workers and what will be required to achieve compliance.
>
> [Haven] Newton [of Fieldcrest Mills] noted that the standard will remain in place while the review proceeds. "It's never bad to say that we want to re-examine something . . . industry needs to be careful not to behave as if this were a battle of competing ideologies."

When the proposed regulation revisions were announced in June 1983, the permissible exposure level of cotton dust remained unchanged. Responses by the interest groups involved indicated that a cotton dust standard permitting high exposure levels did not have the support of most of the textile industry.[4]

The cotton dust standard had some of the same structural impacts on the textile industry that the Coal Mine Health and Safety Act had on the coal industry. The requirements for capital investments to comply with the standard and to keep up with trends in the industry pressured marginal textile manufacturers out of business, while large textile companies which could afford to comply with the ventilation requirements took advantage of the change in market conditions to purchase these smaller companies. Curran (1984: 17) suggests that increased regulation and increased dominance of large companies in an industry often interact in this way, and that the larger companies may support regulation for this reason. Pearson (1975: 78—87) notes that in the uranium industry, regulation was not proposed despite knowledge of the dangers of radon exposure until after the larger operations began to take over mining. These companies were easier for the government to regulate and were more amenable to regulation. Certainly increased regulation increases the resources which are necessary to successfully negotiate business interactions in an industry.

While safety regulations had a major impact on the textile industry, compensation for brown lung disease did not result in the same broad economic gains for textile workers which the coal miners experienced. In comparison to the 450,000 miners and dependents receiving monthly black lung benefits in 1980, 1500 textile workers in North Carolina had received lump sum or monthly payments for byssinosis as of March 1985, and the number was much smaller in other textile states. Eligibility criteria for byssinosis compensation were strict, although the legal precedents set in brown lung cases may have made it easier for victims of other occupational diseases to win compensation. Because compensation benefits for byssinosis were paid by employers through their insurance companies, textile workers who won compensation could feel grateful to the textile mills. This is an interesting contrast to the situation of miners, who credit the government and their union for assuring them of them right to compensation. The workers compensation system is a new arena for interaction between textile workers and their employers, and the rules involved do give the workers some new leverage in negotiations.

Conclusion

There is some danger in treating the issue of workplace risk in terms of the meaning of the symbols for interacting groups. As Gusfield (1981) points out in his study of the social construction of drunk driving, the "sociological irony" used to examine public problems as symbolic interactions leaves the impression that the observer doubts that the problem is real.

The problem of occupational lung disease, like the problem of drunk driving, is quite real. There is broad scientific consensus that exposure to the dusts in coal mines and cotton mills is a factor in obstructive lung disease, although the extent of occupationally related disease in the industries' workforces is still being debated. There is also a strong indication that at the time black lung and brown lung came to public attention in the late 1960s and mid 1970s, there were thousands of miners and textile mill workers who were disabled by lung disease and who had not been informed that they might have an occupational health problem.

However, the emphasis of this chapter, and indeed of this book, is that the verified existence of a risk is not sufficient for the danger to be publicly recognized. Rather, danger is a meaning which an object takes on in certain circumstances. The scope of the definition of the danger is shaped through the use of the symbol in social interactions. The public understanding of danger bears only a tangential relationship to the objective evaluation of the riskiness of the substance, but a close relationship to the political impact of identifying the risk.

I have suggested here that the public redefinitions of coal mine and cotton dusts as dangerous were part of attempts to challenge the legitimacy of established institutions, including government agencies, mine operators and mill owners, and the miners' union. The new danger definitions were created by groups demanding changes in the existing distribution of power and resources in the Appalachian coalfields and the southern textile states. Redefinitions of danger have the capacity to challenge powerful groups because the powerful rely on their positions as mediators between purity and danger to maintain their authority over others. When miners, mill workers, and public support groups stopped believing that government and industry were protecting employees' health, government and industry lost their ability to define and control interactive situations. They could no longer trade on the good will they had generated from their beneficent roles. I have termed the public redefinition of danger the first stage in the social construction of risk.

In the second stage of the social construction of risk, the authorities recognize the newly defined danger and assume responsibility for solving the problem. Thus Congress quickly passed the Coal Mine Health and Safety Act in 1969 to right the wrongs done to coal miners by previous failures to address the black lung issue. Federal and state governments and the textile industry announced their intentions to regulate and improve conditions in cotton mills and to compensate brown lung victims. Once the public redefinition of mines and mills as dangerous had occurred, these authorities moved to take charge of the issue and to regain the credibility they had previously established as protectors of public welfare.

The third stage in the social construction of risk institutionalizes the new definition of danger through the problem solutions managed by the authorities. The solutions may seem symbolic rather than real, especially to the extent that they fail to visibly alter the status quo. Altering the status quo was the initial intention of those who introduced the new danger definition. However, the use of new symbols in interactions at high levels of power in a society does create structural changes at lower levels. Government recognition of the dangers of both coal mine and cotton dusts generated regulations which altered working conditions and business practices in the two industries. Workers' compensation procedures provided new forums for workers to challenge employers and to obtain new sources of income.

Symbols are a key component of daily social interactions. Through social interactions we maintain and alter the negotiated order, ultimately recreating as well as modifying our culture and social structure. Thus the social construction of risk is a universal and ongoing process, touching our deeply held notions of order and disorder, as well as our material interests in promoting our own definitions of reality.

Notes

1. This discussion of the history of the brown lung problem is based on primary field research conducted in 1981 and 1982. For a more detailed presentation, see Bronstein, Janet M., "Brown Lung in North Carolina: The Social Organization of an Occupational Disease," Ph.D. Dissertation, University of Kentucky, 1984.

2. Quoted from a grant proposal submitted by the Carolina Brown Lung Association to the Campaign for Human Development, November 1975. Archives of Dr. Bennett Judkins, Belmont Abbey College, Belmont, North Carolina. See also Judkins (1986: 116).

3. Quote from interview conducted by Bennett Judkins, January 1981, transcribed in fieldnotes, volume C, page 149.

4. In late 1983, the Amalgamated Clothing and Textile Workers Union and the American Textile Manufacturers Institute submitted a joint post-hearing brief to the Occupational Safety and Health Administration, recommending that the permissible exposure limits for cotton dust remain unchanged in the revised cotton dust standard. As of November 1985, OSHA was still attempting to secure the approval of the revised standard (with unchanged exposure limits) from the Office of Management and Budget.

References

Alston, Chuck. 'Cotton Dust Rule May Cut Costs,' *Greensboro* (N.C.) *Daily News* February 11, 1982.

Ashford, Nicholas. *Crisis in the Workplace*. Cambridge: MIT Press, 1976.

Ball, Richard A. 'Social Change and Power Structure: An Appalachian Case,' in John D. Photiadis and Harry K. Schwarzweller (eds.), *Change in Rural Appalachia. Implications for Action Programs*. Philadelphia: University of Pennsylvania Press, 1971, 147—165.

Batteau, Allen, 'Appalachia and the Concept of Culture: A Theory of Shared Misunderstandings,' *Appalachian Journal*, Autumn/Winter 1978—1979, 9—31.

Calavita, Kitty. 'The Demise of the Occupational Safety and Health Administration: A Case Study in Symbolic Action,' *Social Problems*, 1983, 30: 437—448.

Cohen, Abner, *Two Dimensional Man: An Essay on the Anthropology of Power and Symbolism in Complex Society*. Berkeley: University of California Press, 1974.

Curran, Daniel J. 'Symbolic Solutions for Deadly Dilemmas: An Analysis of Federal Coal Mine Health and Safety Legislation,' *International Journal of Health Services*, 1984, 14, 5—29.

Douglas, Mary. *Purity and Danger: An Analysis of Concepts of Pollution and Taboo.* New York: Praeger, 1966.

Douglas, Mary and Aaron Wildavsky, *Risk and Culture: An Essay on the Selection of Technological and Environmental Dangers.* Berkeley: University of California Press, 1982.

Edelman, Murray. *Politics as Symbolic Action: Mass Arousal and Quiescence.* Chicago: Markham Publishing, 1900.

Estes, Carrol and Beverly C. Edmonds, 'Symbolic Interaction and Social Policy Analysis,' *Symbolic Interaction,* 1981, 4, 74—86.

Finch, Robert H. 'Brown Lung Disease — A Reply from HEW Secretary Finch.' *Congressional Record* (House), October 9, 1969, 115; 29383.

Fisher, Stephen L. 'Victim Blaming in Appalachia: Cultural Theories and the Southern Mountaineer,' in Bruce Ergood and Bruce E. Kuhre (eds.) *Appalachia: Social Context Past and Present.* Dubuque: Kendall/Hunt, 1976, 139—148.

Fox, Daniel M. and Judith F. Stone. 'Black Lung, Miners' Militancy and Medical Uncertainty 1968—1972,' *Bulletin of the History of Medicine,* 1980, 54, 43—63.

Geertz, Clifford. *The Interpretation of Cultures.* New York: Basic Books, 1973.

Good, Paul. 'Kentucky's Coal Beds of Sedition,' *The Nation,* September 4, 1967, 201, 166—169.

Gusfield, Joseph R. *The Culture of Public Problems.* Chicago: University of Chicago Press, 1981.

Hall, Peter M. 'A Symbolic Interactionist Analysis of Politics,' *Sociological Inquiry,* 1972, 42, 35—75.

Hanes, Gordon. 'Management's Responsibility for People,' in David Fraser and Mario Battigelli (eds.). *Transactions of the National Conference on Cotton Dust and Health.* Chapel Hill: University of North Carolina School of Public Health, 1970, 59—61.

Harris, Reginald T., James A. Merchant, Kaye H. Kilburn, and John D. Hamilton. 'Byssinosis and Respiratory Diseases Among Cotton Mill Workers.' *Journal of Occupational Medicine,* 1972, 14, 199—205.

Hewitt, Kenneth. 'The Idea of Calamity in a Technocratic Age,' in Kenneth Hewitt (ed.), *Interpretations of Calamity from the Viewpoint of Human Ecology.* Boston: Allen and Unwin, 1983, 3—32.

Judkins, Bennett M. *We Offer Ourselves as Evidence.* Westport, Connecticut: Greenwood Press, 1986.

Kerr, Lorin E. 'The Occupational Pneumoconiosis of Coal Miners as a Public Health Problem,' *Virginia Medical Monthly,* 1969, 96, 121—126.

Maines, Donald R. 'Social Organization and Social Structure in Symbolic Interactionist Thought,' *Annual Review of Sociology,* 1977, 3, 235—259.

Marschall, Daniel. 'The Miners and the UMW,' *Socialist Review,* 1978, 8, 65—115.

McCaffrey, David P. *OSHA and the Politics of Health Regulation.* New York: Plenum Publishing Co., 1982.

McGillicuddy, Robert D. 'The Legislative History of Black Lung Reform Efforts,' in House Education and Labor Committee, Black Lung Benefits Reform Act and Black Lung Benefits Review Act of 1977. Committee Print, February 1979.

Meiklejohn, Andrew. 'History of Lung Disease in Coal Miners: Part III, 1920—1952,' *British Journal of Industrial Medicine,* 1952, 9, 208—220.

Merry, Sally. *Urban Danger: Life in a Neighborhood of Strangers.* Philadelphia: Temple University Press, 1981.

Mittleman, Eugene. 'A Legislative History,' in National Research Council, Committee on Natural Resources and the Environment. *Supplemental Report: Coal Workers Pneumoconiosis — Medical Considerations, Some Social Implications.* Washington: National Academy Press, 1976, 95—99.

Nader, Ralph. 'Brown Lung in the Textile Industry,' *Congressional Record*, (House) August 11, 1969, 115; 23117—23118.

O'Leary, John. 'An Executive Perspective on the Black Lung Compensation Program' in National Research Council, Committee on Natural Resources and the Environment. *Supplemental Report: Coal Workers Pneumoconiosis — Medical Considerations, Some Social Implications*. Washington: National Academy Press, 1976, 100—102.

Ortner, Sherry B. 'Theory in Anthropology Since the Sixties,' *Comparative Studies in Society and History*, 1984, 26, 126—166.

Page, Joseph A. and Mary Winn-O'Brien. *Bitter Wages*. New York: Grossman, 1973.

Plaut, Thomas, 'Political Alienation and Development: A Perspective from Appalachia,' in Bruce Ergood and Bruce E. Kuhre. (eds.) *Appalachia: Social Context Past and Present*. Dubuque: Kendall/Hunt Publishing Co. 1976, 295—302.

Raleigh (N.C.) *News and Observer* 'Byssinosis Bill Given Approval.' April 28, 1971, 5.

Rosen, George. *The History of Miners' Disease*. New York: Shuman's, 1943.

Schilling, Richard S. F. 'The History of Byssinosis and the British Experience,' in David Fraser and Mario Battigelli (eds.), *Transactions of the National Conference on Cotton Dust and Health*. Chapel Hill: University of North Carolina School of Public Health, 1970, 7—12.

Smith, Barbara E. 'Black Lung: The Social Production of Disease,' *International Journal of Health Services*, 1981, 11, 343—359.

Smith, Barbara E. 'Too Sick to Work, Too Young to Die. An Oral History of the Black Lung Association,' *Southern Exposure*, 1984, 12, 19—29.

Stanley (N.C.) *News and Press*. 'Cannon Offers Retirees Free Brown Lung Tests,' May 15, 1979.

Stephenson, William. 'Administrative Law Problems with Byssinosis,' *Chest*, Supp. April 1981, 79, 132S—133S.

Strauss, Anselm. *Negotiations: Varieties, Contexts, Processes and Social Order*. San Francisco: Jossey-Bass, 1978.

U.S. Department of Labor. *An Interim Report to Congress on Occupational Disease*. Washington: Government Printing Office, June 1980.

U.S. Senate, Subcommittee on Labor, Committee on Labor and Public Welfare. *Hearings on the Occupational Safety and Health Act*, Part One, Greenville, South Carolina, April 28, 1970. Washington, D.C.: U.S. Government Printing Office, 1970.

PART V

Part V

Organizations and the Social Construction of Risk

The focus of this section of the volume is on organizations and the social construction of risk. The three chapters examine cases where specific organizations, such as government agencies and industrial firms, play a major role in risk selection and perception. An important emphasis in each chapter is the contribution of internal and external factors to the construction of risk by organizations.

The first chapter, by Richard Gale, focuses on the role of government agencies in a dispute over an industrial waste site in Seattle, Washington. The case offers an interesting comparison to the Love Canal case discussed in Chapter 3 and Chapter 13. In Love Canal, the debate was largely between state and federal authorities on the one hand and citizens' groups on the other. By comparison, in Seattle the debate was between two government groups — one group consisting of federal officials and one part of the Mayor's advisory committee, and the other group consisting of politicians and another part of the Mayor's advisory committee. Local citizens were, for the most part, excluded from the decision-making process and were unorganized and impotent throughout the debate.

Several important conclusions can be drawn from Gale's study. First, the study illustrates the extent to which the environmental protest movement has become institutionalized. Second, the study suggests that risk selection and risk perceptions may vary at different stages in the life-cycle of an organization. The life-cycle model of organizations — which posits that organizations are born, mature, age, and die — has been extensively debated over the last few decades (Bernstein 1955; Downs 1976; Ripley and Franklin 1975; Plumlee and Meier 1978; Mitnick 1980; Anderson and Zeithaml 1984; Kimberly and Miles 1981). While the results of such studies are still preliminary in nature, the model suggests that the goals of organizations shift over time and that organizations may select different risks for attention and concern at different stages in the life-cycle. Finally, Gale's study graphically illustrates the difficulties experienced by scientists involved in public controversies (see also Part VI of this volume). In the Seattle case, the professional credibility of scientists was continually threatened by media coverage of expert disagreements and scientific uncertainties. Scientists, whose research is usually done far from public

B. B. Johnson and V. T. Covello (eds.), The Social and Cultural Construction of Risk, 229—232.

view, found themselves in a "fish bowl" atmosphere where mistakes and internal differences of opinion were immediately reported by the media. These and other difficulties reflect many of the problems faced by scientists working in centralized bureaucracies (Namer 1984; Scott 1966; Van Maanen and Barley, 1984: 310, 335—336, 344).

The second chapter in this section, by Michael Brown, focuses on the determinants of worker perceptions of occupational risks and the role of management in informing workers about such risks. Brown points out that several factors influence management decisions on whether and how much risk information to communicate to workers. These include the presence or absence of economic incentives to inform workers about risks, the existence of a high level health advocate, the degree of authority held by managers familiar with shop floor conditions, the time horizon of management — e.g., a short- or long-term focus, and management attitudes toward the ability of workers to understand and appropriately act upon risk information. External factors that influence management decisions to communicate include activism by local union officials — which depends in part on levels of concern among union members — and the existence of state and federal worker right-to-know laws and regulations. Brown's empirical examination of these factors suggests that the current system provides only minimal incentives for managers to communicate fully and effectively with workers about risks.

In cases where managers do provide workers with full information about risks, the response by workers is substantially less than optimal. Responses by workers can be categorized into three types: denial, fatalistic acceptance, and activism. The first type of response — denial — is especially common when adverse health effects manifest themselves only after several decades. Other factors that encourage denial are (1) cognitive dissonance — for example, the belief that only an irrational person would work in a dangerous job and so the job must be safe; (2) group norms — for example, a machismo attitude toward risk fostered through peer pressure by fellow workers and supervisors; and (3) the desire to preserve existing social relationships — for example, concerns about the disruptive effects of revelations on relationships with supervisors and fellow workers.

The second type of response — fatalistic acceptance — tends to be more economic in origin. Workers may acknowledge the existence of significant health risks, but believe that these risks are offset by the economic benefits of the job and by the costs of leaving the job in the absence of realistic alternatives.

The third type of response — worker activism — is the most rare. When it does occur, activism appears to be based largely on personality characteristics, levels of information, "transformative experiences" (e.g., the recognition that no changes will occur without somebody taking action) and concerns about the future of one's children. Activism does not appear to be the result of different perceptions of risks between activists and nonactivists.

One clear implication of Brown's study is that psychological factors cannot adequately explain why only some risks and not others are selected for attention and concern by workers and managers. Instead, risk selection and perceptions in organizations appear to be rooted in social relationships and in the range of choices that workers perceive to be available within the specific organization and the wider society.

The last chapter in this section, by W. Bernard Carlson and Andre J. Millard, focuses on the dynamics of risk selection and perception within organizations. The chapter explores this issue through an examination of the debate over the choice between alternating current (a.c.) electricity and direct current (d.c.) electricity in the late nineteenth century. One of the principal figures in this debate was Thomas Edison, who led the fight against the adoption of a.c. technology. At least two important implications can be drawn from the case study. First, the study illustrates the substantial role that subunits within an organization can play in the risk selection process. Carlson and Millard note, for example, that risk selection evolved from a conflict between the research, marketing, and production units in one of the firms that was studied (cf. Gregory 1983). Second, in accordance with the life-cycle model of organizations, the study raises several interesting questions about the importance of safety issues at different stages in the development of an organization.

References

Anderson, Carl R. and Carl P. Zeithaml. 'Stage of the Product Life Cycle, Business Strategy, and Business Performance,' *Academy of Management Journal*, March 1984, 27(1), 5—24.

Bernstein, Marver H. *Regulating Business by Independent Commission*. Princeton, New Jersey: Princeton University Press, 1955.

Downs, Anthony. *Inside Bureaucracy*. Boston: Little, Brown, 1967.

Gregory, Kathleen L. 'Native-View Paradigms: Multiple Cultures and Culture Conflicts in Organizations,' *Administrative Science Quarterly*, September 1983, 28(3), 359—376.

Kimberly, John R. and Robert H. Miles. *The Organizational Life Cycle: Issues in the Creation, Transformation, and Decline of Organizations*. San Francisco: Jossey-Bass, 1981.

Mitnick, Barry M. *The Political Economy of Regulation*. New York: Columbia University Press, 1980.

Namer, Gerard. 'The Triple Legitimation: A Model for a Sociology of Knowledge,' in Nico Stehr and Volker Meja (eds.), *Society and Knowledge: Contemporary Perspectives in the Sociology of Knowledge*. New Brunswick, New Jersey: Transaction Books, 1984, 209—222.

Plumlee, John P. and Kenneth J. Meier. 'Capture and Rigidity in Regulatory Administration: An Empirical Assessment,' in Judith V. May and Aaron Wildavsky (eds.), *The Policy Cycle*. Beverly Hills, California: Sage, 1978, 215—234.

Ripley, Randall B. and Grace A. Franklin. *Policy-Making in the Federal Executive Branch*. New York: Free Press, 1975.

Scott, W. R. 'Professionals in Bureaucracies: Areas of Conflict,' in H. Vollmer and D. Mills (eds), *Professionalization*. Englewood Cliffs, New Jersey: Prentice-Hall, 1966.

Van Maanen, John and Stephen R. Barley. 'Occupational Communities: Culture and Control in Organizations,' in Barry M. Staw and L. L. Cummings (eds.), *Research in Organizational Behavior*, Vol. 6. Greenwich, Connecticut: JAI Press, 1984, 287—365.

Chapter 9

The Environmental Movement Comes to Town: A Case Study of an Urban Hazardous Waste Controversy

Richard P. Gale*

Introduction

Contrary to earlier predictions (Downs 1972), the environmental movement has endured as a strong social movement, and now enjoys a relatively comfortable mid-life maturity. The consequences of this maturity are several. The movement can rely on complex and favorable legislation and a well-developed body of movement-supportive judicial decisions. The movement is less reliant on public adversary protest strategies than in earlier days. It can count on more established review procedures, such as environmental impact statements, to provide itself and the public time to review actions of interest. Maturity also means that not all movement issues will meet with a strong countermovement response. In addition, issues may broaden. For the environmental movement, earlier wilderness and rural area preservation issues have been joined by urban community issues, such as the one under study here. In this sense, then, the mature environmental movement has come to town.

This chapter examines events surrounding the temporary closure in 1984 of a city park (Gas Works Park) in Seattle, Washington, located on the lakefront site of a former coal and oil gasification plant. Following a brief description of the site and an overview of the controversy, we shall examine four elements of the Gas Works case which, it is argued, are associated with mature social movements, and characterize the way in which such movements address social issues.[1] These four elements are (a) movement-generated government agencies, (b) politically inexperienced and diverse movement-generated professions, (c) assumed public constitu-

* Research and preliminary analysis were completed while the author was on sabbatical leave, and a Visiting Professor of Forest Resources, University of Washington. This article was completed while the author was a Visiting Scholar, Institute for Marine Studies, University of Washington. Susan Stiles Gale provided helpful suggestions and editorial assistance. The author is solely responsible for all opinions expressed.

B. B. Johnson and V. T. Covello (eds.), The Social and Cultural Construction of Risk, 233–250.

ency, and (d) governmental and professional "face saving" in response to eventual negative scientific findings.

The general thesis of this chapter is that when mature movements manifest these four characteristic elements in resolving issues, the results may not be optimally beneficial to either the public or the movement.

The data base for this chapter included newspaper articles which appeared in Seattle newspapers from April through August of 1984. The numbers correspond to the citations in the text. Article by-lines have been indicated where appropriate.

I. An Overview of the Gas Works Park Problem

One central feature of Seattle geography is Lake Union, a 640 acre body of water located immediately north of downtown. The lake shore displays conflicting land uses, from marinas, condominiums, and office buildings, to struggling marine-oriented service industries. In contrast to this development is Gas Works Park, the only green space on the shoreline. It is characterized by several immense rusting structures which were part of a coal and oil gasification plant operating from 1906 to 1956. The 20 acre Gas Works Park, completed in 1976, retains several large pieces of equipment, and also includes a picnic and play area, and "Kite Hill," an artificially constructed hill several hundred feet high, which affords both spectacular kite flying and a magnificent view of the downtown skyline.

Although the design of the park, which relied heavily on the retention of former industrial machinery and is what one observer terms an "industrial conversion park" (Richard 1983: 32), generated substantial community controversy, it was a design success for its architect, Richard Haag, and a recreational success for the community.

> Per square foot, Gas Works Park is indisputably the most popular park in the Northwest. The winner of many design and architecture awards, it brings tourist money to Seattle without short-changing the locals. Very simply, it is a point of local pride (Richard 1983: 12).

Early in 1984, the federal Environmental Protection Agency (EPA) tested lake bottom sediments immediately offshore from the park. Finding evidence of polynuclear aromatic hydrocarbons (PNAs), they posted signs warning against swimming or fishing. Then, they began sampling the soil in the park itself. They took soil samples at different depths, as well as preliminary "grab samples" of surface soil. Upon learning that the preliminary samples contained 15 carcinogens within the PNA group, the EPA forwarded these results to Mayor Charles Royer in the early afternoon of Friday, April 20 (Jones 1984a: A-1, A-13.)[2] After consulting with the local health department, the U.S. Center for Disease Control, and Region 10 of the EPA, late that same afternoon Royer ordered the park closed and "Gas Works Park Closed Until EPA Tests Are Completed" signs and wooden barricades placed at the entrance.[3]

The initial prediction was that the park would remain closed until evaluation of the EPA tests; one estimate was that results would be available in June (Hadley 1984d: C1). The Mayor immediately appointed an advisory committee, composed of health experts drawn from government units and the University of Washington, and neighborhood representatives. The advisory committee was co-chaired by a professor of environmental health and an administrator from the combined city-county health department, and met more than a dozen times during the three-month closure. The park was finally reopened to the public in late August.

II. Four Elements Associated with Mature Social Movements

A. *Movement-Generated Agencies*

The state, manifested in government agencies, plays varying roles over the life of a soical movement (Gale 1986). One of the hallmarks of a successful social movement is its ability to create government agencies which represent movement concerns. Although these agencies often deviate from their initial movement-specified aim, and may eventually come under attack from their "sponsoring" movement, their existence is important in the evolution of issues. In the Gas Works case, several environmental movement-generated government entities played key roles.

This section of the chapter describes the six government bodies which played key roles in the controversy.[4] These are the federal Environmental Protection Agency (EPA); Center for Disease Control (CDC); the city-county Department of Health; Seattle Parks and Recreation Department; the University of Washington Department of Environmental Health, a part of the School of Public Health and Community Medicine; and the Mayor's Ad Hoc Gas Works Advisory Committee. To this organizational list, we would add the federal Superfund program, although it was not represented by a separate organization during the controversy.

Two of the entities involved in the Gas Works controversy can be squarely aligned with the environmental movement — the EPA and Seattle Parks and Recreation. Three others are more closely connected to traditional public health concerns — the CDC, Health Department, and the University of Washington. The final body, the Ad Hoc Committee, significantly does not fit into either of these clusters. This group typifies government response to a new, but potentially significant, issue requiring close political control by major actors.

The Gas Works issue would not have emerged without the environmental movement because it was the movement's flagship agency, the Environmental Protection Agency (EPA), that recommended that the

Mayor close the park.[5] Almost immediately another layer of government stepped into the controversy: "Superfund," the toxic waste, cleanup program funded by the federal government (Hadley 1984b: A6; *Seattle Post-Intelligencer*, 1984a: A10, 1984b: A10). The Superfund can be considered as aligned with the environmental movement because one of its tasks is to assign liability to companies responsible for toxic wastes, and to require that they help pay for site cleanup and rehabilitation. The local Parks Department is also aligned with the environmental movement because of its combined concern with public safety and park use, and the fact that park maintenance employees would be most directly exposed to park hazards, particularly those involving contaminated soils (Hedges 1984).

In its initial stages, the issue primarily involved contact between the EPA, and the Mayor and his Ad Hoc Committee. Although other municipal agencies, such as the park and health departments were involved through membership on this committee, some members of the City Council were critical because the closure and the committee of experts were totally under the Mayor's control.[6]

EPA's involvement in the issue was complex and multi-faceted. First, EPA's representatives, and EPA's general role, were such that their ultimate power to supercede the Mayor remained evident. Citizens were warned that if they did not stay out of the loosely fenced park, the EPA would step in and fence the park more securely. (People continued to use the park throughout the closure [*Seattle Post-Intelligencer* 1984c: D1].) Citizens and the press were reminded that even if the Mayor's committee (and the Mayor) wished to reopen the park, the EPA retained the option of keeping it closed (Hadley 1984f: D1, Wilson 1984c: D1).

The relationship between the more routinized tasks of the EPA, and the less defined role of its Superfund, also increased the complexity of the issue. EPA's scientists vascillated as to their primary purpose: investigation of whether the park might qualify for Superfund cleanup, or establishment of the health risk associated with public use of the park. Shortly after the closure, the possibility of Superfund help seemed welcome to some (*Seattle Post-Intelligencer* 1984a: A10), as it was assumed that federal funds would be essential for what was believed to be a massive site rehabilitation project (Jones 1984c: A1). Others, however, feared that Superfund eligibility would also mean continued closure, and years before the park would be available for public use. Still others criticized Superfund officials' interest in the park, arguing that more dangerous, industrial sites demanded immediate action.

The role of the federal Center for Disease Control (CDC) also increased the complexity. The National Center for Disease Control in Atlanta was kept informed of the issue from the time the EPA preliminary findings were available, and their advice was sought repeatedly as the Ad Hoc Committee tried to decide whether to reopen the park. Midway through the work of the committee, the CDC was reported (by a sympathetic public official) as being critical of additional studies to be undertaken by university researchers, who, in turn, were supplementing work done earlier by EPA scientists. Later, the CDC was contacted via conference telephone, and its response to that call contained three recommendations which became the core of the committtee's recommendations. The CDC report was the basis for a July 4th news story headlined "It's safe to open Gas Works, say national health officials" (Hadley 1984g: A3). Advice from the CDC, which was typically endorsed by the medical members of the Ad Hoc Committee, reflected input from the federal health establishment. As such, this advice had the potential for conflicting with the environmentally oriented, and, in this case, more cautious perspective of the EPA.

Finally, this organizational complexity was even manifested in one individual. One committee member, an environmental engineer and attorney, was an EPA employee who had been assigned, through the Intergovernmental Personnel Act program, to the city's Office of Intergovernmental Relations. The multiple "hats" worn by this person were evident — while portraying himself as a city employee, he both took the EPA position in public meetings, and independently contacted the press with suggestions for complex, expensive, "solutions" long before the committee had established the potential risk (Jones 1984c: A1).

B. *Movement-Generated Professions*

Diverse professions also play important roles in social movements. The interests of mature, stable social movements may be manifested in the emergence of new movement-aligned professions, or in the refocus of traditional professions to reflect movement concerns. The process by which movements evolve is thus influenced by the behavior of these professionals. The discussion which follows describes three features of movement-generated and aligned professions — their representation among the total number of professions involved in the issue, how they relate to other professions in collegial and leadership roles, and their experience in operating in highly politicized situations.

Representation of movement-generated and traditional professions

The professions represented on the nine-member committee and by the EPA reflected both environmental movement-generated and traditional public health fields (Jones 1984a: A1, A13). Among the former were occupational medicine, hazardous waste transport, environmental toxicology, environmental engineering, and environmental law. The latter included epidemiology, family medicine, medical education (with an emphasis on poisoning among children), and public health.

Several of these professions had an interdisciplinary component and all but one committee member were employed by public agencies. (The exception was a family practitioner who was one of several neighborhood representatives.) The secure, established positions held by these professionals contrast with what would be expected in a less mature movement in which those professionals who were involved would either be affiliated (as volunteers with a likely university base) with social movement organizations, or employed by mediating government agencies or the countermovement.

Relations between professionals

Despite the general similarity of these fields (all were health related, and included no social sciences), it was clear that each had its own primary interest, and that these disciplinary concerns often did not mesh well. For example, the occupational health specialist stressed the difficulty of defining any minimally "acceptable" dose of carcinogens, the epidemiologist worried about inhalation, and the pediatrician repeatedly reminded the committee that the ingestion of ash would not generally be a problem since most of the substance would pass through the child's digestive system. Disagreements also surfaced over testing procedures: the toxicologist argued for a "vacuum cleaner" soil sampling system, and the epidemiologist responded that such sampling would be meaningless, except for estimating health impacts on people crawling on the ground! In another exchange, the validity of "visual assessment" was debated; an environmental engineer expressed concern about "blackish" dirt, to which the toxicologist responded that the appearance had little to do with actual contamination.

Because of the external role of the EPA and other organizations, and the fact that much committee time was spent either waiting for test results or trying to assess immediate versus mid- or long-term risks, significant factions did not emerge within the committee.

However, it is clear that, almost from the beginning, the environmental toxicologist and the occupational medicine administrator, both from the UW, tried to direct the discussion toward consideration of whether the risks were sufficient to justify continued park closure. For example, only a week following closure, the toxicologist indicated his belief that the worst pollution was probably confined to a few areas, and that other portions of the park could be re-opened during rehabilitation of contaminated areas (Jones 1984b: A26). Several days later, at a committee meeting, he recommended reopening the parking lot, since it was used by cyclists and not the subject of any testing. This suggestion was rejected by others who believed that it would encourage park use (Hadley 1984a: C1).

Both of these professionals persisted in their efforts to translate the potential risk into terms appropriate to the specific case. For example, trying to balance concern about carcinogenic risk and potential exposure from park use, the toxicologist stated, "A single short-term exposure won't produce cancer. It's a question of an incremental increase in risk" (Jones 1984a: A1, A13). He also contrasted chronic occupational exposure to that of the general public:

> With any chemical carcinogen, you're concerned generally about chronic exposure. A single exposure or one or two inadvertent exposures pose little risk relative to the situation we know about in most occupational exposures (Hadley 1974c: C1).

At several points, the occupational medicine administrator made similar contrasts. Stating her opinion that the vast majority of park users probably have "no significant adverse health risk," she also noted that "The risks are inordinately low, but any risk is of concern to us" (Hadley 1984d: C1).

Throughout park closure, these comments seemed to have little impact on the deliberations. When they were made during committee meetings, they were seen by some as reducing the reliance of the committee on the scientific data which some hoped would definitively establish risk levels. The comments of the toxicologist and occupational medicine specialist were similar to those made by members of the medical profession. For example, the pediatrician repeatedly reminded the committee that small amounts of soil ingested by children would quickly pass through the digestive system. The county epidemiologist was reported by the press as believing that he saw no health-related reason for protecting the public from the benzo(*a*)pyrene at the park (Hadley 1984g: A3). However, despite these shared concerns about possible overestimating of the risk, there were few linkages between these professionals, in part because they came from different specialities and were employed by different organizations within the Seattle metropolitan area.

Political experience and leadership styles

The course of the issue was also affected by the differential political experience of the specialists and the different agendas and styles of the committee co-chairs.[7] For example, the key EPA figure was a young environmental engineer who had supervised the initial collecting of grab samples. Shortly after the closing of the park, he led twenty observers and media representatives on a tour of contaminated park sites. He was very nervous about the media's questioning, and responded to the many microphones thrust at his every comment with allusions to the fact that he had never been in such a public position before. His youth and lack of experience in operating in the public eye lent an air of tentativeness and uncertainty in the early stages of park closure. The power he did command appeared to come more from his general position as an EPA employee than from his experience of status within the agency. One reason that the Ad Hoc Committee took a long time to reopen the park after the findings were available was that it was unclear how the EPA might respond to the committee's recommendations. It is also notable that no key EPA official was publically involved throughout the course of the issue.

Dealings with the public and the issue of closed meetings also reflected the political inexperience of professionals on the committee. Washington law requires governmental bodies to hold at least one open meeting prior to commencing a series of closed sessions. The committee held two open meetings, but these concerned mostly organizational details. Citing the need for free discussion among its members, the committee then decided to close its meetings to the public and the press (Wilson 1984a: A3). City Council members were upset at this action. Council President Norm Rice stated:

> It surprises me that they've taken that tack, especially because the public is very concerned about it and they want to be kept informed. While I recognize there might be facts about the chemicals and things that might cause concern, I don't see any need for closing the meetings (Wilson 1984a: A3).

Council member Sam Smith added:

> I think it's a public park, a public property, and I think the public ought to be privy to something that's going on. They're discussing whether it's safe for us to go to the park, and they're closing the meeting. I think it's ridiculous (Wilson 1984a: A3).

This action also upset several interested members of the public who were not on the committee. The response of the City Attorney to their query was that the open meetings law was not violated because the committee had only advisory power.

The issue of open meetings persisted. After several weeks, the public and the media were informed that the committee would have an open meeting. Arriving at the Mayor's Conference Room at noon, they encountered the committee just adjourning from a closed meeting, and preparing for a carefully staged briefing session for the media and public. Eventually, the meetings were opened, and members of the public were given an opportunity to make comments following each meeting.

Finally, the styles and experience of the committee co-chairs also influenced the process. One was a division administrator in the city-county health department, and she appeared far more comfortable assigning duties related to preparation of the final report than answering questions from the public or the media. Her skills were managerial and coordinative, and she responded with a nervous defensiveness to questions. These characteristics influenced handling of the issue when the Ad Hoc Committee was deciding how to report its findings. At a meeting she chaired (the co-chair was out of town) the committee's energies were directed to organizing a complex final report rather than to quickly providing the information needed to open the park.

The style of the university-based environmental engineer sharply contrasted with the health administrator. He responded protectively to public queries, and was quick to reinforce the seriousness of the issue. Clearly, his role was that of scientist and expert, as evidenced in the following portion of a newspaper article (Hadley 1984d: C1).

> Jerry Ongerth, chairman of the panel [Committee], said the group will focus on the risk to public health presented by the park. Consequently, the panel members, who are volunteering their time will not be giving advice on how to clean up the park. That is the concern of state and federal agencies, Ongerth said. . . . However, Ongerth [also] commented: "You could easily spend hundreds of thousands of dollars just defining the problem." Cleaning up such a huge site could run into millions, he said.

The differential roles of the co-chairs are also reflected in the fact that the health administrator was seldom quoted in the press. In consequence, the public portrayal of the issue reflected the viewpoint of an aggressive university-based, research-oriented, environmental engineer, rather than the more cautious, public service orientation of an health administrator.

C. *An Assumed Public Constituency*

One feature of a mature social movement is that emergent issues which are within the established focus of the movement may not generate the level of controversy associated with less mature movements, in which each issue must be fought for, as the movement seeks to establish its "turf" or realm of responsibility. Thus, general categories of issues on which the movement has previously taken a stand operate with an assumed public constituency that is perceived to extend to any relevant, new issues which arise. In these instances, movement-aligned agencies and professions can operate with assumed, rather than direct or active, movement or public support. Further, in some instances the movement's previous stance and public support may be so strong that no general opposition to issue resolution surfaces. Instead, controversy is limited to details such as the relative seriousness of the problem and alternative ways of resolving it.

In the Gas Works Park case, throughout several months of closure and repeated news headlines that the park would soon be reopened (Gates 1984: A8; Hadley 1984f: D1; Schaefer 1984: B1; Wilson 1984b: A1), the issues remained surprisingly free of controversy. Despite the heavy use of the park by diverse users, as well as its physical prominence in the city, the Mayor's sudden park closure did not generate formal comments or positions from established citizens' groups. What did emerge, however, was a core group of citizen observers who attended Ad Hoc Committee meetings and spoke with the media after each meeting. In addition, the controversy provoked a few letters to the editor and newspaper interviews with the "denizens" who continued to use the closed park (*Seattle Post-Intelligencer* 1984c: D1).

Public acquiescence was probably premised on health and cancer fears, as well as the generally liberal, responsive community image of the Mayor (a former TV commentator). Public concerns were probably also quieted because the initial closure announcement suggested a possible early July reopening (Wilson 1984b: A1). (The park was actually reopened in mid-August.) Thus, the lack of public concern over loss of park access, as well as the almost total lack of questioning of the study process or the decisions made during that process, suggest a passive, but supportive, public. The relatively short closure also dampened the public's response. Either a longer projected closure at the start of the process or a protracted delay in reopening would probably have led to organized public pressure to reopen the park.

Environmental organizations also took a passive stance. When this author called the Washington Environmental Council to question the policy of barring the public from meetings of the Mayor's committee, the response was that the organization generally approved of the committee's actions, and did not wish to become involved in the issue. The only other involvement by an environmental organization appeared in the form of a newspaper quote from the "toxic materials director" of Greenpeace Northwest. His comment was very general: "If we don't want to write Lake Union off as a loss, we're going to have to remove that material or it's going to keep leaching into the lake" (Jones 1984b: A26).

This public acquiescence also helped justify the early decision to close committee meetings to the public. This would not have been possible had the environmental "villain" been an active industry, rather than a long-defunct quasi-public utility. That the Mayor and his panel of experts were protecting the public's health, particularly that of children, gave them licence to exclude the public from its initial deliberations.

D. *Face-Saving or What to Do with a "Cold Potato"*

Most people are going to have more exposure breathing the air in downtown Seattle than from visiting Gas Works Park. (Statement from county health department epidemiologist [Hadley 1984h: C1].)

The worst-case scenario — a child eating large amounts of the most high by contaminated dirt over a long time — is extremely unlikely, to say the least. (Editorial titled "Let's start using Gas Works Again" [*Seattle Times* 1984b: A18].)

The general trajectory of environmental and public health concern about Gas Works Park began at a high level, and declined slowly (and unevenly) over the five-month lifespan of the issue. (The issue is not completely dead, however, more than a year later. Superfund assistance is still discussed, and the city continues to undertake rehabilitation projects in the reopened park.)

Initially, then, Gas Works Park contamination was a "hot potato," and this set the stage for a high level of public concern and media interest. The initial, selective EPA grab samples yielded very high levels of chemical carcinogens: "Park's worst chemicals found in children's play area," was the headline (Jones 1984a: A1, A13). (The headline on the continuation portion of the article was even more alarming: "Worst carcinogens found in children's play area" [Jones 1984a: A1, A13].) The highest level of contamination was found inside a metal "dome" located in the play area, and this structure was welded shut by the city even prior to closing the park. High levels of contamination were also found in the "blackish dirt" present in a large sandbox. These initial findings were followed by additional, but significantly lower, levels of benzo(a)pyrene on "Kite Hill," an artificial hill constructed of soil hauled to the park from a construction site several miles away. The headlines which followed the Kite Hill contamination were even more severe: "All of Gas Works Park is polluted, latest tests show" (Hadley 1984a: C1).

The "cooling" of the hot potato occurred nearly two months after the closure. On June 19, the panel reviewed the independent UW team findings (Hadley 1984f: D1; Schaefer 1984: B1), which indicated that the concentrations of benzo(a)pyrene varied from 8.15 to 52.7 ppm dry weight. The report also indicated that the range of PNA concentrations within the park were "significantly elevated" above those found at several reference sites (other city parks located away from Lake Union) (Kalman 1984). These data were presented, however, with no reference to health risk comparisons. Thus the numbers themselves tended to be misleading. The citizen audience was dismayed, as it appeared that toxic levels for many areas within the park far exceeded levels from the reference samples. The prognosis for park reopening looked bleak.

It was thus most surprising when the committee's discussion swung almost immediately to the topic of reopening the park. The consensus appeared to be that the park posed no clear health danger at current use levels. The committee focused on the three or four "hot spots" in the park needing some attention. With estimates of polyaromatic hydrocarbon concentration varying between 10 and 50 ppm, the toxicologist on the committee remarked that a concentration of 30 ppm "wouldn't appear to be much greater than staying in a smoke-filled room for a while" (Schaefer 1984: B1).

This "cold potato" conclusion to weeks of cancer-filled headlines left the committee with several problems. First, how could they reconcile the considerable public concern generated by the sudden closure with their swift and unexpected assessment that the park was not particularly dangerous? Second, how should the committee prepare a complex report which would ultimately recommend only minimal park changes? (They had committed themselves earlier to preparing a complex report by assigning report-writing tasks to each member.) Third, should an anxious, restless park-using public be forced to wait for the final report, or should the committee send its recommendation, in summary form, to the Mayor as soon as possible?

Some committee members believed that the park should open immediately with no change in facilities, other than protecting the public from several "hot spots." The toxicologist's position was summarized in the newspapers as follows:

"We're looking at the park as a whole." Cautioning that to use only the highest estimates of contamination would grossly overestimate the danger, he also observed that "People don't just play in one four-inch square."

Some committee members implied that they were even opposed to minimal remedies:

Helgerson [an epidemiologist with the city-county health department] discounted the dangers of even these hotspots and said that the greatest risk appears to be the public's anxiety about the park (Hadley 1984f: D1).

The landscape architect who designed the park, but was not a committee member, had an even more critical view: "I feel that with a few minor corrections the park will have a clean bill of health and we will stop polluting the public's mind" (Schaefer 1984: B1).

Others were more cautious. The environmental engineer co-chair focused on the comparative danger of different substances, and noted that "acceptability of risk is a personal decision" (Schaefer 1984: B1). His conclusion, however, was as follows: "The public health risk in usage of the park is not unreasonable. The exposure one gets is one that is acceptable" (Sumida 1984: A1).

The position of the occupational medicine administrator reflected some caution. Disputing the position of the epidemiologist quoted above, she argued that it was not too frivolous to be cautious, and was quoted as follows: "There are known areas that can be separated off. It makes sense to be a little cautious" (Sumida 1984: A1). The position of the EPA officials was the most conservative, as their preference was to keep the park closed two additional months for more tests (Wilson 1984c: D1). The EPA engineer indicated, however, that they did not plan to seek an order forcing additional closure (Wilson 1984c: D1).

The vehicle for mitigating their "cold potato" resolution and maintaining some balance between those panelists who wished to immediately reopen the park and those who wished to proceed cautiously was to recommend both immediate short-term "symbolic" changes in the park, and to endorse other longer term site rehabilitation and improvements, some of which, such as installation of a sprinkler system, had been part of the original park plan.

The Executive Summary (City of Seattle 1984) prepared by the committee reflected their continuing ambivalence in establishing park user risks:

> It should be stressed that data which indicate that PNAs are human carcinogens are derived from occupational studies where exposures were greater for longer periods of time than could conceivably occur from recreational use of Gas Works Park. . . . The process of extrapolating studies on occupational exposures to ill-defined, sporadic environmental exposures is inexact and extremely difficult. . . . Ingestion of soil contaminated with PNAs is considered to be the most important route which could lead to significant exposure to PNAs. We consider the likelihood of such exposure remote for all park users except for infants and toddlers who may ingest contaminated soil when placing objects or dirty hands in their mouths.

In the end, the committee recommended reopening the park, contingent upon the following four recommendations: (1) protecting park users from two areas of highest contamination — an old lakefront concrete pier, and large pieces of industrial machinery in the "playbarn"; (2) posting informational signs "that will remind parents to monitor their children's activities and encourage them to follow good personal hygiene practices, such as not placing foreign objects in their mouths and to wash their hands before eating"; (3) following-up sampling and monitoring; (4) reinforcing the ban on swimming in Lake Union from the park's shoreline (City of Seattle 1984).

The park was reopened, initially with the two high contamination areas fenced off. The equipment in the play barn was later cleaned and painted, and a concrete covering was added to the "pier" area. Signs were posted concerning swimming and fishing in Lake Union, and several areas received additional topsoil and new grass. The city had additional tests done on the park, and it is still a possible candidate for Superfund rehabilitation (Hadley 1984j: A1). The park rapidly returned to its high level of use.

III. Conclusion

The case described here emerged at a time of maturity and stability for the environmental movement. It occurred in an area of the United States noted for environmental concern. Yet the issue might never have emerged without the impetus of the EPA.

In reflecting on this case, it is difficult to balance the potential public health dangers which could have resulted from underestimating health hazards with continued closure of a very popular public facility. The latter was a distinct possibility which could have resulted from any combination of factors such as investigative delays, conflict between government agencies, and complex and expensive site rehabilitation.

The case is instructive for citizens interested in hazardous waste issues. The Gas Works case may be substantively satisfying to environmentalists, in that the hazards were defined as real, were investigated with the help of impartial scientists at significant public expense, and were generally handled seriously (Gates 1984: A8). (For a case in which agencies were somewhat less willing to accept responsibility, see Nigg and Cuthbertson 1982.) However, the process for professional-public interaction was more typical of pre-movement days when the public was excluded and decisions were left to the experts. Although the initial decision to close meetings was quickly reversed, public interests external to the Ad Hoc Committee remained unorganized and impotent throughout the issue. The public was clearly placed in a powerless, defensive position, and was forced to wait out the slow decision-making of the committee. Nearly two months prior to reopening, a committee member summarized the public's perception: "The public is just anxious that we hurry up and reopen their park so that they can resume the activities they're accustomed to" (Wilson 1984b: A1).

There are also several possible lessons for environmental scientists. First, each committee member worked hard to present their professional opinion convincingly to a sometimes dubious public. Despite these efforts, jokes about "eating Gas Works dirt" persisted. In addition, the committee found it difficult to agree on how to put the park's hazard risk in the context of other daily health risks. From the perspective of one park jogger:

> This is really dumb! I've been jogging up that hill [Kite Hill] every day for the last year, and now I find out that what I've been doing to make me healthy could be making me sick. Everything seems to be hazardous to your health these days. I might as well start smoking again! (Byrd 1984: 1, 11)

Second, because most environmental issues are interdisciplinary, it is important for environmental professionals to be sensitive to the differing assumptions and perspectives of other fields whose expertise may also bear on the issue under investigation. At times the interaction among the professionals on the committee suggested to this observer that working in an interdisciplinary setting, especially one well covered by the media, was a new experience.

Third, although several Ad Hoc Committee members were designated as public representatives, one missed many meetings, and another identified more with the professionalism of other committee members than with the public. One consequence was that it sometimes appeared that the park had become a research site for several of the disciplines represented on the committee. Public comments offered to the committee at the close of these sessions often were pleas to remember that the park did have a human constituency anxious to once again enjoy a favorite place. Environmental scientists thus need to remember that people are also part of the environment, and must be included both formally and substantively in the resolution of environmental issues.

Fourth, analysis of the case suggests a continuing tension between those supporting closure of the entire park (and adjacent parking lot), and those who felt that a high priority was to reopen those portions with low risk. Those taking the former position were less likely to place the risk in the context of likely exposure by park users. The latter appear to be the better model for environmental scientists operating in the public eye, in part because of the general exposure of the population to a multitude of hazardous substances in daily life.

Finally, with the exception of several closed meetings, the committee was forced to operate literally in the eyes of the public and the media. There was no laboratory or agency office to retreat to — the committee was an extreme example of "public arena" decision-making, contrasting with the more closed "professional arena" settings of most natural resource agencies (see Miller and Gale 1985). While most environmental issues do not reflect the continued public and political exposure characteristic of the Gas Works case, it nonetheless behooves environmental scientists to prepare themselves, intellectually and emotionally, for operating in a highly politicized, public setting.

In conclusion, it was encouraging that the environmental movement did come to Seattle on behalf of Gas Works Park. The general handling of the issue provided a positive, politically acceptable template for handling similar issues, although many of the other sites may involve far more serious health risks and rehabilitation problems (Hadley 1984e: D2). Yet evaluation of the role of movement-linked agencies such as the EPA and Superfund is mixed, and not optimally beneficial. The role of the EPA varied from helpful consultant to a "big brother" who might close the park for a longer term despite recommendations by the Ad Hoc Committee. The public and the Ad Hoc Committee both suffered because of a combination of a high level of EPA concern and that agency's inability, apparently because of workload and lab space limitations (Wilson 1984a: A3), to produce data analysis within what an anxious public considered a reasonable time period.

Relations between the city and the EPA were complex and varied. At times, it was the EPA which challenged the city to increase its level of concern with health hazards. In other instances, it appeared that the park was caught in a power contest between a federal agency and a strong city government. The role of the Superfund was even more problematic, and could have ultimately posed major problems for any timely resolution of the park's hazards.

The Gas Works issue was clearly affected by the maturity and strength of the environmental movement. The substantive interests of the movement were essentially delegated to the EPA. The cautious, "purist" perspective of the EPA resulted in a nonnegotiating stance more typical of a youthful movement. Procedurally, however, a more youthful movement might have stimulated more aggressive public involvement, and, possibly, a swifter park reopening. Public aggressiveness might have been more pronounced, and the EPA employees might have been more likely to reflect the enthusiasm and movement linkage typical of that agency's earlier days (Bachman 1982). Most importantly, though, the issue was resolved, and the park environment improved. Seattle and the Puget Sound area now await their next "Gas Works."

Notes

1. It is interesting to speculate on how this issue might have developed previous to the emergence of the environmental movement in the seventies. Although there was community controversy about the park's design, this was a much different issue than the discovery of potentially dangerous soil (Richard 1983).

2. Newspaper articles in Seattle papers concerning the Gas Works issue, published from April through August, 1984, and the personal observations and informal interviews by the author, constitute the data base for this case study.

3. It is probably significant that the park was not physically sealed. People continued to use the park, although in drastically reduced numbers. Fencing the closed park would have suggested a seriousness to the problem which no one was ready to assert. It might, however, have resulted in propark mobilization or more intense scrutiny of the handling of the issue.

4. Other government organizations with sporadic active interest in the issue included the U.S. Food and Drug Administration, concerned about the catching and selling of possibly contaminated crawfish from Lake Union (Jones 1984b: A26), Seattle Police Department divers operating at a police boat station near the park (Jones 1984b: A26), and the union representing city park maintenance workers.

5. One might argue that public health organizations might have eventually discovered the problem in the absence of environmental movement aligned agencies. The perspective of this paper, however, does not support this view. The three public health organizations were thus cast in the reactive, rather than advocative, role of providing experts.

6. It is also interesting that almost no economic interests were affected by the park closure. Early in the controversy, the Washington Natural Gas company, which had assumed management of the facility in 1955, made it clear that its responsibilities ceased when the site was purchased by the city in 1962. The only active concerned business was "Suspended Elevations," a kite shop located near the park.

7. However, several members of the Ad Hoc Committee were more experienced in a public setting than many of the government- and university-based specialists. The two Mayor's office staff members assigned to the committee were skilled in working with the

public and the media, although they lacked the technical expertise to understand much of the committee's deliberations. One was the Mayor's press secretary and the other an assistant who had previously successfully resolved a downtown land use dispute. The two Ad Hoc Committee members appointed as representing both public and neighborhood interests were politically experienced as they had served on city advisory boards. However, their experience actually made it more difficult for other members of the public to gain access, since these two individuals tended to operate in a "professional public person" mode.

References

Bachman, W. A. 'Interior's Policy Decisions Tilted Toward Environment,' *Oil and Gas Journal*, 1982, 76; 33—38.

Byrd, Michelle. 'Gas Works: Controversial from the Start,' *University of Washington Daily*, May 10, 1984, 1 c. 1—3, 11 c. 1—4.

City of Seattle. *Executive Summary*, Health Advisory Committee to Mayor Charles Royer, 1984.

Downs, Anthony. 'Up and Down with Ecology — the 'Issue-Attention' Cycle,' *Public Interest*, 1972, 28; 38—50.

Gale, Richard P. 'Social Movements and the State: The Environmental Movement, Countermovement, and the Transformation of Government Agencies,' *Sociological Perspectives*, 1986, 29, 202—240.

Gates, Carolyn. 'Planners Then Didn't Know What We Know Now.' (Letter to Editor), *Seattle Times*, June 18, 1984, A8 c. 1.

Hadley, Jane. 'All of Gas Works Park is Polluted, Latest Tests Show,' *Seattle Post-Intelligencer*, May 2, 1984a, C1 c. 3—6.

Hadley, Jane. 'State Wants 14 Sites Added to the Superfund Hazardous List,' *Seattle Post-Intelligencer*, May 3, 1984b, A6 c. 1—4.

Hadley, Jane. 'What's Known About the Dangerous Toxins at Gas Works Park,' *Seattle Post-Intelligencer*, May 4, 1984c, C1 c. 3—6.

Hadley, Jane. 'Gas Works Park Decision Not Due Before Mid-June,' *Seattle Post-Intelligencer*, May 5, 1984d, C1 c. 5—6.

Hadley, Jane. 'EPA Criticized for Rushing Cleanup Plans Past Public,' *Seattle Post-Intelligencer*, June 12, 1984e, D2 c. 3—6.

Hadley, Jane. 'Gas Works Park May Reopen with Little Change,' *Seattle Post-Intelligencer*, June 20, 1984f, D1 c. 5—6.

Hadley, Jane. 'It's Safe to Open Gas Works, say National Health Officials,' *Seattle Post-Intelligencer*, July 4, 1984g, A3 c. 3—6.

Hadley, Jane. 'Gas Works Park May Open Soon with (Fairly) Clean Bill of Health,' *Seattle Post-Intelligencer*, July 13, 1984h, C1 c. 1—4.

Hadley, Jane. 'State Asks EPA Help in Park Cleanup,' *Seattle Post-Intelligencer*, July 14, 1984j, A1 c. 6, A7 c. 1—2.

Hedges, Fritz. Seattle Parks and Recreation Memorandum addressed to Jill Marsden, Chair, Health Advisory Committee, City of Seattle, June 25, 1984.

Jones, Lansing. 'Park's Worst Chemicals Found in Children's Play Area.' *Seattle Times*, April 28, 1984a, A1 c. 2—6, A13 c. 4—6.

Jones, Lansing. 'In 1962, Gas Works Park was a 'Fantastic Deal',' *Seattle Times*, April 29, 1984b, A26 c. 1—6.

Jones, Lansing. 'I-90 Clay May Save Gas Works Park,' *Seattle Times*, June 13, 1984c, A1 c. 1—4.

Kalman, David. Letter from Kalman to Mr. Chuck Kleeberg, City of Seattle. Department of Environmental Health, University of Washington, June 18, 1984.

Miller, Marc L. and Richard P. Gale. 'Professional and Public Natural Resource Management Arena: Forests and Marine Fisheries,' *Environment and Behavior*, 1985, 17, 651–678.

Nigg, Joanne M. and Beverly Ann Cuthbertson. 'Pesticide Applications near Urban Areas: A "Crisis in Confidence" for Public Health Agencies,' *Journal of Health and Human Resources Administration*, 1982, 4, 282–302.

Richard, Michael. *Seattle's Gas Works Park*. Seattle: Tilikum Place Printers, 1983.

Schaefer, David. 'Health Experts May Advise Royer to Reopen Park,' *Seattle Times*, June 20, 1984, B1 c. 1–2.

Seattle Post-Intelligencer. 'Superfund Help' (editorial), May 3, 1984a, A10 c. 1–2.

Seattle Post-Intelligencer. 'Superfund Works its Will in Seattle' (editorial), May 10, 1984b, A10 c. 1–3.

Seattle Post-Intelligencer. 'Gas Works Denizens Don't Wait for Reopening,' August 13, 1984c, D1 c. 1–5.

Seattle Times. 'Gas Works May Open Soon, says Royer,' July 10, 1984a, C2 c. 1.

Seattle Times. 'Let's Start Using Gas Works Again' (editorial), July 29, 1984b, A18 c. 1–2.

Sumida, Midori. 'July Opening of Gas Works Park Unlikely,' *Seattle Times*, July 4, 1984, A1 c. 5–5.

Wilson, Duff. 'City Asks UW to Test Soil at Gas Works,' *Seattle Post-Intelligencer*, May 26, 1984a, A3 c. 1–2.

Wilson, Duff. 'Gas Works Park Could Open in July,' *Seattle Post-Intelligencer*, June 27 1984b, A1 c. 5–6.

Wilson, Duff. 'Royer says Gas Works Park Could Reopen by August 17,' *Seattle Post-Intelligencer*, July 28, 1984c, D1 c. 3–5.

Chapter 10

Communicating Information about Workplace Hazards: Effects on Worker Attitudes Toward Risks

Michael S. Brown[1]

Introduction

Employee access to information about workplace risks has generated intense debates over government policies and growing conflicts on the shopfloor. The success of lobbying and negotiating efforts on the part of workers, their unions, and health professionals is reflected in the proliferation of collective bargaining agreements, state and local laws, and federal regulations which mandate the transfer of hazard information from management to employees (Ashford and Caldart 1983; Brown 1984). Colloquially known as the "right-to-know" movement, interest in educating workers about the dangers of toxic substances on the job follows a long history of employer neglect toward communicating risk information and varying degrees of employee interest in occupational health and safety.

The last several decades have been a time of explosive growth in concern about occupational exposures to hazardous materials. After a long period of relative quiescence about health hazards on the job, rank-and-file workers, labor unions, and health activists campaigned on the shopfloor and in Congress for improved conditions. The Occupational Safety and Health Act of 1970 brought substantial change in the duties and obligations of management and the federal government toward protecting employees.

Yet by the end of the decade, workers and their unions realized that legal language did not always translate into safer conditions. They began to argue that workers had to be informed about hazards on the job. Knowledge, argued a prominent union spokesperson, can lead to action, while ignorance all too often means tragedy (Hancock 1982). The right-to-know campaign grew out of a well-documented history of employer refusals to identify and communicate information about known toxic substances, and efforts to mislead workers into believing their jobs were safe.

Some instances of employer malfeasance have had tragic consequences

B. B. Johnson and V. T. Covello (eds.), The Social and Cultural Construction of Risk, 251–274.

for workers. In the early 1930s at Gauley Bridge, West Virginia, the New Kanawha Power Co., a subsidiary of Union Carbide, blasted a water tunnel through rock that was up to 95 percent silica (U.S. Congress 1936). Congressional inquiry established years later that employees were told that the work was safe at the same time that engineering personnel entered the tunnel while wearing protective masks. Thousands of workers, most of whom were black sharecroppers, were exposed and rapidly developed silicosis. Some estimated that over 1,000 tunnel workers died of the disease.

A more chronic problem manifested itself in the case of asbestos. Court litigation has established that asbestos manufacturers, particularly Johns-Manville, knew about the hazards of breathing the dust since the 1930s and withheld the information from workers for over 30 years (Borel v. Fibreboard Paper Products Co., C. A. 5th, 493 F.2d 1076, 1975 and Beshada v. Johns-Manville Products Corp., 90 N. J. 191, 1982). The company even went so far as to monitor workers' health and lay off or urge early retirement for workers disabled by asbestos without disclosing the findings of their medical examinations (Johns-Manville Products Corp. v. Contra Costa Superior Ct., 612 P.2d 948, 1980).

Such management attitudes are not necessarily relics of past practices. Recently, the State Attorney General for Cook County, Illinois brought murder charges against executives of a firm using cyanide and other highly toxic chemicals to recapture silver from processed film (Nordgren 1985). Management had allegedly painted over warning signs, removed hazard labels, and encouraged its predominantly non-English speaking workers to ignore safety procedures. The subsequent death from cyanide of a recent immigrant to the United States was not simply a case of management neglecting adequate safety precautions, but appeared to be a result of an active campaign to control what workers knew about hazards and consequently their attitudes toward risk.

This paper looks critically at the relationships between the communication of hazard information and workers' responses to risk. What might have happened had the tunnel drillers at Gauley Bridge known about the silica dust; would asbestos workers have avoided the crippling effects of asbestosis and the almost sure death associated with mesothelioma if they were educated about asbestos-related disease and precautions; what actions might the young Polish immigrant have taken if the warning labels had remained on the cans of cyanide and someone had explained them to him? Clearly, not all workers worry about health hazards on the job (see Quinn and Staines 1979). While some employees are active information seekers and express a great deal of concern about workplace risks, others ignore or deny that chemicals are a serious problem on the job.

The research reported here explores the organizational context of these experiences and the efforts of some workers to become better informed about risks. It seeks to embed the issue of hazard communication and worker responses to risk in the social, economic, and political structures surrounding the workplace (see Douglas 1976; Burawoy 1979; Douglas and Wildavsky 1982). Using a set of interviews conducted in the course of a project which explored broad questions of worker attitudes toward risk (Nelkin and Brown 1984), this paper focuses on the responses of workers to the communication of risk information.[2]

The next section discusses the production of hazard information. It looks at the processes by which various parties including management, unions, workers, and the government develop knowledge that is of interest to workers. In the following section, several theories accounting for the patterns of communication and miscommunication between employees, unions, and management are explored along with the legal obligations of employers toward informing employees about job risks. The last two parts look at worker responses to information and analyze the implications of attitudes for public policies.

Producing and Disseminating Hazard Information

Chemical substances are present in most work environments, even those popularly thought to be clean and safe. White-collar employees, especially clericals, may be exposed to office hazards such as fumes from reproduction machines, correction fluids, and poor ventilation. Blue-collar workers encounter chemical substances in nearly every occupation: machinists may use cutting oils, electrical equipment workers are exposed to heavy metals and transformer fluids, and welders liberate toxic fumes in the process of cutting and joining metal. Professionals such as artists and sculptors can create fumes or dusts, doctors and nurses are exposed to anaesthetic gases during surgical operations, and scientists and technicians using laboratory chemicals may be exposed in the process of conducting experiments. Even in the micro-electronics industry, noted for clean-room conditions in the production of computer circuits, worker exposure may be substantial (LaDou 1984). Not all substances encountered in the workplace are hazardous, yet identifying those that are and communicating information about them is a complex and controversial task (see Fisher 1982 for a useful summary). This section discusses some problems in determining workplace hazards workers at risk.

The amount of chemicals used in industry, government, and education is staggering. Although millions of compounds have been synthesized in laboratories, over 58,000 are presently in commercial use (USEPA 1982). In addition, innovative companies submit about 1300 premanufacture notices for new chemical compounds each year to the U.S. Environmental Protection Agency (USEPA 1984). A recent National Academy of Sciences (1984) study found that only a very small percentage of commercially valuable chemicals have been tested for chronic effects; not many more have even been evaluated for acute toxicity.

For those substances that are identified as potentially harmful to health, scientists attempt to match toxic effects with changes in exposure levels. Quantifying this relationship, however, is difficult. Much of the controversy over regulatory standards focuses on defining threshold levels (below which effects are not observed) and dose-response curves for specific chemicals. Given limited scientific tools, disagreement over the number of toxic substances and who is at risk comes as no surprise. Estimates of substances that may be of workplace concern range from the hundreds to the tens of thousands.[3] Millions of workers may be exposed to regulated chemicals; however, the level of exposure and the degree of risk is constantly in dispute.

Workers are less concerned with the scientific debates over testing and dose-response curves than with identifying hazards on the shopfloor. This involves the production and communication of two kinds of information. People want to know what substances are used on the job and the effects associated with them and they want to determine if they are suffering from work-related illnesses and the sources of these health problems.

Efforts to identify substances may be thwarted by management claims for trade secret status or by employer ignorance of chemical identities. In compiling data for its National Occupational Health Survey (USDHEW 1974, 1977a, 1978), NIOSH found that over 70 percent of the exposures recorded in the survey were to trade name products. An estimated 15 million workers worked with a trade name product containing an OSHA-regulated substance. About 615,000 were exposed to trade name products containing an OSHA-regulated carcinogen. The survey led the agency to conclude that many workers and employers were not likely to know about potentially hazardous exposures which should be monitored and controlled (USDHEW 1977b).

When the identity of substances is known, communication of detailed hazard information is possible. Management may inform workers through the use of material safety data sheets, training programs or warning labels (Santodonato 1981). Alternatives to management sources of data include unions for organized workers, COSH groups,[4] health professionals, government agencies, and coworkers. However, unless government agency or union-sponsored inspections occur, workers have no choice but to ask management for information about exposure levels. Without such information, workers cannot determine the extent of risk, the degree of employer compliance with regulatory standards, and their options for action. Lack of adequate exposure monitoring is a key factor in limiting worker knowledge of hazards. It also hinders the development of epidemiological evidence which may identify factors at work affecting employee health.

Organizational Theory and Patterns of Hazards Communication

In the absence of government regulation, what factors might influence management decisions to collect and disseminate hazard information to workers? Further, in the absence of management action, how might unions respond? Most common are explanations which focus on the economic importance of hazard communication. An alternative, or, perhaps, complementary analysis suggests that the organizational consequences of informing workers play an important role in communication practices.

Economic Interpretations

Neoclassical economists posit the efficient operation of market transactions based on a key assumption that buyers and sellers possess perfect information pertinent to decision-making. In the context of labor markets, this implies that managers and workers are aware of hazards present in the workplace and negotiate wage rates according to the willingness of both parties to balance money against risk (Viscusi 1979). Economic analysts point out, however, that information, like all goods and services in society, is not cost-free (Stigler 1961; Nelson 1970; Wilde 1981). Assumptions of perfect knowledge are inappropriate because the development and acquisition of information requires the expenditure of resources. Although management is in the best position to know what substances are present in the work environment and whether they are hazards, they have limited incentives to develop and disseminate risk information.[5] If workers were to either reduce wage demands in exchange for risk information or pay outright, management would have an incentive to provide the desired product (see Smith 1976). The history of employer failures to disclose risk only shows, according to these economists, that employee demand for information has been minimal.

Critics of this avenue of economic analysis question the conclusion that there has been little worker demand for hazard data. They point to repeated cases of employers providing inaccurate or misleading information to workers. In this view, the cost of information collection and dissemination pales in comparison to the costs of cleaning up the workplace and compensation to victims of occupational diseases (see Ashford 1976). Informing workers about job hazards will tend to increase the cost of doing business. Knowledgeable workers may demand safer conditions or removal of hazardous substances or processes. Demands may be backed up with actions affecting productivity such as grievances or job actions, complaints to regulatory agencies which could result in fines and orders to eliminate hazards, or adverse publicity. Workers may quit when they find out about risks and recruitment may be difficult, causing havoc with production schedules. Finally, information about job hazards may enable employees who become ill from occupational exposures to substantiate claims for compensation.

Organizational Interpretations

In addition to economic constraints influencing attitudes towards hazard communication, there are few organizational incentives for dissemination of risk information which would promote a cleaner work environment. Building on Cyert and March's (1963) theory of organizational behavior, McCaffrey (1982) points out that occupational health has no advocate in organizations comparable to managers with sales, production or finance responsibilities. He suggests that occupational health will suffer especially when promotion criteria hinge on short-term feedback reflected in quarterly earnings reports. Neither government inspections nor a large number of compensation claims for work-related illnesses which might induce a change in behavior are likely to occur over the short-term. In sum, organizations do not reward managers for protecting the health of employees.

Within organizations, hierarchical structures further limit promotion of health and safety. Higher level managers with the authority to make investment decisions are removed from day-to-day operations and rarely see shopfloor conditions. Line supervisors are more familiar with conditions, but their primary responsibility is overseeing production. Knowledge may lead to complaints and in the daily interactions between supervisors and shopfloor employees, the less workers complain, the smoother production flows. Thus, there is little incentive to share information with workers.

Traditional analysts of organizations have long maintained that hierarchical structures are necessary for efficient operating conditions. Building on principles of "scientific management" established by Taylor (1911), theorists argue that in the absence of explicit control mechanisms, line workers will not meet the output goals of the organization (Kerr 1954; Kornhauser *et al.* 1954; Behrend 1957). However, as Galbraith (1971) has pointed out, managerial goals need not be limited to maximizing profit, but may also include maintaining organizational stability and hierarchy.

Burawoy (1979b) has shown that workers recognize and act upon their own interests which necessarily contradict those of management. While management seeks to reduce wage costs and control the quantity and quality of production, workers in opposition look to reduce their level of effort for a given wage rate. One means for inducing worker compliance with managerial needs is to control what workers know about the production process. Deskilling comes directly out of Taylor's dictum to use labor for its brawn and reserve for management the right to assert its brains. Though his analysis has been attacked for its selectivity, ahistoricism, and lack of theoretical power, Braverman (1974) is persuasive in arguing for a steady dimunition over time in worker knowledge inherent in craft occupations. Management has made concerted, but, as Storey (1983) points out, not altogether successful, attempts to obtain and control knowledge necessary to limit the influence of line personnel on the tempo and quantity of output. The assumption on the part of management has been that the more workers know, the more they will subvert managerial goals.

In the case of occupational health, control of knowledge helps maintain the "myth" that only supervisory and executive staff are capable of understanding and using technical information about risk. Monopolizing the development and dissemination of hazard data allows management to define adequate safety measures. It follows that it is in workers' interests to allow management to determine appropriate actions. Thus managers argue that much of the information demanded by unions and health activists would "overload" employees and would either confuse them or be useless (see, for example, statement of the Chemical Manufacturers Association in U.S. Congress, 1981: 266—267). The bulk of hazard information is properly the province of "experts" whose loyalty, in most workplaces, rests with the employer (see Berman 1978).

Successful efforts to control information flows can lead to a sense of powerlessness among workers at risk. John Gaventa (1980; 1983) has explored this theme in the contest of labor struggles in the coal towns of southern Appalachia and in the relationship between experts and those struggling for social change. Elites exercise power over subordinates through information control and induce them to take risks they might otherwise avoid. More importantly, it limits the ability of those at risk to raise issues and contest their situation. Without knowledge, workers, for example, are not likely to challenge either hazardous conditions or the social structure which allows dominant elites to determine the level of risk in the workplace. Worker claims of evidence of hazards through personal experience or collective documentation of exposure and effects are ridiculed as lay ignorance. Their experiences must be certified by the management's experts before being considered legitimate. As a consequence, workers are encouraged to remain passive and responsive to management directives. The maintenance of information control thereby promotes among workers a sense that they lack both the ability and the power to become informed and to use what they know.

Unions

Unions currently represent 21.6 percent of the nation's labor force (USDL, 1985). Despite their relatively low representation rate, they have had a substantial effect on occupational health policies (see Ashford 1976; Berman 1978). Some unions have long been active in health and safety (for example, the auto workers, steel workers, chemical workers, coal miners, and rubber workers); others only recently have addressed the problems.

The impetus for much of the current efforts of unions came from the New Directions program established by the Occupational Safety and Health Administration (OSHA) under President Carter. Utilizing federal grants to hire technical staff, generate educational materials, and train local and regional safety representatives, many unions provide alternatives to employer sources of information. Their influence may even extend to nonunion workers through organizing drives, the formation of local committees on occupational safety and health and lobbying for more stringent laws and regulations.

Unions operate in an organizational environment which focuses on the needs of shopfloor workers (see Estey 1981). They are not subject to the same type of market pressures as management. The organizational structure of most unions is divided into local shops who have direct contact with members, and a national or international body. Unions rise and fall on the strength of their local organization, with the national offices performing coordinating, informational, and general policy-setting functions (Estey 1981). Health and safety problems mirror this division of labor. Locals may call on the national staff, but the rank-and-file and their officers are responsible for resolving day-to-day problems. Thus, while the public views the actions of the national offices of labor organizations and the AFL-CIO as the basis for judging unions, members look to the performance of their local and its officers. Their attitudes help shape whether workers look to their union for health and safety information.

Local union officials must respond to several pressures. First, unlike management they are usually democratically elected by the workforce. Electoral success often rests on a willingness to stand up to management and fight for the interests of the rank-and-file. Once elected, constituencies may shift. Some officials maintain their militancy. Others, locked in daily interaction with management, start to identify with the company. They moderate demands and look to deflect rank-and-file activism and avoid confrontations. They may limit action on issues which are not "bread-and-butter" questions of wages, benefits, and job security. If most workers are not very concerned about health and safety, officials can avoid pursuing the complaints of activists, secure in the knowledge that their union position is not threatened. They may even be able to use management threats to shut down the plant or to discipline workers as a means of avoiding action on hazardous conditions.

This does not mean that union officials can safely ignore the needs of their members. Those who identify strongly with management run the risk of generating rank-and-file opposition which could result in their being voted out of office. More militant workers may tap a desire for greater action on nontraditional issues such as health and safety, as well as more traditional concerns such as wages, pursuit of grievances, and protection of job security. Federal New Directions grants have given activists a measure of financial independence within unions to act on occupational health issues without too much intrusion from union hierarchies. The resulting increase in union activists is measured in the levels of contracts containing safety and health language (see BNA 1983), in the petitions, testimony and demands for effective OSHA regulations and standards, and the proliferation of state and local right-to-know laws.

Changing Terrain: Right-to-Know Laws and Regulations

Over the past century, there have been three major periods delimiting the obligations of management toward informing workers of hazards. Prior to the advent of workers compensation legislation in the early part of the twentieth century, relationships between employees and their employers were governed by common law precedents. These obligations were swept away when states adopted compensation schemes as the sole remedy for injuries and illnesses related to the job. The third period is marked by the passage of the OSH Act and the efforts to gain federal, state, and local rules governing disclosure policies.

Few specific legal obligations were imposed on employers before the turn of the century. Common law held that employers had "the duty to give warning of dangers of which the employee might reasonably be expected to remain in ignorance" (Prosser 1971: 526). This did not necessarily mean that an employer had to inform workers of hazards. The courts presumed that experience provided adequate knowledge about routine risks (Moore v. Morse and Mallory Shoe Co., 89 N. H. 332, 1938; Engelking v. City of Spokane 49 WA. 446, 1910). An employer had only to provide warnings for risks which were abnormal or extraordinary. Once informed, employees were assumed to have accepted those risks voluntarily and could not bring suit if injured or made ill by the job. Common law obligations to provide adequate warnings were eliminated when states adopted a workers' compensation system as the sole remedy for workplace injuries. Barring lawsuits, they reduced the incentive for employers to inform workers of extraordinary hazards.[6]

Attitudes toward communicating health and safety to workers remained substantially unchanged until Congress enacted the Occupational Safety and Health Act in 1970. Congress indicated a desire for occupational health programs (Section 2(b) (1)) and joint labor-management efforts to reduce occupational diseases (Section 2(b) (13)). Thus, it directed the Secretary of Labor to establish training and education programs for employers and employees on hazards and their prevention (Section 21(c)(1)).

Some of the 100 or so regulations OSHA has adopted contain training requirements and although most refer to safety hazards, a significant number involve toxic substances. OSHA also has recommended training guidelines that involve employee participation (USDL 1983a). Employers should ask workers about the presence of hazardous conditions and their opinions on existing barriers to knowledge. As a voluntary policy, however, it can be adopted or ignored.

OSHA has made access to company medical records a mandatory rule. Employees and their designated representatives may read and copy records covering themselves and similarly situated employees (USDL 1980). The rule, however, does not require companies to generate any records. Medical records are not necessarily useful in identifying substances present in the workplace, proper procedures, or protective actions. OSHA viewed the rule as complementing a yet-to-be-promulgated hazard information program which would provide workers with extensive knowledge about job hazards.

Frustrated with the lack of comprehensive federal action, unions and health activists have succeeded in securing "right-to-know" laws in over 20 states and localities. Some simply require making lists available to workers, while others require comprehensive training programs. New Jersey enacted a particularly stringent law requiring employers to disclose to the state the identity of hazardous substances and to issue material safety data sheets developed by state officials to workers. Information about toxics is also available to community residents.

Industry, prompted by the proliferation of strict, though sometimes contradictory, rules at the state and local level pressed for enactment of a federal rule preempting local action. Following nearly a decade of work on a federal rule providing for employer communication of workplace hazards, OSHA issued its Hazard Communication standard in November 1983 (USDL 1983b). The new regulation was substantially changed from its initial formulation in the mid-1970s to its final form, reflecting a victory of advocates of management rights to control of hazard communication.

Under the Carter Administration, OSHA proposed a rule which would have identified the chemical name of all substances in all workplaces and allowed employees access to trade secrets (USDL 1981a). Shortly after taking office, the Reagan Administration withdrew the proposal and reissued it as a Hazard Communication Standard (USDL 1981b, 1982). The final rule, which went into full effect in May 1986, applies only to manufacturing employers and is designed to supersede state and local statutes. It requires chemical manufacturers and importers to assess the hazards of chemicals which they produce or import, and manufacturing employers to provide training and education on hazardous substances including their effects, emergency procedures, and proper handling. The rule is extremely flexible, however, allowing an employer to determine what hazards should trigger the new requirements. The rule grants physicians and health professionals access to trade secret information, but does not allow workers the right to specific chemical names. Industry likes the new rule, believing it is workable and will provide needed uniformity across the country. In contrast, organized labor is disturbed, because the OSHA rule is weaker than most state and local laws. Several unions along with three states have sued OSHA hoping to overturn preemption of state and local laws by the new regulation (BNA 1984a, 1984b).

Hazard Information and Worker's Perceptions of Risk

Over the last decade there have been significant changes in workers' demands for information and communication practices among employers. Although shifts in policy have led to increased employee and union requests for sophisticated hazard data, along with the implementation of computerized data bases and training programs in health and safety by a variety of firms, not all workers seek information, nor are all employers forthright about hazards. The discussion of organizational and economic analyses of information control coupled with past practices suggests that hazard communication will not necessarily lead to action on health risks. Indeed, if management is successful in controlling risk knowledge and workers are unable to get information elsewhere, we would expect to see many workers minimally concerned about hazards.

This section explores the responses of workers in a variety of situations to job risks and seeks to embed their responses in the social and economic structures affecting their lives. Given the limited nature of the interviews which form the basis of the discussion,[2] the analysis should best be considered preliminary and indicative of trends deserving further research.

Denial

Work, for most people, means doing their job, social interaction, and a paycheck. No one likes to think about the possibility that a job may not only put food on the table but result in a debilitating illness or affect the health of one's children. Confronted by risk, some refuse to acknowledge its relevance to their lives. Others justify and then ignore risks as part of the job.

Social psychologists have put forward several theories accounting for these responses to risk information (see Weinstein 1979). One approach suggests that people use denial to avoid anxiety and feelings of hopelessness when faced with risks that they cannot control. This may include avoiding information about the seriousness of risk (Bishop 1974). Robert

Beilin (1982) argues that people use denial to preserve existing social relationships that would be upset if they integrated knowledge of potentially tragic outcomes into their lives. In contrast, the theory of cognitive dissonance holds that people filter information in an effort to achieve consistency of knowledge with beliefs and actions (Janis and Mann 1976). If forced to confront information which conflicts with beliefs, the theory suggests they will change their behavior or suffer from stress. Both approaches suggest that people will not be likely to seek information if they believe that risks are minimal or nonexistent and there is little that can be done about them.

In this context, management messages reinforce a pattern of denial. Employers' arguments imply that if workers follow orders, they will be safe. Employees do not have to take an active role in health and safety; management can be trusted to protect them. Many workers want to believe that their employer would not create conditions that will lead to harm. They have no methods of coping with the long-term implications of continued exposures. Oriented toward short-term goals such as avoiding accidents and collecting a paycheck, workers can have attitudes summed up by a chemical worker:

> You walk into the lunch room and it sounds like there are a bunch of horses in there. Everybody's coughing. If you ask why, they say, "Well, I work at —, that's why I got the cough." If somebody's telling you, "You shouldn't drink that because it's got saccharin in it," they say, "Well, what the hell do I care if it's got saccharin in it? I work at —." There's always jokes about chemicals. It's something real and people know it. But let's face it. Most just get it out of their mind. They don't think about it. People are more worried about getting burned or inhaling something than about any long-term sickness. Since it's not happening now, it's not something to worry about. You always think, "I ain't gonna be here long. I'm only here for a couple years." When I first came here, I thought five years, that's all I'll be here. I'll be here five years next month and I have no plans to leave. Most everybody that I talk to says the same thing, but 95 percent of us stay.

Interviews suggest that supervisors or senior workers who care little about hazards and are more interested in getting the job done with a minimum of conflict set the tone for the entire workplace. A maintenance worker in an egg-processing plant described how he was ridiculed by his supervisor and coworkers for wearing gloves when using a powerful cleaner. Attributing the lackadaisical tone to his supervisor's attitude, he described the reaction of the company president to improper practices:

> The company president told B. and this other guy "Don't stick your bare hands in the Orbit [an acid-based cleaner]." "Well," B. says, "I've been sticking my hands in there for years." You know, they're resigned to the fact that they're living with these chemicals.

Similar attitudes surface when a minority of workers decide to use personal protection equipment. A mold maker recounted his experiences:

The easy way out for me is to wear a mask. I get kidded about that: "Hey, doctor." Here I am with my hands looking like I've just come out of a coal mine with my mask on. But most people don't even bother to wear masks.

In these cases ridicule becomes an enforcement mechanism by which the group avoids coming to terms with the implications of the presence of hazards in the workplace (see Haas 1977).

Even when group norms are not enforced or do not exist, individuals may cope with risk by denying its significance or lapsing into fatalistic attitudes. Much of this is connected to feeling that there is little one can do to change either the situation or to adopt protective measures. The psychological reaction some have to risk was a source of speculation for a pipefitter at a chemical plant:

Over a course of a month's time, we must be exposed to a hundred different chemicals ranging from caustics to cyanide. If you really sat down and thought about what you're exposing yourself to and what it can do to you, I don't think anyone would work there. But the human mind is strange. It blocks stuff like that out and you don't really think about it.

Some simply do not see hazards as requiring special care. Work comes first as a laboratory technician observed:

We'd never get any work done if we had to wait around to find out if things were safe. This is carcinogenic, that stuff grows toes on your feet or something . . . . After a while they'll say you can't breathe the air because it's carcinogenic.

Denying risk may extend to the point of refusing to become more knowledgeable about hazards. The mold maker recalled the reaction of his coworkers when he tried to educate them about hazards in their shop:

I've passed the right-to-know stuff around the shop. Two or three guys looked at it and threw it down. They say, "Hey, you're scaring me. I can't come to work anymore." They don't want to accept the idea that coming to work could kill them.

These sentiments also may appear in shops with a union active in health and safety. The union's message suggests that ignorance and complacency contribute to ill health; that management through neglect or willful intent may expose them to dangerous conditions; and that the situation will improve only if they become activists. Avoiding information from the union and thus avoiding the implications of working in a risky environment enables some workers to sidestep difficult decisions about their health, their pocketbook or both.

Acceptance

Some of the workers interviewed for this study believed that the hazards they face are an acceptable part of the job. This attitude reflects several different types of situations. Some believe that risks have been reduced as low as possible and the benefits of the job compensate in some way for the residual. Others, cognizant of the risks of continued employment, are skeptical of finding alternative sources of income in a troubled economy.

Some workers find substantial satisfaction in their chosen occupation despite significant hazards. Firefighting, for example, is a job with enormous short-term and long-term risks, yet there is a heroic dimension to their work:

The risks are always in the back of everybody's mind. If they weren't we wouldn't be alive. At the same time, they're put in the background because of the nature of the job. The primary job is rescue, saving people's lives and their property. That outweighs the hazards. It really does. The hazards are there, but they're not as important as some little four year old who's off in a closet someplace scared to death because he set his house on fire and now he doesn't know what to do. I guess if there's a chance that you could save somebody's life, your own life kind of gets put in the background.

Benefits are not always evaluated in life and death terms. A gardener who sprayed pesticides in the course of caring for the roses at a botanical garden spoke in terms of combining natural beauty with social interaction:

I love what I do. Probably would do it even if I wasn't getting paid for it. So I don't really equate the pay with the risks that I take. I don't really feel that I am taking that many risks. I work for the beauty of the gardens. I enjoy doing this type of work, making things look good so that other people can enjoy them.

In contrast, social isolation and the quality of life were important to a woman who worked on ocean-going freighters as a deckhand:

In any job, there's certain things you have to put up with. I'd rather stand on an icy deck than be a waitress and smile at businessmen who say, "Hey, sweetie, come over here." It's a question of what sort of things make you uncomfortable and what you are willing to tolerate. Some people would rather accept risk than spend their whole life walking behind the rear end of a donkey with a plow. If you're worried about the quality of your life as well as the length of it, you may be willing to incur a calculated risk.

And some, such as a furniture refinisher, see their work in a larger historical context:

The furniture I work with is stuff that I could never hope to own. But I love the stuff. I get a buzz out of doing a nice job on a nice piece of furniture. I'd go so far as to say that I'd rather die 10 years before my time and have had some job satisfaction than work at GM and be bored to death. I don't actively wish to use things that are bad for me. I've looked around at possible alternatives, but they don't produce the job. I got some work in a museum. It's there because I've used materials which are, if you like, carcinogenic; there's no way I could have done those jobs using totally safe materials. So I could say, if I really went out on a limb, that it's worth the cancer problem to conserve pieces of historic significance.

Yet other workers do not think of their job as intrinsically worthwhile.

They work for money, not self-fulfillment, and resign themselves to risks they cannot ignore. Wages are the reason they put up with hazards. Yet, as a chemical worker in a food-processing plant pointed out, people avoid making the classical economic choice of more risk for more pay. The reality is much more brutal:

A guy [I worked with] had a perforated septum. He showed it to me and says, "it's the fumes." I asked, "Why don't you quit?" He says, "Well, I'm 46 years old and you're 18. I've got a third grade education, three kids . . . where the hell am I gonna go?" I found out he was right. Because when I quit there and went to the food plant, there was no difference. It's like going from one horrible place to another horrible place. You're kind of locked.

Economic hard times compound difficult decisions about risk. The fewer the job opportunities, the more workers are apt to stay with hazardous jobs. They are forced to perceive their situation as their job *versus* their health. A repair technician, describing the situation in his plant, spoke eloquently of the plight of the blue-collar worker who struggles to maintain a middle-class life:

They asked me to work in the lime kiln and I got stuff hanging all over me: white, drippy, slimy stuff all over my hands, my clothes and everything else. Then I said to myself, "It's not worth my $9.50 an hour to do that." Most guys, however, won't tell their foreman, "I'm not going to do it," because they just got hired and they'll lose their job. Most of them have been working for $4.00 an hour and now they're making $9.00, and they've got good benefits and they're not going to lose that for anything. If you have a family, two kids, a house, a car, and bills to pay, you forget your personal feelings about certain things. Nobody wants to do things that they feel are wrong, but then again they have the responsibility to provide for their family. We really don't have a choice. I can't refuse to work knowing that tomorrow I can get another job. I can't look for a year and half for a job. I'd lose everything.

Activism

Although workers may avoid hazard information as too threatening, they still may wonder about the consequences of exposures to hazards. The mold maker's coworkers, despite their refusal to read about the right-to-know laws, still wondered about risk:

"He just retired and he's dead," or "He was only 56," or "Does everyone who gets out of here die early?"

Some do not simply ask the question, but seek out answers and struggle for change on the shopfloor. Thus, there may not be a great deal of difference in perceptions of risk between those who end up accepting their situation and workers who become activists. The interviews give little support to the notion that access to hazard information generates demands for change. What seems to be the case is that people become activists for other reasons, but information *seeking* is critical to their efforts for change.

Our respondents indicated that activists, if they were present in their workplaces, tended to be a very small minority of the workforce with distinctive personalities and social characteristics. Personalities, transformative experiences, concerns about the future, or some combination lead a few individuals to taking on the status quo. In some situations, job risks are another example of injustices they believe should be corrected; others were moved by deplorable conditions to become activists on the shopfloor.

Some workers put health and safety issues into a larger perspective. They are unwilling to accept a status quo they perceive as unfair. As an aspect of labor-management relations, risks on the job are an issue that should be fought for just like wage rates, seniority, benefits, and dignity of working people. Thus people described their personalities as questioning the basic assumptions of hierarchy typical in most organizational settings. As a machine operator in a university print shop put it:

> I go along with the bumper sticker that says, "Question Authority." I'm a radical thinker . . . . I believe that no one should accept the final word on anything. I don't agree with Reagan that business is the backbone of the country. I believe the average worker is the backbone of this country . . . .

Several others ascribed their activism to an unwillingness to meekly accept conditions. More than that, they believe that everyone has a duty to speak out, yet recognize that not everyone does and that they are outspoken. Several workers commented:

> I call [the people I work with] jellyfish or ostriches . . . . They don't want to get involved . . . . It's because people are like that they have to work in the conditions they do. In our family, if you're going to gripe, do something about it. If you ain't going to do something, shut up. If I don't like where I have to work, I say so. Why shouldn't I?

> I hate to see people get walked on. Somebody just has to take charge. With some people — if somebody does something to them, they just lay back and take it . . . . Somebody has to take a firmer stand on hazards.

> People come in to do their work with the attitude, "don't bother me. I'm here to do a job." There's a selected few who carry the ball for everybody else. You're talking probably 50 out of 800. The ones who do the most complaining are the ones who won't come forward and say anything in a safety meeting. I'll speak up because I just don't like taking risks. I don't like doing things that will hurt myself or other people.

Surprisingly, few workers mentioned experiences with hazards as the events which triggered beliefs that they had to get involved. Experiencing risk does not seem to be as significant an impetus to action as much as a recognition that conditions are not likely to change unless an individual worker chooses to seek changes. Currently a union organizer and a health and safety representative, a former chemical operator described his transformation:

> After being out of work for six months, and a high school dropout, I realized there weren't many jobs around. You either took a job at minimum wage in a warehouse where your health deteriorated because you couldn't afford to feed yourself, or you took a job [as a chemical operator] at a wage you could live on but where you subject yourself to all kinds of hazards. I told myself, "You know, this is reality, so you better fight to change this place and do something about it."

An important motivation among many of the activists was their orientation to the future. Some saw activism in a personal light: their children should not have to put up with the conditions they endure every day. This was the concern of an orchard worker exposed to pesticides on the job who led an organizing drive for union representation:

> I've been active in getting a union, hopefully to make some jobs less hazardous. It might make things a little better for my kid so he won't have to deal with that kind of stuff. He'll have millions more chemicals when he's old enough to work . . . . Chemicals are here to stay; they're a threat, but they're also a boon. You got to deal with them. If pressure could be put on employers to not go with the cheapest but the safest, that will be to my kid's benefit.

Interestingly, while everyone wanted a better life for their kids, which for some meant a white-collar professional job, many saw the younger generation following in their footsteps. In an unspoken recognition of the class nature of American society, they recognized that whatever ambitions they might harbor for their own children, someone's children will be working their jobs after they are gone:

> I used to be one of those guys who think, "Well if I get it, I get it. Too damn bad. If my number's up, my number's up. I found out that's not the way to think. If there's s problem that can be eliminated, eliminate it for the next guy. If we don't, before too long there won't be a next time. I feel that my father and his forefathers kept it going up to my time: now it's my responsibility and yours and anybody else out there to keep it going for the kids.

Common to most of the workers who became involved in health and safety issues was the role of hazard information in perpetuating their activism. While several noted that their union was a good source of information, many of those interviewed were disappointed in the efforts of management or government agencies. Those who became activists tended to first become better informed about hazards. As they became aware of the extent of risk they and their coworkers faced, they became more determined to act as a focal point for complaints. Coworkers who were reluctant to approach management instead asked activists for information and identified hazards throughout the workplace. Often, activists became more committed to changing their conditions of employment when they began to see both themselves and their coworkers at risk. Identifying with their coworkers as a class whose interests clash with management, they are reluctant to abandon activism even when they are jeopardizing their own health.

In one interview, a chemical plant worker told of suffering from mercury poisoning on the job and unexplained seizures suffered by his baby daughter which he suspected were related to his chemical exposures. He became very knowledgeable about the hazards he and his coworkers faced and turned into a fierce advocate for change. Active in his union and the leader of a wildcat strike over hazards, he is intensively involved in educating coworkers and, through a local COSH organization, workers throughout his community about job risks. He explained his dedication, at some personal costs, as if it were a crusade:

I guess I'll never be able to relax and enjoy myself, like some people can, until the issue of workers' safety and health is resolved. I see workers themselves learning about the problems faced by workers. I see unity, I see a coalition of workers for workers . . . . If you don't do something for yourself then nobody's going to do it for you. And that holds true in safety and health. My wife is very upset about my working with all these chemicals. She would like me to quit. But if I were to do that and obtain another job, I would be just throwing in the towel, giving up that commitment to a safe and healthy workplace. I'd be reducing my own risk of occupational disease or accidents, but doing nothing for the workers who remain.

Discussion

The evidence from this research suggests the need to dig deep into the relationship between hazard communication and worker responses to risk. Simple constructs such as an economic bargain over wages and conditions or personal attitudes toward risk-taking versus risk avoidance do not adequately explain what is likely to happen when right-to-know policies are fully implemented. Greater recognition of the effects of the source and content of risk information and the political implications of hazard communication efforts is needed.

In the interviews conducted for this study, beliefs about limited job choices and ability to control risk permeated the views of workers who denied the seriousness of hazard exposures. They avoided hazard information because it seemed to serve no purpose other than to generate greater emotional conflict. Knowledge forces workers to confront the rationality of believing that either risks are minimal and they are personally immune, or that the short-term benefits justify the long-term risk of ill health. When workers believe that risks cannot be changed, information is perceived as useless and a potential source of either friction with management or emotional stress, as suggested by an apprentice painter:

It's funny how few of the journeymen know what the dangers are, what the chemicals will do to you or what's really in it. If you refuse to use it, you might get laid off, fired or suspended, or at least the boss will remember who you are. So most guys don't want to rock the boat. I guess they figure ignorance is bliss.

Management communication about risks reinforces feelings that workers have little control. Claims that known risks are minimal or that conditions are being investigated suggest to workers that immediate action is unnecessary and that management is acting in a responsible manner. This is a very effective argument with workers who want to believe that someone is taking care of them.[7] Explicit threats to shut down a plant or claims that a cleanup is too costly speak to the financial concerns of many workers. These communication practices are particularly effective in the absence of any intrinsic job satisfaction which might predispose some workers to conclude that existing risk levels are acceptable. Moreover, if appropriate alternative messages from trusted sources such as a union do not exist, workers tend to do their jobs and not think about the risks.

This contrasts with workers who perceive that change is possible and that they should speak up about problems. They discount management claims, especially when they run counter to their experiences on the shopfloor or when their union argues that exposures should be reduced. The existence of uncertainties provokes an argument that prudence dictates exposure reductions. In their view it is better to avoid potential health problems than wait for unequivocal data. For these workers, identification of hazards in their workplace is a call to action.

These findings suggest several implications for current policies. First, the enactment of state and local right-to-know laws, and the recent OSHA hazard communication rule designed to supersede non-Federal action, may not have the consequences intended by their proponents. For example, OSHA has argued that its new rule will "reduce the incidence of chemical source illnesses and injuries . . ." (USDL 1983b: 53281). However, some of the laws, particularly the OSHA rule, allow a great deal of employer flexibility in deciding what workers will be told and how to tell them. As flexibility increases, firms are in a position to manipulate the content of information in a manner which reinforces tendencies among workers to deny or passively accept hazardous conditions.

In addition, the focus of current policies on management obligations to inform workers neglects development of alternative sources of hazard information. In situations where employees do not trust their employer to provide adequate information and there is no union active in health and safety, there may be few places workers can turn for adequate data. OSHA's New Directions program provided a start for a variety of organizations involved in educating workers about occupational health. However, recent cutbacks and a lack of alternative funding sources have diminished the ability of these groups to continue their work. New funding and possible expansion of services may be necessary to provide a counterbalance to employer-oriented programs.

This suggests that even if right-to-know laws are strengthened, they are not likely to have a substantial effect on working conditions without a militant workforce. Policies which improve communication of hazard information, although important and necessary for workers who are inclined to participate in decisions about risk, do not by themselves alter risk attitudes or labor-management relations. Other conditions which influence workers' concerns about health hazards such as employment alternatives and job control are unaffected by right-to-know programs. Consequently, for these policies to be effective, more attention should be given to increasing workers' ability to control occupational risks.

The research reported here provides the basis for more extended work in understanding what prompts varying responses to risk. Future studies should look more closely at the effects of the type and degree of hazard, organizational structure, and management responses to worker initiatives, as well as hazard communication practices. A next step might be a prospective study which investigates the effects of different information programs on worker actions in a variety of occupational settings.

Notes

1. The material for this paper was gathered while the author was at the Program on Science, Technology, and Society at Cornell University, Ithaca, New York. It is based on work supported by the National Science Foundation Program on Ethics and Values in Science and Technology, under grant no. ISP 8112920. Any opinions, findings, and conclusions expressed here are those of the author and do not necessarily reflect the views of the National Science Foundation. The author is currently at the Massachusetts Department of Environmental Management, Boston, Massachusetts.

2. The interviews with workers occurred over a 10-month period in 1981 and 1982. Respondents were recruited in a network fashion which involved identifying initial contacts and seeking additional workers from them. No effort was made to gain a random sample, although a range of occupations was sought; the sole criterion for entry into the study was exposure to chemicals due to the job. Approximately 80 interviews were conducted covering a variety of topics including attitudes toward work, self-reported health problems, observations of health hazards, and knowledge of workplace risks. The interviews were recorded, transcribed, and edited for clarity. A more extensive description may be found in Nelkin and Brown (1984) and Brown (1984).

3. The National Institute for Occupational Safety and Health (NIOSH) lists over 59,000 suspected toxic substances, including 3,200 potential carcinogens (USDHHS 1983). Another federal agency included 88 substances in its 1982 report on human carcinogens (USDHHS 1983). Several standard reference works also vary: Sax (1981) lists 2,400 entries for chemical carcinogens, while the American Conference of Governmental Industrial Hygienists included some 450 substances, of which 50 were considered carcinogens, in its 1983 compilation of significant airborne contaminants (ACGIH 1983).

4. COSH groups or committees on safety and health are local or regional affiliations of rank-and-file workers, unions, and health professionals who develop and disseminate information materials for workers and employers, provide training, and lobby for better shopfloor conditions (see Levenstein *et al.* 1984).

5. The discussion that follows is also applicable to the nonprofit and government sectors, especially when scarce resources force cost-conscious decision-making.

6. In some states, workers' compensation laws impose an obligation to warn workers about certain hazards. A California court upheld a claim of a worker with an asbestos-related disease, stating that management must inform an employee of any work-related illnesses uncovered during medical examinations (Johns-Manville Products Corp. v. Contra Costa County Superior Ct., 612 P.2d 948, 1980). Failure to do so demonstrates a willful intent to do harm.

7. Slovic *et al.* (1980) contend that, given a choice, people do not want to confront risks, but prefer to be told a competent expert is taking care of them and that risks are small.

References

American Conference of Governmental and Industrial Hygienists (ACGIH) *TLVs: Threshold Limit Values for Chemical Substances and Physical Agents in the Work Environment with Intended Changes for 1983—1984.* Cincinnati, Ohio: ACGIH, 1983.

Ashford, Nicholas A. *Crisis in the Workplace: Occupational Disease and Injury.* Cambridge: MIT Press, 1976.

Ashford, Nicholas A. and Charles, C. Caldart. 'Framework Provides Path Through Right-To-Know Law,' *Occupational Health and Safety*, October, 1983, 11—19ff.

Baram, Michael S. 'The Right to Know and the Duty to Disclose Hazard Information,' *American Journal of Public Health*, 1984, 74(4), 385—390.

Behrend, Hilde. 'The Effort Bargain,' *Industrial and Labor Relations Review*, 1957, 10, 503—515.

Beilin, Robert. 'Social Functions of Denial of Death,' *Omega: Journal of Death and Dying*, 1982, 12(1), 25—35.

Berman, Daniel. *Death on the Job: Occupational Health and Safety Struggles in the United States.* New York: Monthly Review Press, 1978.

Bishop, Richard L. 'Anxiety and Readership of Health Information,' *Journalism Quarterly*, 1974, 51, 40—46.

Braverman, Harry. *Labor and Monopoly Capital: Degradation of Work in the Twentieth Century.* New York: Monthly Review Press, 1974.

Brown, Michael S. *The Right To Know: Hazard Knowledge and the Control of Occupational Health Risks* (Ph.D. dissertation). Ithaca, New York: Cornell University, Department of City and Regional Planning, June 1984.

Bureau of National Affairs (BNA). *Basic Patterns in Union Contracts* (10th ed.). Washington, D.C.: Bureau of National Affairs, 1983.

Bureau of National Affairs (BNA). 'Three States Join in the Steelworker Suit on Hazard Communication; Fourth May File,' *Occupational Safety and Health Reporter*, January 5, 1984a, 811—812.

Bureau of National Affairs (BNA). 'OSHA Rule would Pre-Empt State Laws, Auchter Tells Pennsylvania Senate Group,' *Occupational Safety and Health Reporter*, February 2, 1984b, 948.

Burawoy, Michael. 'The Anthropology of Industrial Work,' *Annual Review of Anthropology*, 1979a, 8, 231—266.

Burawoy, Michael. *Manufacturing Consent: Changes in the Labor Process Under Monopoly Capitalism*. Chicago: University of Chicago Press, 1979b.

Cyert, Richard and James March. *A Behavioral Theory of the Firm*. Englewood Cliffs, New Jersey: Prentice-Hall, 1963.

Douglas, Mary. *Purity and Danger: An Analysis of the Concepts of Pollution and Taboos*. New York: Routledge and Kegan Paul, 1976.

Douglas, Mary and Aaron Wildavsky. *Risk and Culture*. Berkeley: University of California Press, 1982.

Estey, Marten. *The Unions: Structure, Development, and Management* (3rd edn.). New York: Harcourt, Brace, Jovanovich, 1981.

Fisher, Ann. 'The Scientific Bases for Relating Health Effects to Exposure Level,' *Environmental Impact Assessment Review*, 1982, 3(1), 27—42.

Galbraith, John K. *The New Industrial State* (2nd rev edn.). Boston: Houghton Mifflin, 1971.

Gaventa, John. *Power and Powerlessness: Quiescence and Rebellion in an Appalachia Valley*. Urbana: University of Illinois Press, 1980.

Gaventa, John. 'The Powerful, the Powerless and the Experts: Knowledge Struggles in an Information Age,' unpublished, 1983.

Haas, Jack. 'Learning Real Feelings: A Study of High Steel Ironworkers' Reactions to Fear and Danger,' *Sociology of Work and Occupations*, 1977, 4(2), 147—170.

Hancock, Nolan W. Statement in United States Congress, House, Committee on Education and Labor, Subcommittee on Health and Safety. *OSHA Oversight Hearings on Proposed Rules on Hazard Identification*, 97th Cong., 1st Sess., 1982.

Janis, Irving L. and Leon Mann. 'Coping with Decisional Conflict,' *American Scientist*, 1976, 27, 261—275.

Kerr, Clark. 'Industrial Conflict and its Mediation', *American Journal of Sociology*, 1954, 60, 230—245.

Kornhauser, Arthur, Robert Dubin and Arthur M. Ross (eds.). *Industrial Conflict*. New York: McGraw-Hill, 1954.

LaDou, Joseph. 'The Not-So-Clean Business of Making Chips,' *Technology Review*, 1984, 87(4), 22—36.

Levenstein, Charles, Leslie Boden, and David Wegman. 'COSH: A Grass-Roots Public Health Movement,' *American Journal of Public Health*, 1984, 74(9), 964—965.

McCaffrey, David P. *OSHA and the Politics of Health Regulation*. New York: Plenum Press, 1982.

McDevitt, George. Interview, February 22, 1982.

National Academy of Sciences, Steering Committee on Identification of Toxic and Potentially Toxic Chemicals for Consideration by the National Toxicology Program, Board on Toxicology and Environmental Health. *Toxicity Testing: Strategies to Determine Needs and Priorities*. Washington, D.C.: National Academy Press, 1984.

Nelkin, Dorothy and Michael Brown. *Workers at Risk: Voices from the Workplace*. Chicago: University of Chicago Press, 1984.

Nelson, Phillip. 'Information and Consumer Behavior,' *Journal of Political Economy*, 1970, 78, 311—329.

Nordgren, Sarah. 'Cyanide Deaths Lead to Murder Indictment of Corporate Officials,' *Boston Globe*, April 14, 1985, 27.

Prosser, William. *Handbook on the Law of Torts* (4th d.n.). Saint Paul, Minnesota: West Publishers, 1971.

Quinn, Robert P. and Graham L. Staines. *The 1977 Quality of Employment Survey: Descriptive Statistics with Comparison Data from the 1969—70 and the 1972—73*

Surveys. Ann Arbor, Michigan: University of Michigan Institute for Social Research, 1979.

Santodonato, Joseph. 'Design and Implementation of a Workers' Right-to-Know Program,' *American Industrial Health Association Journal*, 1981, 42, 666—670.

Sax, N. Irving. *Cancer Causing Chemicals*. New York: Van Nostrand Reinhold, 1981.

Slovic, Paul, Baruch Fischhoff, and Sarah Lichtenstein. 'Informing People About Risk,' *Banbury Report 6: Product Labeling and Health Risks*. Cold Spring Harbor, New York: Cold Spring Harbor Laboratory, 1980.

Smith, Robert S. *The Occupational Safety and Health Act: Its Goals and Achievements*. Washington, D.C.: American Enterprise Institute, 1976.

Stigler, George. 'The Economics of Information,' *Journal of Political Economy*, 1961, 69, 213—225.

Storey, John. *Managerial Perrogatives and the Question of Control*. London: Routledge and Kegan Paul, 1983.

Taylor, Frederick W. *Principles of Scientific Management*. New York: Harper, 1911.

United States Congress, House, Special Subcommittee of the Committee on Labor. *An Investigation Relating to Health Conditions of Workers Employed in the Construction and Maintenance of Public Utilities*. 1936.

United States Congress, House, Committee on Education and Labor, Subcommittee on Health and Safety, *OSHA Oversight Hearings on Proposed Rules on Hazards Identification*, 97th Cong., 1st Sess., 1981.

United States Department of Health and Human Services (USDHHS), National Institute for Occupational Safety and Health. Registry of Toxic Effects of Chemical Substances, 1981—1982. DHHS (NIOSH) Publication No. 83—107. Washington, D.C.: Government Printing Office, 1983.

United States Department of Health and Human Services (USDHHS), National Toxicology Program. *Third Annual Report on Carcinogens*. Washington, D.C.: Government Printing Office, 1982.

United States Department of Health, Education and Welfare (USDHEW), National Institute for Occupational Safety and Health. *National Occupational Hazard Survey. Volume I: Survey Manual*. Washington, D.C.: Government Printing Office, 1974.

United States Department of Health, Education and Welfare (USDHEW), National Institute for Occupational Safety and Health. *National Occupational Hazard Survey. Volume II. Data Editing and Data Base Development*. Washington, D.C.: Government Printing Office, 1977a.

United States Department of Health, Education and Welfare (USDHEW), National Institute for Occupational Safety and Health. 'The Right to Know,' July 1977. Reprinted in U.S. Congress, Senate, Committee on Human Resources, Subcommittee on Labor. *Hearings on Monitoring of Industrial Workers Exposed to Carcinogens, 1977*. 95th Cong., 1st. Sess., 1977b.

United States Department of Health, Education and Welfare (USDHEW), National Institute for Occupational Safety and Health. *National Occupational Hazard Survey. Volume III: Survey Analysis and Supplemental Tables*. Washington, D.C.: Government Printing Office, 1978.

United States Department of Labor (USDL), Bureau of Labor Statistics. *Employment and Earnings*. January 1985.

United States Department of Labor (USDL), Occupational Safety and Health Administration. 'Access to Employee Exposure and Medical Records: Final Rules and Proposed Rulemaking,' *Federal Register*, 1980, 45, 35212—35303.

United States Department of Labor (USDL), Occupational Safety and Health Administration. 'Hazards Identification: Notice of Proposed Rulemaking and Public Hearings,' *Federal Register*, 1981a, 46, 4412—4453.

United States Department of Labor (USDL), Occupational Safety and Health Administration. 'Hazards Identification,' *Federal Register*, 1981b, 46, 12214.

United States Department of Labor (USDL), Occupational Safety and Health Administration. 'Hazards Communication,' *Federal Register*, 1982, 47, 12092—12124.

United States Department of Labor (USDL), Occupational Safety and Health Administration. 'Training Guidelines: Request for Comments and Information,' *Federal Register*, 1983a, 48, 39317—39322.

United States Department of Labor (USDL), Occupational Safety and Health Administration. 'Hazard Communication,' *Federal Register*, 1983b, 48, 53280—53348.

United States Environmental Protection Agency (USEPA), Office of Toxic Substances. *Toxic Substances Control Act Chemical Substances Inventory. Cumulative Supplement II.* May 1982.

United States Environmental Protection Agency (USEPA), Office of Toxic Substances. TSCA Assistance Office. Personal communication with Carol Hatfield, April 16, 1984.

Viscusi, W. Kip. 'Job Hazards and Worker Quit Rates,' *International Economic Review*, 1979, 20, 29—59.

Weinstein, Neil. 'Seeking Reassuring or Threatening Information About Environmental Cancer,' *Journal of Behavioral Medicine*, 1979, 2(2), 125—139.

Wilde, Louis L. 'Information Costs, Duration of Search, and Turnover: Theory and Applications,' *Journal of Political Economy*, 1981, 89, 1122—1141.

Chapter 11

Defining Risk within a Business Context: Thomas A. Edison, Elihu Thomson, and the a.c.—d.c. Controversy, 1885—1900

W. Bernard Carlson and A. J. Millard

Introduction

This chapter examines the identification and assessment of the risk construction of electrical power systems, an essential foundation of modern industrial society. When electric power systems were first introduced in the 1880s, they presented a threat of physical injury and fire. Electricity was an invisible, mysterious force, and the risks associated with its use were not fully understood at the time. The dangers were identified by electrical pioneers who were constructing the first electrical power systems. Men like Thomas Edison and Elihu Thomson made the first risk assessments and then quickly took the issue of risk into a broader environment; the debate was then carried into the public domain and the eventual management of electrical risk was carried out by inventors, engineers, businessmen, and legislators.

This chapter will show how the risk from electricity was assessed and managed. It will describe the motivation and actions of leading electricians who participated in this process. Thomas Edison and Elihu Thomson were the electrical pioneers who first identified a risk with high-voltage electricity, and each had a different response to the risk arising from new technology. While Thomson tried to resolve the problem with a technological solution, Edison used his reputation to draw attention to the risk and initiate a public discussion. The debate over the risks of electricity reached a climax in the so-called "battle of the systems," a controversy centered on replacement of small-scale direct current (d.c.) with large-scale, alternating current (a.c.) power systems. Lasting from approximately 1885 to 1900, this battle brought the issue of risk into the public eye and greatly influenced the development of the modern technology of electricity supply. While this debate was a public discussion about the deployment of new technology, it was based on the fierce competition of electrical manufacturing companies in the market place. The economic interests of

B. B. Johnson and V. T. Covello (eds.), The Social and Cultural Construction of Risk, 275—293.

these companies, and the interaction of different interest groups within them, shaped the resolution of the struggle. The risk of electrical power systems was addressed within a mixed matrix of public and business organizations and this set an important precedent for the future management of risks associated with new technologies.

Thomas Edison and the Identification of the Risk from Electricity

The serious study of electricity began in the eighteenth century, but it was not until the nineteenth century, with the discovery of the phenomenon of electromagnetic induction by Michael Faraday in the 1820s, that the study had practical consequences. Faraday invented a primitive generator and electric motor, but no economically feasible means of generating power in large quantities was available until the 1870s. Lighthouses began to utilize electric arc lamps in the late 1850s, and by the 1870s they were being used in a number of factories, theatres, and public buildings. Arc lamps produced illumination by having a high-voltage electric current jump a narrow gap between two carbon electrodes. While they produced a brilliant light, making them well-suited for illuminating streets and large buildings, they were too powerful for use in homes and shops. Consequently, it soon became obvious that large pecuniary rewards awaited the inventor who could devise another method of electrical illumination consisting of smaller or "subdivided" units.[1]

Thomas Edison, one of America's great inventive geniuses, responded to the opportunity of electric lighting. In 1879, he "subdivided" the light by inventing an incandescent lamp which produced illumination when a low-voltage current passed through a high-resistance filament in an evacuated glass bulb. One enthusiastic press report, possibly inspired by Edison himself, emphasized the new invention's safety and hailed it as "a lamp that cannot leak and fill the house with vile odors or combustible vapors, that cannot explode and that does not need to be filled or trimmed."[2] Edison followed up on the lamp by developing and building a complete system of lighting. He installed his first major system in 1882 at Pearl Street in New York and this was the first central electric lighting system in the world. This power station supplied customers within a half-mile radius with electricity to illuminate incandescent lamps.[3]

From the beginning, Edison had three priorities in creating his central station system: it had to be simple, safe, and economical. Simplicity was obtained by using direct current (d.c.) which kept the complexity of the distribution circuits and the number of system components to a minimum. Edison secured safety by keeping the voltages low (100—120 volts), by including safety accessories such as fuses and cut-outs, and by installing his distribution wires underground. In pursuit of economy, Edison devised his three-wire distribution scheme to reduce the size of costly copper mains and he carefully sited his stations in densely populated urban centers (such as downtown New York, London, and Berlin) where there were large numbers of customers to share the costs of the system.

Edison believed that his incandescent lighting system would be used universally. He dreamed of lights in every home and business, and to accomplish this he did his utmost to create public confidence in the safety of his system. He insisted that his lighting system should offer no danger at any point. As the Edison companies emphasized in their sales brochure, "there is no danger to life, health, or person, in the current generated by any of the Edison dynamos."[4] Edison knew that maximum economic and technical success would come only with general adoption of his invention which in turn was dependent on the safety of his system.

Despite Edison's efforts to render his system safe, not everyone was convinced. In particular, his major competitors in the urban illumination market, the gas companies, retaliated by drawing attention to the dangers of electricity. With the support of the popular press, the gas companies were able to publicize accidents involving electricity and this alerted the public, legislators, and insurance companies to the risk from electric lighting systems. The struggle between the gas lighting industry and its newest competitors often took the form of a propaganda campaign in which each side criticized the economy, reliability, and risk of the opposing technology. Undaunted, Edison fought back vigorously and sought public acceptance of his new technology by regularly stressing its safety and reliability while reminding the public of the dangers of gas explosions.

Although attacked by the gas companies as being unsafe, incandescent lighting grew in popularity during the mid-1880s. In particular, many small cities and towns were anxious to secure electric lighting. Unfortunately for Edison, his companies were unable to penetrate the new market since his central station system had been designed for use in large, densely populated cities where there were numerous customers within a short radius who could absorb the high cost of the distribution network. Many of the municipalities desiring the light lacked the population density needed to offset the expensive copper mains. Edison and his engineers responded by developing more efficient system components, such as a better lamp and a five-wire network, which were intended to reduce the overall cost of the central station system and permit it to be installed with a more diffuse customer base.[5] Nevertheless, this research failed to address the central problem of supplying electric lights to more people over a larger geographical area. Just as Edison had invented his incandescent system to supplant arc lighting, so other electrical inventors resorted to a new technology to overcome the limitations of Edison's system.

The technology to which electrical inventors now turned was high-voltage alternating current. They chose high voltage because it permitted electric power to be transmitted over longer distances using a significantly smaller conductor. Alternating rather than direct current was desirable because inventors soon found that the voltage of an alternating current could be raised or lowered by using a converter or transformer. By combining high voltages with a.c., electricians in the mid-1880s discovered that they could build a system in which high-voltage a.c. could be transmitted and distributed over long distances yet stepped down by transformers so that it could be used by customers to run incandescent lights.

The technology of a.c. was initially developed in England by John Gibbs and Lucien Gaulard and in Hungary by Zipernowsky, Blathy, and Deri (ZBD). However, it was brought to commercial fruition by the Westinghouse Company in America. With the encouragement of George Westinghouse, Jr., William Stanley and other engineers designed a practical system which was first installed in Buffalo, New York, in November 1886. Further refinements were soon provided by Elihu Thomson and the Thomson-Houston Electric Company who installed their first a.c. plant in Lynn, Massachusetts, in the spring of 1887.[6]

Edison followed the development of a.c. with interest because his companies had considered adopting the system while it was being developed in Europe in the early 1880s. The Edison companies even purchased the American rights to the ZBD patents, but never utilized them. Edison commissioned a survey of all research in this field, and he carefully studied the evaluations of the first a.c. plants. In 1888 and 1889, Edison and his assistants also tried designing their own high voltage systems. After experimenting with a.c., Edison chose to concentrate on a high-voltage d.c. system; he may have chosen to investigate such a system because it had the advantages of long-distance transmission while avoiding the problems Westinghouse and others were having with power losses in their transformers.[7]

As Edison studied high-voltage a.c., he came to the conclusion that it was dangerous on the grounds of its complexity, poor reliability, and threat to public safety. Edison was particularly concerned that the Westinghouse Company frequently installed their system using overhead wires in order to reduce installation costs; in contrast, Edison insisted that his utilities install more costly underground conductors in the interests of safety. Similarly, Edison noticed that linemen working with a.c. failed to understand the hazards of the new technology. By experiment, Edison further demonstrated that the high voltages used in a.c. systems could kill those coming into contact with it. All of these observations led Edison to quip that one way to eliminate criminals would be to "Hire [them] . . . out as linemen to some of the New York electric light companies."[8]

Thus, drawing on careful research, experiment, and his own extensive experience in electric lighting, Edison assessed the risk of a.c. and decided that it was too dangerous for widespread use. Even though he could have designed and deployed his own a.c. system in order to remain competitive in the market for small cities and towns, he chose not to pursue this option. Instead of doing what was economically expedient over the short term, Edison chose to promote his safe d.c. system, believing that safety was essential to his goal of having electricity adopted universally.

The Battle of the Systems: Creating a Larger Environment for the Issue of Risk

Edison followed his negative assessment of a.c. technology by taking steps to limit its use in the Edison companies by insisting that they should not exercise their option on the ZBD patents. Since any accidents caused by a.c. would harm the public image of electricity, he felt it was essential to discourage this new technology on all fronts. Significantly, he was able to enforce this policy for two reasons. First while Edison was neither president nor majority stockholder of his key company, the Edison Electric Light Company (EELC), he had gained effective control of this firm by building a coalition of stockholders who elected the officers he wanted. His enormous prestige in the organization gave Edison great informal power. Second, Edison was able to shape the technology of this and his other companies because he was their chief source of technological innovation. The EELC as well as other Edison companies had contracted exclusively with Edison in order to secure his latest inventions and improvements; consequently, if he refused to develop certain products, then the companies were forced to do without.[9] As we shall see, in contrast to Elihu Thomson, Edison dealt with the management of his companies from a position of strength and thus was able to influence technological policy.

Not only did Edison discourage his own companies from developing a.c., but he sought to stop his competitors from installing a.c. lighting systems. Just as the gas companies had first attacked their d.c. system as unsafe and unreliable, so now Edison and his companies now turned on their a.c. rivals. Both Westinghouse and Thomson-Houston were doing a brisk business selling a.c. equipment to small towns and cities and Edison mounted a campaign to show that their a.c. equipment was unsafe. In part, he undertook this campaign because his companies were finding it difficult to compete in this segment of the market but more importantly, he engaged in this struggle because he feared that poorly designed and installed a.c. systems would impede the broad adoption of electric power. Vigorously debated in both public and professional forums, this struggle between rival technological systems has been called the "battle of the systems."[10]

Thomas Edison played a leading part in attracting public attention to this battle; as the greatest electrician of his day, his technological achievements stood foremost in the public eye and he was acknowledged in the press as "the best informed man in America, regarding electric currents and their destructive power."[11] Edison was just the man to bring the issue of safety of high-voltage current to a very wide audience. In other countries the "battle of the systems" remained primarily a concern of professional engineers and the electrical industry, but in the United States the "death current" quickly became a public issue after Edison publicized its risk.

Edison was a shrewd enough businessman to recognize the marketing tool he had in his worldwide fame and he assiduously cultivated his image in the press. As the best known inventor in America, Edison was good press and he welcomed newspapermen to his famous laboratory in West Orange. In addition, several important figures in the electrical press had either worked for Edison or were among his ardent admirers, making it easy for him to place articles in key journals. Newspaper reports of his activities provided free advertising for new products he was developing. He would often announce inventions well before he had accomplished them, in order to scare off the competition. He also used the media to shape the history of an invention, giving himself the prominent position in its development. Thus Edison capitalized on the media to bring his views about the danger of high-voltage a.c. to a wide segment of the population.

A good example of Edison's media campaign against a.c. is the article he published in the *North American Review*, a popular magazine which often dealt with science and technology. Taking advantage of his friendship with the editor, Edison submitted an article in 1889 in which he summarized his reasons for opposing a.c. In "The Dangers of Electric Lighting," the great electrical pioneer based his risk assessment of high-voltage technology on his view of the future. Edison foresaw a time when every American city would be covered in a web of electrical wires, some of them carrying high voltages. Mass use of electricity depended on mass acceptance and Edison knew keenly that the risk of accident and fire weighed heavily against its acceptance by the public. To him, high-voltage current was a technological time bomb; its full danger would not be felt until sometime in the future when it was in common use. As he observed:

> With the increase of electric lighting (which today is used to only a very limited extent as compared with its inevitable use) and the multiplication of wires, these dangers which exist in a thousand different parts of the city will be manifolded many times.[12]

These were considerations that drove him to oppose alternating current. It was not, as many of his biographers have stated, personal economic motivation.[13] While his companies took the short-term economic view that much would be lost if they did not enter the market for small cities and towns, Edison was much more concerned with the long-term impact of a.c. Furthermore, at the time he opposed a.c. he was rapidly divesting himself of his financial interest in the electrical industry, having decided that other fields such as manufacturing phonographs and ore-milling offered greater technical challenges and monetary rewards. Consequently, it was not money that Edison had at stake in "the battle of the systems"; it was his reputation as an engineer and inventor. No other person was as closely associated with electricity in the public mind. His name had become synonymous with high technology and practical engineering. Edison feared that a major accident, a nineteenth-century Three Mile Island, would not only hinder the diffusion of electricity, but also reflect back on him and the companies that bore his name. He therefore had reason enough to make the public aware of the risk of alternating current and to educate them on the differences between the d.c. system marketed by Edison and the a.c. systems marketed by Westinghouse.[14]

To support his "media blitz," Edison had his assistants follow up with a series of demonstrations and experiments to show a.c. technology as uneconomical and unsafe. His West Orange laboratory became the site of experiments in which high-voltage electric current was used to kill animals, proving that a.c. could kill. The Edison Electric Light Company stood behind this campaign; they organized and paid for it. They also supported the activities of Harold P. Brown, an independent electrical engineer who waged his own campaign against the high-voltage technology marketed by Westinghouse. Brown staged several exhibitions of the killing power of a.c. and even challenged George Westinghouse to a duel with electricity, each receiving progressively larger shocks — Brown with direct current and Westinghouse with alternating. The immediate goal of Brown's personal campaign was to convince the state of New York to use a.c. as an agent of electrocution. As a result of his agitation, a Westinghouse a.c. generator was used in the first electrocution in 1890. This attracted a great deal of attention, increased public awareness of the risk of high voltages, and delivered a severe blow to Westinghouse and a.c. technology.[15]

The publicity campaign was the first step in Edison and his companies' campaign against a.c. The next was to turn this public apprehension about the dangers of high voltage into concrete legislation. The Edison interests initiated and promoted legislation that would limit the risk from a.c. power systems. This took several forms: it proscribed the use of overhead lines, it encumbered transmission systems with safety devices, it limited the voltage of transmission cables, and it outlawed alternating current in dwellings and offices. These conditions favored the existing d.c. technology while eliminating the economic advantages of a.c. systems. This kind of stringent regulation was not without precedent in industrial society and Edison cited the regulation of high-pressure boilers when he suggested that electrical systems be restricted to a safe voltage (600 volts) and be regularly inspected.[16]

Despite extensive promotion and technical support from the West Orange laboratory, no state government actually adopted any legislation to remove the risk of alternating current. The Edison interests came very close to success in Virginia and Ohio, but eventually other political considerations outweighed the risk of high voltages.[17] Urban and municipal governments took the strongest stand against electrical risks: several large cities, particularly New York, outlawed the overhead cable and restricted voltages in dwellings.[18] Yet, overall, the legislative program to manage electrical risk failed in the United States. Consequently, the management of the risk of high voltages was carried out within the context of the corporations who were developing and installing power systems. Let us turn now and review how one inventor, Elihu Thomson, coped with the risk of a.c. within the context of a business firm.

Coping with the Risk of Alternating Current: Elihu Thomson and the Thomson-Houston Electric Company

Like Edison, Elihu Thomson responded to the opportunity of electric lighting in the late 1870s. However, unlike the "Wizard of Menlo Park," Thomson did not create a radically new technology but instead chose to perfect the existing technology of arc lighting. Collaborating with Edwin J. Houston, Thomson constructed several arc-lighting systems in the late 1870s. To solve the problem of subdividing the electric light, Thomson designed an automatic regulator which permitted the generator to adjust its output so as to match the number of lights consumers turned on and off. This feature proved to be highly practical and appealing to businessmen, and in 1882 Thomson contracted with a group of shoemakers in Lynn, Massachusetts, to manufacture and market his system.[19]

Together, Thomson and the Lynn shoemakers organized the Thomson-Houston Electric Company. In contrast to Edison who often played a prominent role in the management of companies bearing his name, Thomson chose not to take an active part in management and instead concentrated on invention and engineering for the firm. To pursue his research, Thomson organized his own workshop and office within the factory which he called his Model Room. Meanwhile, other individuals and groups within Thomson-Houston took responsibility for the key functions of the firm. Charles A. Coffin, a former shoe manufacturer and salesman, became executive vice-president in charge of marketing and finance. Edwin Wilbur Rice, who had been Thomson's research assistant, became factory superintendent and supervised a team of engineers who solved the problems of manufacturing electrical machinery on a large scale. Thanks to Thomson's inventiveness, Coffin's marketing skills, and Rice's ability to solve production problems, the Thomson-Houston company grew rapidly; to cite one set of statistics, total sales for the company grew from $400,000 in 1883 to over $10,000,000 in 1891.[20] In 1892 Thomson-Houston merged with its major competitor, Edison General Electric, to form the General Electric Company.

To understand how Thomson went about developing a.c. equipment and how the company chose to address the risk associated with these new products, it is helpful to view the company's management as a set of competing interest groups, each with its own business-technological mindset. This mindset consisted of different ideas about how new technology and products were to be used to capture markets and hence insure the growth and prosperity of the firm. Because each mindset was a general perception of how products would be used by consumers, each mindset also embodied certain assumptions about risk or safety.[21] As head of the Model Room, Thomson's viewpoint was that of a technological perfectionist and systems builder: design the best possible systems and over time they will capture the largest market share. The best system, Thomson argued, was a safe system, since safety features would make it easier for customers to take up the new technology and adapt it to their needs. A second mindset found within Thomson-Houston was that of the Coffin and the marketing group. In contrast to Thomson, Coffin and the other managers took a short-term view: respond to the market quickly and give people the products they want. Implicitly, they viewed safety in commercial terms: if safety helped sell lighting systems without raising the price, then they favored it. However, beyond that, the marketing people embraced the notion of *caveat emptor*: let the customer be ultimately responsible for the safe use of the product. Finally, the third group, Rice and the production engineers, evinced little interest in safety matters. From their perspective, profits would be increased if manufacturing processes were standardized and streamlined. By reducing the cost of producing electrical machinery, the engineers believed that Thomson-Houston would be able to reduce its prices and thus capture a larger market share. In short, the management of Thomson-Houston was hardly monolithic in terms of strategy and goals; rather, the firm contained a diversity of opinions and this diversity shaped how the firm introduced new products and set policy.[22]

It was within this complex organizational context that Thomson developed an a.c. system for incandescent lighting. In March 1885, Thomson sketched out a new a.c. system in which he placed the primary windings of the transformers in parallel in order to step the voltage down from 1,000 to 110 volts. Recognizing the risk of using such a high voltage, he included several safety devices in order to minimize the danger of electrocution. On the primary or high-voltage side of the transformer he inserted a fuse while on the secondary or low-voltage side he added a ground connection which would conduct the high voltage away from the electric lights in the case of a short circuit between the primary and secondary coils. By including such devices it is clear that Thomson had recognized the risk of a.c. and that he coped with it by using foolproof safety devices. From the outset, he was confident that he could design a system that balanced the risk of using higher voltages with the benefits of broader distribution.[23]

Curiously, after sketching out this system, Thomson set it aside and pursued other projects. While he must have viewed it as an interesting possibility, he did not recognize any immediate commercial potential. Instead, Coffin was the first to perceive how Thomson-Houston might use a.c. to enter new markets. During the summer of 1885, Coffin visited Europe and was deeply impressed by the ZBD a.c. system. Although Coffin always claimed that he knew nothing of the intricacies of electrical technology, he sensed that the ZBD system pointed the way to a new and profitable product line. Upon his return, he urged Thomson to pursue his work with transformers coils and to file additional patent applications as soon as possible.[24]

Thomson responded to Coffin's request by filing new patent applications and by devoting most of 1886 to designing and testing an a.c. system. By June of that year, an experimental a.c. power line was run between two buildings at the Lynn plant to deliver current to incandescent lamps. By year's end, Thomson-Houston was advertising a.c. generators in its catalogue, but it had yet to install any system.[25]

In the meantime, the Westinghouse Electric Company was moving quickly to develop their a.c. lighting system and they installed their first commercial system in Buffalo in November 1886. To insure a strong patent position for themselves, the Westinghouse Company secured an American patent on behalf of Gaulard and Gibbs. Significantly, this patent gave Westinghouse broad control over any distribution system which had the primaries of transformers connected in parallel. In contrast to Westinghouse's patent success, all of Thomson's a.c. distribution patent applications were rejected or placed in interference, thus putting Thomson-Houston in a defensive position. If they were going to survive in the a.c. field, Thomson-Houston would have to either contest the Westinghouse patent in court or devise equipment that bypassed the broad patent.[26]

While Thomson responded to the Westinghouse patent by filing a new set of applications, the marketing group grew concerned. By early 1887 Coffin began to wonder if development was perhaps not working fast enough. Why was Westinghouse already selling a.c. equipment but Thomson-Houston was not? In February 1887, Thomson admitted to Coffin that the development work was proceeding slowly. Thomson explained that although he had been working as rapidly as possible he was determined to introduce a complete a.c. system with generators, lamps, transformers, and safety devices all matched to each other and that designing such a system took time. Thomson insisted on taking a systems approach because he believed that an a.c. system designed as a single entity would be the safest and most reliable; to quote him, "when we enter this field we wish to . . . be sure of success from the start with a complete and economical system, and the preparatory work that we have done will . . . tell in the end."[27]

Rather than opposing Coffin, Thomson took several steps in 1887 and 1888 to help the company's short-term marketing position. First, to reduce the threat of patent litigation, Thomson helped the company arrange a patent sharing agreement with Westinghouse which permitted Thomson-Houston to manufacture a.c. systems without infringing the a.c. distribution patent. Second, knowing that the patent agreement could only be a temporary expedient, Thomson introduced new products. To make his system more appealing to potential customers, the utility companies, Thomson designed a recording wattmeter which permitted utilities to accurately measure and bill consumers for the electricity they used. Thomson also worked on an a.c. motor which would have allowed utilities to sell electricity for both lighting and power purposes. Third, Thomson initiated a new patent strategy; since Westinghouse had control of the broad principle of using transformers for distribution, Thomson decided that he would secure patents on the design of the most efficient transformers. Thus, while the Westinghouse company might control the principle, they would be unable to apply it since all of the best transformer designs would be owned by Thomson-Houston.[28]

As a result of these steps, Thomson-Houston shipped its first a.c. machines in May 1887 to the Lynn Electric Lighting Company and by 1889 it had installed 27 a.c. systems.[29] However, in return for accommodating Coffin and the marketing group, Thomson proposed that they help implement his ideas about safety. Thomson was anxious that the company minimize the risk of a.c. by promoting the use of his safety features. Maintaining that the best system was a safe system, Thomson remarked ". . . I am a believer in the establishment of all safeguards which conduce to the good working of a system, especially when they do not add greatly to the cost of making the installation."[30] Notably, Thomson thought that his safety inventions should be part of the marketing strategy used to fight Westinghouse. In December 1888, he proposed that all power plants using Thomson-Houston a.c. equipment be fitted with the latest safety devices and that Westinghouse then be forced by means of publicity to install safety equipment as well. Because Thomson-Houston controlled the patents for the best safety features, Thomson felt that this strategy would block Westinghouse from acquiring a larger share of the market for a.c. systems.

As Thomson soon learned, the chief difficulty with this strategy was that it presumed that Thomson-Houston would encourage and even insist that its customers install safety devices. Unfortunately, this was hardly the case in 1889. Instead, their customers frequently ignored safety equipment and careful installation procedures in the rush to get "on line." Undoubtedly pushed by investors to begin selling electricity as soon as possible in order to make a return on the capital invested, utility operators were forced to keep their construction and installation costs to a minimum. Just as Edison had previously noticed, utilities did not purchase the necessary safety accessories, they used poorly insulated wire, and their linemen were often sloppy and indifferent to the special requirements of alternating current. Although Thomson realized that such poor work was influenced by the competitiveness of the electric lighting field, it nonetheless offended him. "I do not believe in this kind of economy," he warned Coffin, "Rather it would be better not to have the business than incur the risks which are thus involved."[31] As an inventor, Thomson would have liked to have control of his creations from the first idea to the final installation but he soon realized that the marketing group was unwilling or unable to see to this. In this situation, while Thomson had adjusted his efforts to suit marketing's short-term interests, marketing failed to reciprocate by promoting his ideas about safety.

In failing to endorse Thomson's ideas about safety, marketing soon discovered that they had lost a key ally. In the fall of 1889 Coffin and the marketing group were anxious to involve the company in the "battle of the systems." In all likelihood, they saw this as an opportunity to challenge and surpass the Edison and Westinghouse companies. Defensively, they may have also been concerned that the negative publicity about a.c. might harm sales and they wanted to reassure customers that it was safe to buy Thomson-Houston equipment. Since the company was already manufacturing safety equipment, all that would be necessary for a strong position in the debate was to have a leading authority in the electrical industry promote the general use of alternating current. Given his established interest in electrical safety, Thomson was the obvious man for this public relations effort.

In October 1889, Coffin asked Thomson to write an article entitled "How to Make Electricity Safe." Along with addressing the general issue of the safety of a.c., Coffin wanted this paper to offset the negative publicity produced by an accident which had recently occurred at a utility in New York City. The utility had been using a Thomson-Houston a.c. system which was short-circuited when a telephone wire fell across a 1,000 volt power line. Much to Coffin's chagrin, Thomson turned down the assignment. The utility had done a poor job of installing its a.c. equipment, and Thomson felt that he could not personally defend such work. Unsafe installation was unacceptable to his mindset and what is more, Thomson-Houston would not be in this unfortunate defensive position had they listened to Thomson some months earlier.[32]

At first, Thomson simply suggested that Thomson-Houston promote the use of safety devices and that he should remain quiet rather than be "a stumbling block in the way of the company's business transactions." However, when Coffin pressed for a general endorsement of a.c., Thomson exploded. In anger, he informed Coffin that he felt like quitting since "my position with the company has no attractions for me if my ideas of what is needed to constitute good substantial work are not followed but personally neglected." With respect to the safety of a.c., he refused to give a blanket endorsement. "I have no method," he wrote,

> I have no panacea — for all the ills which may follow the use of high potential currents under conditions usually found in large cities. I can no more say how to make electricity safe in such cases I can say how to make railroad travel safe, or how to make steamship travel safe, or how to make the use of illuminating gas safe, nor the use of steam boilers safe. No improvement of our modern civilization has ever been introduced but that involved considerable risk.[33]

Because he had designed an a.c. system, Thomson knew well the risks involved in using a.c. He had made it one of his principles that the best system was a safe system. But when he found that the marketing group was interested only in promoting safety as a short-term, defensive measure, he refused to cooperate. They had not been willing to adopt a marketing policy which reflected his technological goals and so he was unwilling to compromise his principles about safety. As a result, the Thomson-Houston Company took no public position in the a.c.—d.c. controversy, leaving the "battle of the systems" to be fought by Edison and Westinghouse.

One might well ask why Thomson took a strong stand and refused to compromise his ideas about safety. Had he not cooperated with Coffin in 1887 when it was necessary to speed up the development of the a.c. system? The answer lies in how his role within the company changed during this period. During the first six years of the company's existence, Thomson perceived himself as the chief inventor. Most of the firm's products were his handiwork and his engineering had reaped a handsome profit for Thomson-Houston. Yet, there were signs that Thomson might be losing his position as leading innovator. Beginning in 1888, Coffin and the marketing group embarked on a merger campaign during which they quickly bought up seven arc-light manufacturing firms. While this campaign was undertaken primarily to improve the marketing position of the firm by eliminating rivals, it also brought other inventors into the company. Within a short time, Thomson was joined at Lynn by several other inventors, all of whom were soon filing patents for the company. While Thomson was nominally in charge of their work, he may have been worried that his work would no longer be as highly valued by management. In addition, Coffin and the marketing group secured another source of innovation by initiating a policy of buying patents in 1888. Ostensibly, the goal was to secure patents that might be otherwise purchased by competitors and used to block Thomson-Houston's efforts. Thomson intensely opposed this policy, probably because he feared that it jeopardized his control of innovation within the firm. Worried about his position as chief innovator, then, Thomson was in no mood to compromise his business-technological mindset in the fall of 1889, especially as it related to safety. He had cooperated fully in the rapid introduction of the alternating current system only to find that other groups did not respect his efforts and authority. Given this situation, it is hardly surprising that Thomson refused to participate publicly in the a.c.—d.c. controversy when he was so requested by Coffin.[34]

In the long run, as a result of Thomson's forcefulness, the Thomson-Houston Company took no public position in the "battle of the systems." In some ways, this was fortunate because they thus avoided the negative publicity of debate and instead concentrated on improving and marketing their a.c. equipment. Although the firm took no public position, over the next few years Thomson presented several papers to professional engineering societies outlining how the dangers associated with electric lighting could be eliminated. In giving these papers Thomson was careful to differentiate his opinion from that of the company and he discussed the safety problems with both a.c. and d.c. systems. Again and again, he urged professional engineers to develop insulation for high-voltage power lines, to demand a high level of workmanship in installing a.c. equipment, and to install safety features such as fuses, cut-outs, and grounded secondaries. To illustrate how a.c. could be used safely, Thomson even had his own home wired for a.c. lighting, complete with all of the available safety accessories.[35] Just as Edison had brought the question of the risks associated with a.c. before the public, so Thomson provided his professional colleagues with answers to minimize the risk. Together, these two electrical pioneers helped both the public and the electrical industry not only identify but address the hazards associated with this new technology.

Conclusion

What lessons and observations can we draw from comparing Edison's and Thomson's roles in the "battle of the systems"? What do their experiences tell us about the social construction of risk?

First, while it would be easy to explain the risk issue in terms of economic considerations, an economic explanation does not do justice to the complexity of the story. Previous historians, especially Harold Passer, have argued that Edison and his companies turned to the safety issue only when they were unable to compete with Westinghouse in terms of price and product.[36] Unwilling or unable to design a.c. equipment in order to secure a share of the lighting market for small cities and towns, Edison and his companies resorted to the unorthodox tactics of attacking the safety of their competitor's product. Following this line of argument, the "battle of the systems" represents an aberration in which competition surpassed the boundaries of the marketplace and had to be resolved in the public and political arena. For the participants, the safety issue was simply an incidental matter to which they resorted when all other issues failed in the course of the struggle. Significantly, this traditional economic perspective says nothing about how the safety issue was resolved, other than to presume that Westinghouse was obliged to make their a.c. products safe after they had come to dominate the market.

For our analysis, the economic struggle for market share is important but is part of the background rather than the determining force. Instead we would emphasize that the issue of risk was identified and addressed by historical actors working within specific organizational contexts. One of the many things with which they had to cope was the short-term problem of maintaining market share, but other ideas and problems were far more significant. For both Edison and Thomson, the most important issue was the long-term viability of their technological creations. Both inventors wanted to see electricity used extensively to provide light, power, and heat to homes and businesses. While they shared this goal, they nonetheless diverged in significant ways. While Edison preferred to pursue this goal using the existing technology (his d.c. system), Thomson was willing to design new equipment (his a.c. system). Clearly, Edison believed that the risks of a.c. could only be addressed by the public actions of publicity and regulation whereas Thomson was confident that the risk could be contained by using technology, namely incorporating safety devices into the system. Finally, while Edison insisted on taking advantage of his media contacts to make the "battle of the systems" into a public affair, Thomson preferred to work within his company and professional organizations to implement change.

So, in a first sense, we would argue that risk in the electrical industry is best not viewed simply as a component of noneconomic competition but instead as an issue handled by real actors responding to a single goal within a constellation of personal values. More than this, however, we would suggest that the issue of risk was shaped by organizational factors. Both Edison and Thomson dealt with the question of designing and deploying a.c. technology within the framework of the business firm. For Edison, this framework reinforced his position on a.c.; through his coalition of stockholders and his control of the sources of innovation, Edison was able to insist that the Edison Electric Light Company oppose the development of a.c. Anxious to sell their existing product line and maintain their share of the big city market, EELC exuberantly pursued an anti-a.c. campaign, going so far as encouraging Harold Brown and his efforts at establishing electrocution as a form of capital punishment.

In contrast, the organization of the Thomson-Houston Company came to interfere with Thomson's position on the safety of a.c. In particular, the dynamics of groups within the firm led to an impasse wherein the company was unable to participate in the "battle of the systems." While Thomson developed an a.c. system at the request of Coffin and the marketing group, they did not reciprocate by promoting his views on safety. Furthermore, Thomson felt that Coffin and others were attacking his position of authority undercutting his position as the key source of innovation within the firm. By 1889, Thomson was clearly in a position of weakness and his only defense was to refuse to cooperate with Coffin on the safety issue. As a result, Thomson-Houston was unable to take a strong position in the debate and Thomson's efforts at building safety into the a.c. system went unnoticed by the public at large.

In the final analysis, then, we would argue that each man constructed the issue of risk within a different social context. Edison, with his established public image and his familiarity with manipulating the press, chose to make safety an issue outside the firm and within a public arena. Thomson was more concerned with his role intrafirm and consequently chose to make safety an issue within the organization. In both cases, however, we see that these men were neither indifferent to the matter of risk nor driven by simply short-term economic necessity. Rather, they were men of vision, anxious to see the new technology of electricity become widely accepted. Through their experiences we are reminded that while economics and organizational factors do influence the issue of risk, ultimately individuals identify, control, and mitigate the hazards of technology.

Notes and References

1. On the general history of electric lighting, see Thomas P. Hughes, *Networks of Power: Electrification in Western Society* (Baltimore: Johns Hopkins University Press, 1983); Harold C. Passer, *The Electrical Manufacturers, 1875—1900* (Cambridge: Harvard University Press, 1953); and Percy Dunsheath, *A History of Electrical Power Engineering* (Cambridge: MIT Press, 1962).

2. Marshall Fox, 'Edison's Electric Light. The Great Inventor's Triumph in Electrical Illumination,' New York *Herald*, December 21, 1879.

3. Robert Friedel and Paul Israel, *Edison's Electric Light: Biography of an Invention* (New Brunswick: Rutgers University Press, 1986).

4. Edison Electric Light Company, 'The Edison Light' [sales brochure], circa early 1880s, Edison National Historic Site, West Orange, New Jersey (hereafter these archives will be cited as ENHS).

5. Edison's efforts to improve his lighting system are recorded in a Scrapbook, Cat. 1151, 1885—1886 and Charles Batchelor, Journal, Cat. 1336, 1886—1887, ENHS.

6. Hughes, *Networks of Power*, 79—105; Bernard Drew and Gerard Chapman, *William Stanley* (Pittfield, Massachusetts: Berkshire County Historical Society, 1986); for an early description of the ZBD system, see 'Alternating Electric-Current Machines,' *Electrical World*, May 31, 1884, 3, 173—174.

7. Edison was able to use his European contacts to provide information on the first a.c. systems in use. For example, Edison received in November 1886 a report on a.c. technology from Siemens & Halske, the leading German electrical manufacturing company and an Edison licensee. This report can be found in 1886 Electric Light, Edison Electric Light Company file, ENHS. For Edison's high-voltage d.c. system, see 'A New Edison System of Electrical Distribution,' *Scientific American*, July 23, 1887, 10, 42.

8. W.K.L. and Antonia Dickson, *The Life and Inventions of Thomas Alva Edison* (New York: Thomas Crowell, 1894), 326.

9. Thomas Edison, 'The Dangers of Electric Lighting,' *North American Review*, November 1889, 49, 632.

10. Hughes, *Networks of Power*, 106—139.

11. Albany *Journal*, July 24, 1889.

12. Edison, 'The Dangers of Electric Lighting,' 625.

13. Robert Conot, *A Streak of Luck: The Life and Legend of Thomas Alva Edison* (New York: Seaview Books, 1979), 253—255; and Forrest McDonald, *Insull* (Chicago: University of Chicago Press, 1962), 45.

14. The Edison companies advertised their incandescent lighting system as "the only one that is not absolutely injurious to the people using it." See untitled sales brochure, 1888 Electricity file, ENHS.

15. Thomas P. Hughes. 'Harold P. Brown and the Executioner's Current: An Incident in the AC—DC Controversy', *Business History Review*, 1958, 32, 143—165.

16. Edison, 'The Dangers of Electric Lighting,' 634.

17. Hughes, 'Harold P. Brown,' 155.

18. The removal of overhead wires was carried out in New York in 1889 under court order. The process of removal caused a number of accidents and injuries, thus reinforcing the belief that overhead wires were extremely dangerous. See 'Cutting Down the Poles and Wires in New York City,' *Scientific American*, April 27, 1889, 13, 250.

19. For background on Thomson's career, see W. Bernard Carlson, 'Invention, Science, and Business: The Professional Career of Elihu Thomson, 1870—1900,' Ph.D dissertation, History and Sociology of Science, University of Pennsylvania, 1984.

20. Passer, *The Electrical Manufacturer*, 30.

21. This model of the business firm is drawn from two different sources. First, it is based on the behavioral model of the firm developed by Richard Cyert and James G. March in the 1960s. Their model was a conscious rejection of the belief that firms make rational decisions based exclusively on economic considerations. Instead, they asserted that firms come to decisions by "satisficing," that is, by negotiating and compromising the needs and goals of different groups. See their book, *A Behavioral Model of the Firm* (Englewood Cliffs, New Jersey: Prentice-Hall, 1963). For an application of their model, consult Leonard H. Lynn, *How Japan Innovates: A Comparison with the U.S. in the Case of Oxygen Steelmaking* (Boulder, Colorado: Westview, 1982). While Cyert's and March's model implied that each group has a particular viewpoint, they said little about how and why each group chooses to articulate different goals, needs, and perceptions. Consequently, a second idea informing this view of the firm is Reese V. Jenkin's concept of the business-technological mindset. In his study of the photographic industry, Jenkins observed that major changes took place in the industry when entrepreneurs succeeded in matching new technology with new marketing techniques, thus creating what he called a business-technological mindset. See his book, *Images and Enterprise: Technology and the American Photographic Industry, 1839 to 1925* (Baltimore: Johns Hopkins University Press, 1975).

22. For a detailed discussion of these different groups and mindsets within Thomson-Houston, see Carlson, 'Invention, Science, and Business,' 373—404 and 428—430.

23. W. J. Foster, 'Early Days in Alternator Design,' *General Electric Review*, February 1920, 23, 80—90 on 81; 'System of Electric Distribution by means of Induction Coils, March 5, 85,' Exhibit No. 12, 'Testimony on behalf of Elihu Thomson and Edwin J. Houston . . . ,' Interference No. 13,761, Elihu Thomson Collection, GE Hall of History, Schenectady; Elihu Thomson (cited as ET) to H. C. Townsend, April 10, 1885, LB 5/83—8/85, 65—67 and 110—112; ET to J. A. Fleming, October, 12 1885, LB 8/85—3/86, 82—83; ET to Townsend & MacArthur, January 6, 1887, LB 9/86—3/87, 223, Elihu Thomson Papers, Library of the American Philosophical Society, Philadelphia (hereafter cited as TP).

24. ET to G. Cutter, October 7, 1885, LB 8/85—3/86, 76—80 and ET to H. C. Townsend, October 23, 1885, LB 83—11/85, 249—250, TP.

25. ET, 'Electric Induction Apparatus,' June 11, 1886, LB 3/86—2/89, 168—172; ET to Coffin, February 5, 1887, LB 9/86—3/87, 334—337; ET to Townsend & MacArthur, January 6, 1887, LB 9/86—3/87, 223; ET to Townsend & MacArthur, March 1 and 19, 1888, LB 3/87—4/88, 884 and 921—922, TP; 'Extracts (or summaries) of testimony in patent infringement suit of General Electric Co vs Butler Company,' John W. Hammond File, J457—459, General Electric Company, Schenectady, New York; and 'Catalogue of Parts of Apparatus Manufactured by the Thomson-Houston Electric Co. . . .' 1886, Elihu Thomson Collection, Massachusetts Institute of Technology Archives and Special Collections, Cambridge, Massachusetts.

26. Frank L. Pope in Discussion, 'The Distribution of Electricity by Secondary Generators,' *Telegraphic Journal and Electrical Review*, April 15, 1887, 19, 349—354 on 349; E. W. Rice, Jr., 'Missionaries of Science,' *General Electric Review*, July 1929, 32, 355—361; L. Gaulard and J. D. Gibbs, ['System of Electrical Distribution'], U.S. Patent 351,589 (October 26, 1886); Rankin Kennedy, 'Electrical Distribution by Alternating Currents and Transformers,' *Telegraphic Journal and Electrical Review*, April 15, 1887, 19, 346—347; 'Specification of Elihu Thomson, Lynn, Mass. Distribution of Electric Currents,' October 9, 1886, LB 3/86—2/89, 232—240; ET to Thomson-Houston Elec. Co., September 28,

1886, ET to Townsend & MacArthur, November 19, 1886, ET to H. N. Batchelder, December 8, 1886, and ET to Townsend & MacArthur, December 10, 1886, LB 9/86—3/87, 18—19, 95—96, 133—134, and 141, TP.

27. ET to Coffin, February 5, 1887 and ET to E. F. Peck, March 22, 1887, LB 9/86—3/87, 334—337 and 470—471, TP.

28. Frank L. Pope to ET, March 23, 1887, Collected Letters; ET to C. A. Coffin, March 24, 1887, LB 3/87—4/88, 1, TP; and Passer, *The Electrical Manufacturers*, 145—147; ET to Townsend & MacArthur, January 18, 1887, LB 9/86—3/87, 265—266; ET to Coffin, November 20, 1889, LB 4/89—1/90, 729—730, TP; Walter S. Moody to J. W. Hammond, April 15, 1927, Hammond File, L2598—2599.

29. 'Exhibit. The Following List of Thomson-Houston Plants . . . ,' circa 1888, Notebooks, TP.

30. ET to T. F. Gaynor, March 7, 1888, LB 3/87—4/88, 900—901; see also ET to Coffin, December 11, 1888, LB 4/88—4/89, 547—550, TP.

31. Quote is from ET to Coffin, February 13, 1889, LB 4/88—4/89, 761—766. See also ET to Coffin, February 16, 1889, LB 4/88—4/89, 780—782; ET to John J. Moore, October 19, 1889, LB 4/89—1/90, 594—596; ET to Narragansett Elec. Light Co., April 22, 1890, LB 1/90—11/90, 443—444, TP; ET, 'Insulation and Installation of Wires and Construction of Plant,' *Electrical Engineer*, March 1888, 7, 90—91.

32. [A. C. Bernheim] to Coffin, October 16, 1889 and ET to Coffin, November 6, 1889, LB 4/89—1/90, 589 and 658—661, TP.

33. First quote is from ET to Coffin, October 19, 1889, the second from ET to Coffin, December 20, 1889, and the third from ET to Coffin, December 24, 1889 in LB 4/89—1/90, 579—583, 867—878, 903—908, TP.

34. J. W. Gibboney to Bentley & Knight, June 6, 1890, LB 1/90—11/90, 597; ET to Robert C. Clapp, October 23, 1889, LB 4/89—1/90, 602—603; ET to Fish, October 1, 1889 and ET to Capt. E. Griffin February 21, 1890, LB 1/90—11/90, 185—187, TP.

35. ET, 'Safety and Safety Devices in Electrical Installations,' *Electrical World*, February 22, 1890, 15, 145—146; ET to Chas. C. Fry, October 7, 1887 and ET to Prof. S. W. Holman, October 21, 1887, LB 3/87—4/88, 522—523 and 572—575; ET to Lynn Elec. Lighting Co., May 8, 1888, LB 4/88—4/89, 13—14, TP; ET, 'Systems of Electric Distribution,' *Scientific American Supplement*, No. 603, July 23, 1887, 9632—9633.

36. Passer, *The Electrical Manufacturers*, 167—174.

PART VI

Part VI

Experts and the Social Construction of Risk

Experts play a major role in risk selection and evaluating the significance and acceptability of technological risks. In performing this role, experts are often perceived by themselves and by large segments of the public to be unbiased, objective, and value-neutral. The chapters in this section, however, argue that experts are themselves culture bound. More specifically experts share with nonexperts two basic characteristics. First, experts act within the constraints of particular organizations, communities, and societies. Second, the risk judgments of experts are strongly influenced by their social networks and social interactions. At the most fundamental level, the core of the scientific enterprise itself — expert knowledge — can be viewed as an "agreed-upon reality." In reaching agreements about the nature of reality, experts use criteria for the acceptance or rejection of new scientific knowledge that are themselves social constructs (see Berger and Luckmann 1967).

Among the various social and cultural factors that influence expert judgments, perhaps the most important is the occupational community or communities to which an expert belongs. Occupational communities can be defined broadly as groups of people who see themselves as doing the same kind of work, who identify positively with their occupation, and who share ideas, values, and viewpoints (Van Maanen and Barley 1984). Solidarity within such occupational communities is promoted by several factors, one of the most important being the shared possession of specialized, esoteric, scarce, and socially valued skills. A central characteristic of such occupational communities — especially those that seek to attain or maintain the status of a "profession" — is that they often attempt to rigidly control the contents and "agreed reality" of the profession's specialized knowledge and skills. In turn, this body of specialized knowledge and skills is used by the profession to justify and legitimize its autonomy, elite status, and opinions on matters relevant to the profession (Wynne 1982). The success of this effort ultimately depends, however, on the wider society, which exercises control over professions through its power to confer legitimacy and resources (Namer 1984: 211).

The influence of occupational communities on experts is addressed by several chapters in this section of the book. These chapters form part of

B. B. Johnson and V. T. Covello (eds.), The Social and Cultural Construction of Risk, 297—300.

the growing literature in the sociology of science on the influence of occupational environments, institutional setting, social networks, and professional training on expert judgment (e.g., McCaffrey 1982: 58; Davies 1983; Scott 1966; Hofstede 1980; Kopp 1979; Mazur 1981). Several of these themes are specifically discussed in the first chapter. Written by Eugene Ferguson, the chapter describes the role of engineers in the development of boiler safety codes and industrial safety standards in the late nineteenth and early twentieth centuries. Ferguson points out that engineers took little initiative and showed little concern about an obvious safety problem of the time — explosions of steam boilers. He argues that this lack of activism can in part be explained by concerns among engineers about their economic and organizational status. The failure of a technology — such as the frequent explosion of steam boilers — represented an important but not sufficient condition for a safety hazard to be identified and for action to be taken. Instead, a "focusing device" was needed. In the steam boiler case, this focusing device was public opinion, which forced the attention of engineers and their employers on safety problems. Concerns about economic and organizational status also undermined attempts by experts to develop more general industrial safety codes. The industrial safety movement was initiated not by engineers or their employers, but by the insurance industry. When employers did take action (e.g., U.S. Steel), it either was in response to public pressure or reflected the personal beliefs and goals of an isolated, enlightened individual. Employers, as a group, became active in the industrial safety movement only when the new workmen's compensation legislation made support for the movement financially advantageous.

The second chapter in this section, by Joel Tarr and Charles Jacobson, examines three cases that illustrate the internal dynamics of risk selection among experts and the effects of external influences on risk perceptions. The first case deals with the late nineteenth-century clash between sanitary engineers and public health physicians over different approaches to wastewater treatment. Tarr and Jacobson argue that the issues in this specific debate went well beyond the problem at hand and represented an attempt by each profession to impose its own standards and values on the emerging field of public health. These different standards and values had their roots in different types of professional training and in different constituencies — cities for the engineers and the "public" for the public health physicians. The eventual victory of the engineers and the adoption of their approach to wastewater treatment can be traced to several factors. One of the most significant was the greater ease with which engineers could mobilize the concentrated interests of cities, as compared to the difficulties encountered by public health physicians in mobilizing the diffuse interests of citizens (cf. Olson 1965).

Tarr and Jacobson's second case examines the role of experts in identifying the health risks of industrial water pollution. Focusing on the period 1920—1950, Tarr and Jacobson note that experts were initially more concerned about the difficulty of treating such wastes than about possible health risks. This lack of concern stemmed from a variety of factors, including the great diversity of industrial pollutants, uncertainties about the adverse health or environmental effects of industrial pollutants, the lack of generalizable indicators for pollution, and the lack of clear agency jurisdiction over water pollution issues. It was only after World War II that experts expressed strong interest and concern about the health risks of industrial water pollution. Factors contributing to this change included: (a) the development of new chemical sampling and analytic methods, (b) the introduction of new synthetic chemicals with potential adverse effects, (c) the rise of uncontrolled industrial activity, and (d) the increased relative importance of chronic (as opposed to infectious) disease.

Tarr and Jacobson's third case examines the role of experts in identifying the health risks of industrial wastes at Love Canal, New York. This same case has already been examined in Chapter 3 from a different perspective. The two chapters are different in that Chapter 3 focused on the role of social networks whereas Tarr and Jacobson focus on the role of experts and government officials. Tarr and Jacobson note, for example, that inadequacies in methods, equipment, theories, and data severely hampered the functioning of experts and health officials at Love Canal. During the crisis, the most that experts and health officials could say with confidence was that adverse health effects were possible. This lack of certainty and definitiveness, combined with government inaction and the failure to recognize the extent of public concern, exacerbated the problem and thrust a new item — "hazardous waste " — onto the nation's public health agenda.

The third chapter in this section, by Frances Lynn, examines the influence of organizational affiliation on the beliefs and risk judgments of scientific experts. Specifically, the chapter examines the beliefs and risk judgments of physicians and industrial hygienists working in industry, government, and academia. Despite their similar training, Lynn points out that scientists working in these different organizational settings diverge greatly in their beliefs about fundamental issues in risk assessment, including (a) the appropriateness of using animal data for identifying and estimating human risks, and (b) the existence of thresholds for human exposures to carcinogenic substances. One highly provocative implication of the study is that biases in the beliefs and risk judgments of experts are strongly linked to organizational affiliation. Given the available data, however, it is not clear whether and to what extent scientists choose employment in an organization that holds beliefs similar to their own, or whether and to what extent an organization influences the beliefs of scientists working in the organization.

The last chapter in this section, by Sheila Jasanoff, examines differences between the United States and Great Britain in how experts assess the risks of several toxic substances — including asbestos, formaldehyde, 2,4,5-T, and airborne lead. One conclusion of the study is that scientists in the United States place much greater emphasis on risk quantification than scientists in Great Britain. Jasanoff traces this difference in part to fundemental differences in the assumptions adopted by scientists in the U.S. and Great Britain. Specifically, American scientists typically assume, and the British do not, that (1) a false positive is preferable to a false negative, i.e., it is better to err on the side of safety; and (2) the existence of a significant health risk can be established by risk assessment and does not need to be based on proof of actual harm (e.g., actual deaths or injuries).

References

Berger, Peter L. and Thomas Luckmann. *The Social Construction of Reality: A Treatise in the Sociology of Knowledge*. New York: Anchor Books, 1967.

Davies, Celia. 'Professionals in Bureaucracies: The Conflict Thesis Revisited,' pp. 177—194 in Robert Dingwall and Philip Lewis (eds.), *The Sociology of the Professions: Lawyers, Doctors and Others*. New York: St. Martin's Press, 1983.

Hofstede, Geert. *Culture's Consequences: International Differences in Work-Related Values*. Beverly Hills, California: Sage, 1980.

Kopp, Carolyn. 'The Origins of the American Scientific Debate over Fallout Hazards,' *Social Studies of Science*, November 1979, 9(4), 403—422.

McCaffrey, David P. *OSHA and the Politics of Health Regulation*. New York: Plenum, 1982.

Mazur, Allan. *The Dynamics of Technical Controversy*. Washington, D.C.: Communications Press, 1981.

Namer, Gerard. 'The Triple Legitimation: A Model for a Sociology of Knowledge,' pp. 209—222 in Nico Stehr and Volker Meja (eds.), *Society and Knowledge: Contemporary Perspectives in the Sociology of Knowledge*. New Brunswick, New Jersey: Transaction Books, 1984.

Olson, Mancur. *The Logic of Collective Action: Public Goods and the Theory of Groups*. Cambridge, Massachusetts: Harvard University Press, 1965.

Scott, W. R. 'Professionals in Bureaucracies: Areas of Conflict,' in H. Vollmer and D. Mills (eds.), *Professionalization*. Englewood Cliffs, New Jersey: Prentice-Hall, 1966.

Van Maanen, John and Stephen R. Barley. 'Occupational Communities: Culture and Control in Organizations,' pp. 287—365 in Barry M. Staw and L. L. Cummings (eds.), *Research in Organizational Behavior*, Vol. 6. Greenwich, Connecticut: JAI Press, 1984.

Wynne, Brian. 'Institutional Methodologies and Dual Societies in the Management of Risk,' pp. 127—143 in Howard C. Kunreuther and Eryl V. Ley (eds.), *The Risk Analysis Controversy: An Institutional Perspective*. Berlin: Springer-Verlag, 1982.

Chapter 12

Risk and the American Engineering Profession: The ASME Boiler Code and American Industrial Safety Standards

Eugene S. Ferguson

Introduction

Risk and failure have always been inseparable companions of technological change. Bridges have fallen down, boilers have exploded, machines have maimed and killed their users. Furthermore, surprise is an inevitable element of every new project. No technical device or system performs exactly the way its designers expect it to. Surprises and failures have added immeasurably to our stock of knowledge, but the costs in pain and treasure have been great.

Engineers, like the conscientious artisans who preceded them, have always tried hard to build structures and machines that would not endanger those who used them. Unfortunately, their successes have always been partial. Invariably, the new devices have hazards built into them, anomalies as surprising to the engineers as to the public. Nevertheless, in the nearly unanimous judgment of those involved in the new technology, advantages always outweigh hazards. Almost invariably, the public is also attracted by the apparent benefits and, like the engineers, underestimates the risks and side effects. One need only mention the automobile to recognize that despite the risks and side effects, a sufficiently attractive technology will enjoy wide popularity; nor is it likely to be discarded when the full roster of hazards and side effects — accidents and exhaust-induced smog, to name but two — unfolds. When a popular innovation appears, its disadvantages are neither apparent nor likely to be sought out. When disadvantages become evident, there are too many vested interests to permit turning back.

For more than a hundred years, American engineers both as individuals and in organizations have worked to reduce the hazards posed by things engineers have built, but the pace of that work has been excruciatingly slow. The first systematic effort to understand and eliminate steam boiler explosions was undertaken in 1830; the ASME (American Society of

B. B. Johnson and V. T. Covello (eds.), The Social and Cultural Construction of Risk, 301—316.

Mechanical Engineers) Boiler Code, an authoritative set of rules for building safe boilers, was not completed until 1914, 84 years later. Hazards in the workplace were serious enough in 1877 to call forth in Massachusetts a factory act, a political attempt to force improvements in working conditions by requiring the guarding of machinery and state inspections of workplaces.[1] It was not until more than thirty years later, when workmen's compensation laws forced the attention of employers to workplace hazards, that engineering societies began to work on safety standards that would make a substantial improvement in workplace safety.

In following the tortuous paths that led eventually to safe boilers and socially-acceptable levels of safety in industrial workplaces, it will be observed that problems of safety and risk are solved not by engineers alone but by nearly the whole community. The process is one of challenge and response. The challenge is posed by a technical innovation. Society is confronted with a new situation, a frustrating anomaly with which it must deal. The response is not merely one of a quick technological fix, but often must involve social and political change. The brain that catches the first glimpse of a possible way out of the intolerable situation may be in an engineer's head, or it may not. Common sense rather than technical expertise may be the key ingredient. The examples developed in this chapter suggest that a vigorous focusing device is necessary to prompt the engineering profession — that is, the leading engineering societies — to do anything to help relieve those frustrations. In any case, an acceptable response to a technical challenge involves far more than the solution of an engineering problem. Engineers produce the innovations, but the resulting problems involve social choice not merely technical fine-tuning.

Because engineers are not free agents, being bound by patrons who supply their salaries and other resources for doing their work, they must trim their sails to the interests of their patrons. And because their patrons, usually corporations, effectively set the limits of action in engineering societies, it is particularly difficult to carry out a scheme of amelioration that threatens the status quo.

One of the earliest committees of the American Society of Civil Engineers was formed in 1873 to report to the Society "On the Means of Averting Bridge Accidents." When the committee made its report in 1875, its members were in substantial agreement on the standards to follow in order to design a safe bridge. They were divided, however, on whether to press for laws that would embody their standards or simply to let the standards serve as information and advice to engineers who wanted such guidance. Committee member Charles Shaler Smith, one of the foremost bridge designers of his time, thought that because of the frequent shocking failure of bridges, public opinion would soon force the enactment of laws of some kind. Therefore, he recommended that "this Society should take time by the forelock, dictate a law which will be just and equitable [and lobby for its enactment rather than] stand in the background until an aroused public opinion compels legislation which may be injurious to the profession." The matter was discussed for a while, but time's forelock was not disturbed, and the Society took no public stand on the issue.[2]

Within two years of its founding in 1880, the American Society of Mechanical Engineers had published and discussed papers on how to prevent fires in mill buildings and how to keep boilers from exploding. In 1883, leaders of the Society argued unsuccessfully for a government program of testing materials, so that an engineer might predict with some assurance the strength of iron and steel and of other materials used in machines and structures.[3] In 1890, a senior member read a paper on "Accident-Preventing Devices Applied to Machines" at the annual meeting of the Society. He tried to convince his fellows that an engineer's "whole duty" consisted not only in designing a machine that would do its intended task but that included safeguards to protect its operator from the gears, cutters, and other components that accomplished the machine's purpose.[4]

Despite the manifest concern of individual engineers for problems of safety, their influence did not extend much beyond the meeting rooms of their societies because their societies — the ASME, the ASCE, and the American Institute of Mining Engineers — were extremely cautious about expressing an opinion on any subject. The societies published the papers that were given by their members, but they did not endorse what their members wrote or said. They feared that the endorsement of findings or recommendations might at some time embarrass a society or subject it to being sued.

The ASME Boiler Code

Almost as soon as steamboats appeared on the Mississippi River, boiler explosions in the boats began to exact a toll of lives — a toll that some people considered intolerable. The reasons why boilers exploded is perfectly clear, by hindsight. Designs were inadequate for the pressures at which the boilers were operated; the iron used in boilers was not uniformly strong; the workmanship was often marginal; and reasonable care was not exercised in operating the boilers. Yet none of these reasons was self-evident in the early nineteenth century, and theories about boiler explosions confused the issue by invoking certain obscure phenomena that were plausible enough to be difficult to disprove. For example, because boilers sometimes exploded when cold feedwater was being pumped into a hot boiler, it was supposed that the water, coming into contact with the overheated iron shell of the boiler, evolved hydrogen as a gas, which subsequently exploded.

In the 1830s, the Franklin Institute in Philadelphia undertook to find out (a) why boilers exploded, (b) how they might be made safe, and (c) how to promulgate and enforce such rules as the investigation might develop. The federal government, through the Secretary of the Treasury, made a small grant of money to support the investigation. After exhaustive studies and experiments, including a number of experiments designed to refute false notions regarding causes of explosions, a substantial report was completed in 1836, containing rules regarding materials, design, construction, and safe operation of boilers. Between 1838 and 1851, the federal Congress enacted a succession of steamboat laws, but with negligible penalties for unsafe boilers. Steamboat boilers continued to explode, and from 1841 to 1851 over a thousand lives were lost in such explosions. In 1852, despite the warning of Senator Stockton that further regulation of steamboat operators would reduce them, so far as their liberty was concerned, to the status of Russian serfs, the Congress wrote and passed a steamboat inspection law that decreased substantially the loss of life from boiler explosions. In the course of responding to the challenge of explosions, the public view of private rights was changing. A small minority in the Senate still agreed with Stockton, but the majority now considered public safety more important than the traditional sanctity and irresponsibility of private property. After 20 years of rising pressure of public opinion, the public had at last a measure of protection from steamboat boiler explosions.[5]

On the other hand, there were no regulations at all for stationary steam boilers, and by the 1860s such boilers were exploding at an alarming rate. The first and for a generation the strongest check to stationary boiler explosions was an insurance company. In 1866, the Hartford Steam Boiler Inspection and Insurance Company was formed by Jeremiah M. Allen, a young civil engineer. Allen was impressed by the success in England of the Manchester Steam Users' Union, which since 1855 had reduced explosions by inspecting and certifying the boilers of its members. The Hartford company proposed to insure selected clients against damage to persons and property that occurred as a result of boiler explosions. The company selected its clients by inspecting their boilers and deciding that their equipment and operators constituted an acceptable risk.[6]

The success of the Hartford Steam Boiler Company reflected a growing public concern about such disastrous accidents. The concern was reflected also in a series of state and local laws that attempted, through inspection or otherwise, to control what was perceived as a serious but controllable hazard. Yet changes came slowly. As we shall see, in the decades leading to workmen's compensation laws, the courts favored defendants in liability cases, and most of those who lost relatives in boiler explosions or were themselves injured probably fared no better than workers who were injured or killed in the course of their employment. Almost no data on liability cases have been published, but a detailed study of an 1867 boiler explosion in Lancaster (Pennsylvania) underlines the plight of the injured and aggrieved. The widow of the boiler operator, after selling her house to pay her lawyer, sued the boiler's owners for $10,000 for the death of her husband. She persisted for nearly ten years, and in 1876 the court awarded her a total of $526. Hers was the only suit pressed by the eight people injured or by the survivors of the six dead.[7]

State and local laws that spelled out minimum design specifications were annoying to boiler manufacturers — even to those who tried conscientiously to build entirely safe boilers. Regulations were conflicting and often contradictory, so that a particular boiler design might be acceptable in one city or state but be illegal in another. Before 1900, the American Boiler Manufacturers Association attempted without success to develop uniform standards for boiler construction. The chairman of the association's Uniform Boiler Specification Committee was Edward D. Meier, founder and president of the Heine Safety Boiler Company. He was also a member of the American Society of Mechanical Engineers. He hoped that his committee might produce a uniform set of standard specifications that could be adopted as the basis for all state and city boiler laws. Unfortunately, several members of the Association would not endorse the standards written by its committee, so the notion of uniform laws among the states was set aside.[8]

A set of standards, even if endorsed by the Association, might encounter opposition to its adoption as law because of a suspected conspiracy by the boiler manufacturers against the public. The adequacy of any set of standards, by their nature lengthy (the first ASME Boiler Code filled 140 pages), arcane, and subject to differing interpretations, could be verified only by experts; it could not be expected that public officials would interpret the ambiguities and conditional statements in the same way as experts employed by boiler manufacturers. The problem then, as now, in producing an enforceable code was that the experts who knew the most about the subject at hand were those who had invented, developed, and promoted the very apparatus that was perceived by the public as requiring regulation. To find a body that in fact represented the public interest was and is very difficult, because each of us, expert or not, has a particular view of the public interest that may, but probably will not, agree with anyone else's view. What was needed was a set of standards, or code, produced by a group of technologically knowledgeable people whose integrity and sense of fairness was acceptable to the builders of boilers and to the public alike. The American Society of Mechanical Engineers eventually was recognized as such a group, but not until its willingness to deal generously as well as fairly with all of the requirements and viewpoints of the manufacturers had been rigorously tested.

In 1905, a boiler exploded in a shoe factory in Lynn, Massachusetts, killing 58 people and injuring 117. As the public outcry began to subside, a boiler exploded in another Massachusetts shoe factory, killing only one person but reminding the public that the hazard was still present. The reminder was very timely. Within a month, the newly elected governor of Massachusetts in his inaugural address called for new boiler laws; the legislature authorized him to appoint a Board of Boiler Rules; and within less than two years a set of rules, which specified minimum safe standards for boiler construction, had been hammered out and adopted as law in Massachusetts. In 1911, the Massachusetts rules were adopted by the State of Ohio as the basis of its own boiler laws. Ultimately, the Massachusetts rules would furnish the framework of the definitive boiler code of the American Society of Mechanical Engineers. One of the members of the Massachusetts board was John A. Stevens, chief engineer of the Merrimack Manufacturing Company, perhaps the largest textile producer in the country, with mills in Lowell, Massachusetts and North Carolina. Stevens represented boiler users on the board; the other members were selected to represent the interests of boiler manufacturers, boiler inspectors, and boiler operators.

In 1911, Edward D. Meier, still president of Heine Boiler Company, was elected president of the American Society of Mechanical Engineers. He saw in ASME the opportunity to pursue his quest for uniform boiler laws. The Executive Council of the Society authorized him to appoint a committee to write yet another code of standards for boiler construction. He was successful in persuading John A. Stevens to be chairman of the committee that became known as the Boiler Code Committee, a committee, by the way, that is still very active, keeping the code current as changing techniques and materials are introduced into the making of boilers and other pressure vessels.[9]

Stevens's committee borrowed heavily from the Massachusetts rules that Stevens had helped write. After several meetings, the committee was ready to submit its report to interested persons, assuming tacitly if not explicitly that it had written an impartial and objective set of standards. Two thousand copies were sent to individuals in April 1914. The committee solicited suggestions for modification of the code, but rather expected that changes would be minor and that the final code might be approved at the Society's June meeting in Minneapolis.

The committee was dismayed by the storm of protest that its actions provoked. The meeting in Minneapolis was besieged by angry accusations by those whose toes had been stepped on or who thought the committee had ignored their suggestions. A public hearing, open to all interested parties, was hastily scheduled, to be held three months later in New York City. The initial objection to the code was less to its general substance than to the way it was being railroaded through the ASME without discussing all the ideas of those who thought changes were desirable. One member of the ASME Executive Council said explicitly that the committee had not given sufficient attention to the wishes of the boiler manufacturing industry. He wanted to remove Stevens, a boiler user not a boilermaker, from the committee and to start over. Cooler heads prevailed at the Minneapolis meeting, and Stevens remained chairman until 1925.[10]

The committee now took very seriously the public hearing. It was clear that any general support the code might enjoy would rest upon the confidence of interested parties — builders, users, insurers, and all the rest — that their interests had been given adequate consideration. The notion of an impartial committee of experts was dismissed by the whole population of experts who would be affected by the boiler code. They recognized, implicitly at least, that a set of technical rules has a political as well as a technical component. Because it must be written by individuals, and because all individuals start with certain assumptions, have certain loyalties and antipathies, and incorporate their own purposes into seemingly objective technical requirements — because of all this, a continuously-correcting give-and-take must be part of the process of putting together a public statement that will have the support of those in position either to support or sabotage it. The air had been cleared by September; the public hearing was attended by 150 persons who were by that time anxious to bring minds together rather than force them apart. The first day of the hearings lasted from mid-morning to 8.30 in the evening. After the second day, certain special groups, such as boiler tube manufacturers, held further meetings to work out a concensus of specifications to be offered to the Boiler Code Committee. Finally, in mid-December 1914 an advisory committee of 18 men to work with the code committee was assembled, each representing a different interest (and including only four members of ASME); the code committee met with the advisory committee, and the final draft of the ASME Boiler Code was written as both committees worked continuously, six days a week from 10 a.m. to 11 p.m., for almost seven weeks. The final draft was accepted by the executive Council of ASME on February 13, 1915.[11]

The ASME Boiler Code was eventually written into the laws of many states and other jurisdictions.[12] It has been continuously reviewed, interpreted when the intention of specific passages has been questioned, and augmented as new materials and techniques have been introduced. Members of ASME have contributed countless hours and enormous energy to the code, and the Society is justifiably proud of the status that the code has given it. On the other hand, a considerable part of the effort expended on the code should be classed as advocacy for a particular view that is of advantage to an individual or, more likely, to his employer. An engineering society that undertook the construction of a technical code or standard had to depend almost entirely upon voluntary labor; because few engineers were independently wealthy, support for that voluntary labor had to be supplied by the engineers' employers. Furthermore, the time was long past when the ASME could be classed as pioneers in solving the problem of boiler explosions. The death toll from boiler explosions had been declining for many years, due largely to the efforts of others. In 1914, about half as many people were killed in boiler explosions as had been killed each year for several years around the turn of the century.[13]

The success of the code as a safety measure is not diminished by recognizing that it works only because it is the product of a political process extending over a period of many years. The saga of the Boiler Code demonstrates the essential complexity of an easily-stated technical problem — how to keep boilers from exploding — when it is imbedded in a social matrix, as all real technical problems are. An engineering student absorbs in technical courses the notion that he or she is learning a method of solving problems that is logical, direct, and infallible. Refuting that notion, the problem addressed by the Boiler Code was solved because public opinion forced the engineering community's serious attention to the problem; because those who would be affected by the code in an economic way, through profits or jobs, were persistent in making sure that their interests were given full consideration in an involved political and technical process of debate and bargaining; because a few members of ASME, in taking on the Boiler Code, followed a vision of altruistic service to the public good; and because some other members of ASME were more expert than their colleagues in making necessary compromises palatable to those who must give up something in order to gain a consensus. The history of the ASME Boiler Code should encourage humility and a healthy uncertainty in those who would solve problems that involve technical expertise but that also involve social and political dimensions.

Industrial Safety in America

Until the twentieth century, the common law of Great Britain and the United States had little patience with the notion that one is one's brother's keeper. Until common law was modified by the enactment of employers' liability laws, it was virtually impossible for a worker, injured in a factory or other workplace, to recover damages from his or her employer, no matter how dangerous the workplace or negligent the employer in providing reasonable safeguards. Usually a presumption of negligence on the part of the worker would settle the matter; if not, the "fellow servant" doctrine could be invoked, holding the employer blameless if the negligence of one worker injured another. Undoubtedly there were employers here and there that assumed responsibility for their workers' safety while in the workplace; but until employers were made legally and economically responsible for their workers' injuries, there existed a vast number of workplace hazards that might with modest effort have been removed or reduced.[14]

The first employers' liability law was enacted in 1880 in Great Britain. Great Britain's Factory Act of 1844 had called for the fencing or guarding of dangerous machinery, but the Act was difficult to enforce because penalties were not great. A rising concern with the plight of workers led to the 1880 employers' liability law. Under that law, an injured worker still had to sue in order to be compensated. The near impossibility of a worker's recovering in the courts a fair amount from a powerful and determined employer gradually became so obvious that a "workmen's compensation" law, providing compensation through equitable rules rather than through adversary proceedings, was enacted in 1897. Similar laws were made in Austria in 1897, in France and Italy in 1898. Most of the other European countries had workmen's compensation laws by the time the Russian law was enacted in 1903.[15] The American Congress was unwilling to consider laws governing labor unless the labor was involved in interstate commerce; therefore, the individual states had to write their own labor laws. In 1877, under pressure from the Knights of Labor, Massachusetts passed one of the first American factory acts, which required the guarding of machinery.[16] Employers' liability laws were passed by most of the states, starting with Alabama in 1885. Yet a great deal of pressure still had to be brought upon legislatures to have them consider the idea of a workmen's compensation law, which would form a radical break with the traditions of common law. President Theodore Roosevelt encouraged the change by declaring the injustice of forcing the individual worker and his or her family to suffer "the whole penalty" of being injured. Roosevelt pointed out that an employer who compensated an injured worker might count the compensation as a legitimate cost of doing business and therefore payable by the ultimate customers.[17] By hindsight it is clear that an effective industrial safety movement had to be buttressed by workmen's compensation laws. Before those laws were passed, however, a number of other steps were taken toward improving safety in the workplace.

Because employers objected to workmen's compensation laws as unnecessary and as threats to their freedom, engineers could not be expected to volunteer opinions about actions that should be taken to safeguard the workplace. The acceptance by employers of engineering initiative in matters affecting industrial safety was slow to develop and, as in so many other matters, the path leading eventually to grudging acceptance by industrial managers of "safety engineers" started in an unexpected place. The "Factory Mutuals," a group of insurance companies specializing since mid-nineteenth century in industrial fire insurance, limited their risks by accepting only those factories that appeared least likely to suffer major fire damage. Furthermore, their inspectors and staff engineers actively developed procedures and devices that would prevent fires or minimize damage if fires were started. For example, in the 1880s Charles J. H.

Woodbury, a staff engineer, made systematic tests of sprinkler systems; an engineering instructor at the Massachusetts Institute of Technology was retained as a consultant. The sprinkler systems, permanently installed in buildings, were intended to drown a fire before it could get started; eventually, automatic sprinkler systems incorporated a fire alarm to alert owners while extinguishing the fire, hopefully before the owners arrived. Standards for the installation of electrical wiring, lighting, and machinery had been worked out by the Factory Mutuals even before the American Institute of Electrical Engineers was established in 1884. Members of the AIEE later contributed expertise and experience to the writing of a definitive National Electrical Code, a set of rules analogous to the ASME Boiler Code. The Factory Mutuals also sponsored the Underwriters Laboratory, which for nearly a hundred years has tested and approved or rejected electrical and other equipment submitted to it. The UL tag on an appliance certifies its fire safety if used as intended by its maker.[18]

The engineers working as inspectors for the Factory Mutuals kept meticulous records of insured properties and their surroundings. In 1886, John R. Freeman, an 1876 graduate of MIT, became a Factory Mutual inspector. Four years later he took charge of the "plan department." In that capacity, he took a rudimentary system of mapping, intended to show the locations of buildings, hazardous work areas, and potential surrounding hazards, and developed it into an effective instrument of control. A perfected insurance map would show, for example, the type of building construction and the materials used, the types of industrial processes being carried on and their specific locations, the locations of water for fire fighting and its quantity and pressure. The maps supplied graphic records of a large number of characteristics of individual industrial properties; furthermore, as the number of plans increased, a valuable comparative record became available to engineers and others who were trying to piece together a general approach to the protection of property from fire. The plan department is an example of a successful engineering approach to a complex but well-focused problem.[19]

At length, when workmen's compensation laws were enacted, casualty insurance companies wrote policies to insure the ability of a company to pay workmen's compensation when required. Taking ideas and practices from boiler insurance and fire insurance companies, the casualty companies also functioned as engineering research organization, spotting hazardous devices and operations and working out ways to reduce the risk to workers and the number of claims that the insurance company would have to pay. We are ahead of our story, however, and must now return to the belated American response to industrial workplace hazards that eventually made possible the passage of workmen's compensation laws.

An early and unexpected contributor to the industrial safety movement was the United States Steel Company. The forces that were directly responsible for the aggressive safety campaign that developed in U.S. Steel after 1906 are difficult to pinpoint, but the evidence suggests a combination of public opinion, offended by the brutal picture of steelmaking being broadcast by muckrakers; the zeal of a handful of individuals in the mills; and the support given to the campaign by Elbert H. Gary, chief executive officer of the corporation.

In 1906, the first organized safety effort was made at the South Chicago works of Illinois Steel Company, under the leadership of R. J. Young. Young was described by an acquaintance as "a sort of genius who suddenly discovered that accidents were preventable — that it was no longer necessary to kill and main men as a part of the making of steel."[20] There were 47 fatal accidents in that plant in 1906; by 1913 the number had dropped to 7. In 1907, at the first meeting of the Association of Iron and Steel Electrical Engineers, a safety committee was formed.[21] In 1908, a safety committee was also formed in a general meeting of "casualty managers" of the several U.S. Steel plants, men whose job it was to settle claims made against the corporation by injured employees. "Judge" Gary — he was a lawyer before becoming a leader in the steel industry — supported the safety efforts with his promise that "expenditures necessary for such purposes will be authorized."[22]

The safety committee of the Iron and Steel Electrical Engineers was guided by Lewis R. Palmer.[23] His vision and genius for organization led in 1912 to the First Cooperative Safety Congress, attended by a diverse group of individuals whose jobs were involved in some way with industrial safety. Palmer insured the success of the meeting by visiting railroad executives, insurance people, members of the National Association of Manufacturers, federal and state officials in bureaus of labor and mines, and others who might support his agenda. He traveled over four thousand miles to explain in person his vision of a national organization concerned with the promotion of safety; he invited an influential group of advocates to his congress. The 1912 Safety Congress appointed a committee to "organize and to create a permanent body devoted to the promotion of safety to human life in the industries of the United States." The end result of Palmer's zeal was the formation, in 1915, of the National Safety Council, which grew directly out of the deliberations of his Safety Congress's committee.[24]

The basic motivating force of the safety movement, while not entirely clear in 1906, had become perfectly evident by 1912. By that time, the inexorable pressure within the states for workmen's compensation laws was beginning to yield results. A temporary setback occurred in 1911, when the New York law of 1910 was declared unconstitutional by the State Court of Appeals, but by 1915 some thirty states had passed workmen's compensation laws. In 1917, the federal Supreme Court settled the question of constitutionality; by 1920 all the states but Missouri and Mississippi had enacted such laws.[25]

Except for the United Association of Casualty Inspectors, formed in

1911 by a group of engineers who were industrial inspectors for insurance companies (and which association became the American Society of Safety Engineers), and the Association of Iron and Steel Electrical Engineers, the engineering societies did not press for workmen's compensation laws. Those laws came from outside the centers of power in the engineering profession, and the response of individual engineers depended generally upon the support that their employers were willing to give to safety matters. As the National Association of Manufacturers and other trade groups began to accept the inevitability of workmen's compensation laws, they discovered that a safe workplace would reduce their compensation insurance premiums. In fact, those individuals and groups advocating safety in the workplace seized upon the economic argument, that it was cheaper to install safety devices than it was to pay workmen's compensation, as a rallying call to enroll employers in the crusade for safety.[26]

When workmen's compensation laws were accepted publicly by organizations as safe and staid as the National Association of Manufacturers, the laws could then be readily accepted by engineers, and their societies could sponsor with enthusiasm the development of safety standards. For many years, the leading engineering societies had had standards committees, concerned with such standards as those for screw threads, for wire and pipe sizes, electrical connectors, and the like. It was found that safety standards could be worked out by safety subcommittees of the standards committees. In 1916 and 1917, for example, safety codes for cranes, industrial ladders, and guards for woodworking machinery were written by the ASME Subcommittee on Protection of Industrial Workers; a code for the guarding of power-transmission machinery was prepared by a member of the ASME Subcommittee "under the direction of and with the approval of the Committee on Health and Safety of the National Association of Manufacturers."[27] After a few years of uncertainty about who should have jurisdiction over what kinds of safety standards, a conference of interested parties to settle the issue was held in 1919 at the National Bureau of Standards. Called at the request of the National Safety Council, the conference decided that the work of safety subcommittees should be supervised and coordinated by the American Engineering Standards Committee, initially a joint venture of five national engineering societies and the Departments of War, Navy, and Commerce.[28] That committee, which was quickly broadened to include trade associations, gave safety standards sufficient status to insure that they would be generally accepted by government authorities at all levels. Trade associations had an incentive to support the writing of the standards because that support would give them a voice in the substance of the standards, usually through having one of the industry's engineers on the subcommittee responsible for a particular standard. As in the later stages of the ASME Boiler Code, the safety subcommittees included all interests that might be affected by the Standards. For example, the committee that developed a safety standard for the protection of workers using grinding wheels included technical representatives of trade associations and other organizations of grinding wheel makers, state regulatory commissions and commissions to administer workmen's compensation, manufacturers using grinding wheels, casualty insurance companies, and workmen being protected, the last-named through representatives sponsored by the U.S. Department of Labor. Also on the committee, which had 30 members, were national engineering societies, interested technical bureaus of the federal government, and independent specialists. After two years of work by the committee, a safety standard was agreed upon to cover the construction, care, and use of grinding wheels. It was recognized as the industry standard and was adopted by the various state regulatory agencies. In 1926, the Secretary of the American Engineering Standards Committee reported that, since 1919, 14 national safety codes had been completed and that 26 were still being written.[29]

Absent from the agenda of code and standards committees was any sustained inquiry into the problems of industrial poisons. Workmen's compensation laws provided the incentives for attention to hazards that were immediate and recognizable. There was no comparable concern for the more subtle hazards of continuous exposure to poisonous substances that made certain trades particularly susceptible to industrial diseases, such as lead or mercury poisoning, silicosis, or the "phossy jaw" of women who worked with phosphorus in match factories. A few pioneers, such as the remarkable Alice Hamilton, M.D. (1869—1970), identified the hazards of what she called the "dangerous trades" and mounted a lifelong campaign to bring the necessary pressure to bear on those who were in a position to improve the situation. The technical problems of reducing the long-term hazards were and are difficult and often frustrating. Until after the 1950s, however, the focussing devices available to the public were insufficient to direct the attention of engineering societies to an effective program for making safe the dangerous trades.

Conclusions

The ASME Boiler Code, the National Electrical Code, and industrial safety codes have been a continuing source of pride to engineering societies. Shortly after World War II, a president of ASME closed a luncheon address with the observation that ASME had every reason to be proud of its involvement in codes and standards. "In few endeavors," he said, "is there greater opportunity for the exercise of enlightened self-interest, coupled with the performance of an outstanding public service."[30] In 1971, when the "hippies" were attacking many of the basic assumptions of engineers and their societies, an article entitled "Standards — the Evidence of Concern" appeared in the *IEEE Spectrum*, the monthly magazine of the Institute of Electrical and Electronic Engineers. An exasperated author scolded the critics of the engineering profession who had "conveniently overlooked the fact that, through the formulation of standards, engineers over a long period of time have made valuable contributions to the well-being of the public." He was willing to see engineers "increase their involvement in our changing society," as the hippies demanded, but he wanted it understood that "engineers have protected the public in the past by the formulation of standards and it is likely that they will respond with characteristic vigor to the new challenges that society presents to them."[31] As we have seen, the challenges originate in technological hazards, but the responses have been forced and focused by public opinion and the pressures exerted by the entire community.

As noted by the president of ASME, the codes have always been influenced by self-interest. Whether that self-interest is enlightened or otherwise depends upon one's point of view. From the viewpoint of Admiral Rickover, testifying before the Congressional Joint Committee on Atomic Energy, the concensus standards of the engineering societies provide only a "minimum level" of protection. Because the standards have been incorporated into state and local laws, the law-making bodies have in effect turned over to the standards committees their own responsibility for protecting the public.[32]

Given the nature of the engineering profession, anything beyond routine attention will be paid to problems of hazards and side effects only when the engineers' patrons require them (and thus pay them) to do so. The public must continue to be the instigators of action, getting first the attention of patrons and then insisting that the engineers assigned to the problems will provide something more than a "minimum level" of protection from hazards.

Engineers, for their part, must begin to understand that a socially acceptable solution of an engineering problem of any significance must always include a political component as well as a technical component.

Notes

1. Somers and Somers, p. 200.

2. ASCE, 1875, pp. 122—135, 208—222.

3. ASME, 1881-83.

4. Cooper,

5. Burke; Sinclair, 1966.

6. Woodward, pp. 112—117.

7. Winpenny.

8. Greene, p. 7.

9. Greene, pp. 7—8.

10. Greene, pp. 10, v.

11. Greene, pp. 14—15.

12. Blackall, p. 979.

13. Sinclair, 1980, p. 146; Smith.

14. Somers and Somers, chapter 2.

15. *Encyclopaedia Britannica*, s.v. 'Labour Legislation'; Heinrich, p. 380.

16. Somers and Somers, p. 200.

17. *Engineering News*.

18. Manufacturers Mutual; Heinrich, pp. 379—80.

19. Manufacturers Mutual.

20. Young.

21. Close.

22. Tarbell, p. 225.

23. Alexander, 1916.

24. Chaney; Palmer.

25. Somers and Somers, chapter 2; Nat. Indust. Conf. Bd., p. 5, table 1.

26. Blackall; Alexander, 1926.

27. Hansen and Hicks.

28. Agnew, 1928.

29. Agnew, 1926.

30. Blackall, p. 981.

31. Middendorf.

32. Hammer, p. 62.

References

Agnew, P. G. 'The National Safety Code Program,' *The Annals*, 1926, 123, 51—54 (Philadelphia: American Academy of Political and Social Science).

Agnew, P. G. 'Work of the American Engineering Standards Committee,' *The Annals*, 1928, 137, 13—16.

Alexander, M. W. *Addresses* [collection of bound pamphlets, c. 1906—16] in Hagley Museum and Library, Wilmington, Delaware, 1916.

Alexander, M. W. 'Need of Safety from the Employer's Point of View,' *The Annals*, 1926, 123, 6—8.

American Society of Civil Engineers (ASCE). *Transactions*, 1875, Vol. 4.

American Society of Mechanical Engineers (ASME). *Transactions*, 1881—83, Vols. 2—4.

Blackall, F. S., Jr. 'ASME Standards Save Lives and Dollars,' *Mechanical Engineering*, 1953, 75, 979—981.

Burke, J. G. 'Bursting Boilers and Federal Power,' *Technology and Culture*, 1966, 7, 1—23.

Chaney, L. W. and H. S. Hanna. 'The Safety Movement in the Iron and Steel Industry 1907 to 1917.' U.S. Dept. of Labor, Bureau of Labor Statistics, *Bulletin 234*, 1918.

Close, C. L. 'Safety in the Steel Industry,' *The Annals*, 1926. 123, 86—92.

Cooper, J. H. 'Accident Preventing Devices Applied to Machines,' *Transactions ASME*, 1890—91 12, 249—264.

Encyclopaedia Britannica (11th edn.). Cambridge, England: The University Press, 1910—11.

Engineering News. 'President Roosevelt on Employers' Liability Legislation,' 1907, Vol. 57(24), 657.

Greene, A. M., Jr. *History of the ASME Boiler Code*. New York: ASME, c. 1955.

Heinrich, H. W. *Industrial Accident Prevention* (2nd edn.). New York: McGraw-Hill, 1941.

Hammer, W. *Occupational Safety Management and Engineering*. Englewood Cliffs: Prentice-Hall, 1976.

Hansen, C. M. and R. W. Hicks. 'A Code of Safety Standards for Power-Transmission Machinery,' *Transactions ASME*, 1971, 39, 399—409.

Manufacturers Mutual Fire Insurance Co. *The Factory Mutuals 1835—1935*. Providence, R.I., 1935.

Middendorf, W. H. 'Standards — the Evidence of Concern,' *IEEE Spectrum*, 1971, 8, 70—73.

National Industrial Conference Board. *The Workmen's Compensation Problem in New York State*. New York: NICB, 1927.

Palmer, L. R. 'History of the Safety Movement,' *The Annals*, 1926, 123.

Sinclair, B. *Early Research at the Franklin Institute*. Philadelphia: Franklin Institute, 1966.

Sinclair, B. *A Centennial History of the American Society of Mechanical Engineers 1880—1980*. Toronto: University of Toronto Press, 1980. [In this book one finds the only critical (as opposed to promotional or ceremonial) historical inquiry that I know of into codes and standards. See index, "codes and standards."]

Smith, C. O., Jr. *Making Steam Safe*. [Professor Smith, of Drexel University, has carefully compared the experiences in France, Germany, Great Britain, and U.S.A in protecting the public from boiler explosions. I look forward to the publication of his work.]

Somers, H. M. and Somers, A. R. *Workmen's Compensation*. New York: Wiley, 1954.

Tarbell, I. M. *The Life of Elbert H. Gary*. New York: Publ., 1925.

Winpenny, T. R. 1981. 'Mill Hands and Boilers: The Anatomy of a Disaster,' *Journal*, 1981, 84, 110—124 (Lancaster County Historical Society).

Woodward, P. H. *Insurance in Connecticut*. Boston: D. H. Hurd, 1897.

Young, A. H. 'Industrial Personnel Relations,' *Transactions ASME*, 1919, 41, 145—162.

Chapter 13

Environmental Risk in Historical Perspective

Joel A. Tarr and Charles Jacobson

This essay is concerned with the question of conflict among experts in regard to perceptions of risk in environmental decision making. In order to illuminate this theme, it will explore three cases of environmental decision-making over time. Two of these cases involve water pollution questions and are primarily historical while the third focuses on the more contemporary problem of the disposal of wastes in the land. The cases also differ in that the water pollution studies deal with broad changes over time while the land pollution study involves one site. All however, include issues of agenda setting, disciplinary conflicts, and the uses and limitations of scientific knowledge in problem solving in a political setting. Thus, in spite of their differences, they all serve to illustrate similar issues of applied science in a policy-making context.

I. Professional Cultures in Conflict; The Case of Protecting Drinking Water Supplies from Pollution, 1900—1920

Professional judgments concerning problem-solving in the environmental area often involve assumptions about the amount and acceptability of risk a particular approach may involve. Experts from different disciplines working on the same problem may disagree in their approach because of the different values imparted by their disciplinary training. This case study is concerned with a critical division that occurred during the first quarter of this century between sanitary engineers and physicians in public health over how to reduce hazard in regard to protecting drinking water supplies. What makes the case particularly interesting is the extent to which the division grew out of the different value systems imparted by professional training as well as by territorial conflict over control of the field of public health.

The belief that government has an obligation to protect its citizens from hazard in regard to drinking water evolved gradually in the nineteenth century as a response to the visible deterioration of the quality of local supplies, epidemics of infectious disease such as typhoid fever, and the nuisances caused by the disposal of untreated human and industrial wastes in waterways. Most important in promoting change in this century was the

B. B. Johnson and V. T. Covello (eds.), The Social and Cultural Construction of Risk, 317—344.

Sanitary Movement, which resulted in the creation of a number of local and state boards of health and statutes regulating a range of activities relating to sanitation and health. The Sanitary Movement also gave rise to two important professions — public health and sanitary engineering. These two groups agreed on the broad outlines of sanitary policy but disagreed over questions of technical design in relationship to public health, priority and standard setting, and professional competence (Tarr 1984: 226–263). These disagreements were especially marked in a controversy over water quality policy that occurred in the years from approximately 1900 to 1914.

During the late nineteenth and early twentieth centuries a number of states enacted legislation to control the pollution of waterways. Massachusetts led the way in 1886 with an Act to protect the purity of inland water and, by 1905, 36 states had statutes protecting drinking water supplies. Eight of these states, according to the U.S. Geological Survey, had "unusual and stringent" laws (Goodell 1906). The laws were badly needed: as of 1904, inland municipalities with a total population of over 20 million discharged raw sewage into neighboring lakes or streams; seacoast municipalities with a population totaling 6.5 million discharged raw sewage into harbors or tidal estuaries; while municipalities with a population totaling just over 1 million treated their sewage (Fuller 1905: 105).

Supervision of water quality, whether through advisory powers or stricter enforcement provision, was usually delegated to state boards of health. Massachusetts created the first of these bodies in 1869 and all states had them by 1909. State boards of health had a wide range of duties, including the collection of vital statistics, the supervision of food marketing and dwelling sanitation, as well as drinking water quality (Chapin 1921: 138–139). Reflecting the health-related nature of many of these responsibilities, 31 states in 1912 required that the head of the state board of health be a physician; 30 states also required that a majority of the board be physicians. Only seven states in 1912 required the presence of a civil or sanitary engineer on the board (Kerr and Moll 1912). The trend, however, especially in the larger and more urban states, was to include a sanitary engineer as a member of the board and to utilize engineering advice, even though physicians held the ultimate responsibility for decisions. The editorial pages of the leading civil and sanitary engineering journals advocated the formation of state boards of health and generally supported the extension of their powers in regard to mitigating water pollution hazards. They also vigorously demanded the inclusion of sanitary engineers on the boards (*Engineering News* 1899: 169).

During the 1890s, two critical scientific and technological breakthroughs occurred in regard to water pollution control. One was the clarification of the etiology of typhoid fever and its relationship to sewage-polluted waterways through bacteriologic research. At the beginning of the century, methods for determining the presence of coliform bacteria in water were developed as indicators for sewage pollution. The second development was the formulation of an effective means of sewage treatment called intermittent filtration. Both these advances were made by interdisciplinary teams of bacteriologists, biologists, chemists, and sanitary engineers (Rosenkrantz 1974: 103—106). The clarification of the etiology of waterborne infectious disease in regard to sewage pollution and the success of intermittent filtration in treating sewage suggested that this method of treatment offered a sound approach to protecting water supplies from pollution. In 1900 the *Engineering Record* boasted, "the resources of the sanitary engineer are sufficient to bring about the purification of sewage to any reasonable degree. This costs money . . . , but not so much as is often believed" (*Engineering Record* 1900: 73).

A second technological option for municipalities seeking to protect their water supplies from infectious disease was filtration. Techniques of slow sand and mechanical filtration were successfully applied for the first time to the water supply of large cities in the late 1890s. (Chlorination was developed in 1908.) The confirmation that these methods could reduce the risk of contracting infectious diseases from sewage pollution led many inland cities to install filters in the decades after the turn of the century (Whipple 1921).

By 1900, therefore, sanitary engineers and health officers found themselves faced by new alternatives in regard to reducing the risk incurred by sewage pollution of drinking water supplies. The options available for municipalities in obtaining a potable water supply were to both filter water and treat their sewage so as to protect both their own and the water supply of downstream cities, or to filter the water alone, leaving it to downstream users to discover a means to guarantee the safety of their water supplies. Debate over which of these options was preferable from the perspective of protecting both the public health and municipal finances caused a major division within the public health movement between physicians and sanitary engineers.

The Policy Conflict

As experience with filtration methods conclusively demonstrated that sewage-polluted waters could be treated and utilized with greatly reduced risk from infectious disease, engineering opinion shifted on the preferred means to protect drinking water. In the 1890s the effectiveness of intermittent filtration in producing a relatively "pure" effluent had convinced many engineers and health officers that sewage "purification" was feasible at a reasonable cost. By approximately 1900, however, leading sanitary engineers, impressed by the success of water filtration in reducing waterborne disease, were questioning whether upstream municipalities should bear the expense of sewage treatment just because downstream communities drew their water supplies from the stream used for disposal. The *Engineering Record* summarized this position in 1903 when it noted,

> it is often more equitable to all concerned for an upper riparian city to discharge its sewage into a stream and a lower riparian city to filter the water of the same stream for a domestic supply, than for the former city to be forced to put in sewage treatment works (*Engineering Record* 1903: 117).

Or, as sanitary engineer Allen Hazen observed in his influential 1907 book, *Clean Water and How to Get It*, when the question of water pollution was examined as a "great engineering problem," it was apparent that "it is clearly and unmistakably better to purify the water supplies taken from the rivers than to purify the sewage before it is discharged into them." Sewage, he argued, should only be treated in cases of extreme nuisance and not to protect drinking water supplies. The utilization of the natural assimilative capacity of streams for sewage disposal would save scarce municipal dollars for needs more urgent than reducing the sewage pollution nuisance (Hazen 1907: 34—37, 68—75).

As the sanitary engineering community adopted the position on water quality policy enunciated by Hazen, considerable public and professional opinion moved in a different direction. The years from approximately 1900 to 1914 are known as the Progressive Period in American history, a period when a number of reform-oriented and structural changes took place in American society and government (Wiebe 1967). A key component of the Progressive Movement was the thrust toward conservation of the nation's natural resources, public lands, and waterways, often through the utilization of scientific knowledge and expertise. Waterway improvement, particularly in regard to navigation, was a major part of the movement, and concern over the growing sewage pollution of waterways as both a health hazard and a threat to shipping was a natural corollary. Conservation and environmental protection received the support of important politicians such as President Theodore Roosevelt, citizens' groups and business associations, and influential journals of opinion and newspapers (Hays 1959: 122—146). Speaking in Buffalo in 1910, for example, ex-President Roosevelt urged the protection of the purity of the Great Lakes, and declared that "civilized people should be able to dispose of sewage in a better way than by putting it into drinking water." He called for state and national legislation to end water pollution (*The Outlook* 1910: 1).

Aside from congressional legislation in 1912 giving the Public Health Service power to conduct research into the pollution of interstate streams, legislative authority remained at the state level. During the period after the severe typhoid epidemics of 1903, a number of eastern and middle western states passed statutes giving state boards of health increased power to control sewage disposal in streams (*Engineering News* 1904: 129—130). The most activist of the boards were headed by physicians who had been influenced by the New Public Health, a movement generated by the bacteriological advances of the 1890s. The advocates of the New Public Health stressed the need to control the diseased individual, or carrier. In the case of waterborne infectious disease, the New Public Health spokesmen argued for restricting or abandoning the use of streams for sewage disposal, especially if they were also the source of a water supply. According to the Committee on the Pollution of Streams of the Conference of State and Provincial Boards of Health, both water filtration and sewage treatment were required in order to provide a double safeguard for the public health (*Engineering Record* 1909: 157—159). Saving life was a physician's highest obligation, and now science and technology had combined to provide the opportunity to greatly reduce risk in regard to drinking water supplies.

In several key cases during the years from 1905 to 1914, the most important of which involved the cities of Pittsburgh, Pennsylvania and Rochester, New York, the state boards of health attempted to compel the municipalities to cease discharging untreated sewage into neighboring streams because of high typhoid death rates. These cities, they maintained, must treat their sewage before discharge and convert combined sewer systems into separate ones (*Engineering News* 1910a: 154—155; 1910b: 179—180; 1910c: 502—503). In these cases the state health commissioners were confronted by the opposing opinions of leading sanitary engineers acting as consultants to the municipalities. Essentially, the sanitary engineers took the position that the orders of the state boards of health were both bad engineering and bad economics, and promised little in the way of health benefits. Thus the confrontation involved experts vs. experts in a situation where the question was the extent of exposure to risk the public should be exposed to in its drinking water.

In Pittsburgh, for instance, the distinguished team of sanitary engineers, Allen Hazen and George C. Whipple, maintained that there was no precedent for a great city to build sewage treatment facilities to protect the drinking water supplies of a downstream city. Such a facility, they said, would tax the city's financial resources but provide only limited benefits for downstream communities which would still have to filter their water. They recommended that rather than construct a sewage treatment plant, the city should continue to discharge its raw sewage into the adjacent rivers where it would be treated by dilution. Downstream cities should protect their water supplies by constructing water filtration plants (Gregory 1974: 25—42).

The Pennsylvania Health Commissioner, Dr. Samuel G. Dixon, called the Hazen and Whipple report an insufficient response to his original instructions requesting Pittsburgh to develop a comprehensive sewerage plan based on long-range planning with regional implications. The problem of water pollution, he argued, had to be viewed from a health rather than a nuisance abatement perspective; the immediate costs of sewage treatment would be outweighed by the long-term health benefits. The time had arrived, Dixon stated, "to start a campaign in order that the streams shall not become stinking sewers and culture beds for pathogenic organisms" (*Engineering News* 1912a: 548—552).

Professional Conflict and Public Health

In the cases of both Pittsburgh and Rochester, as well as in several other cities with similar pollution problems in this period, the position of the sanitary engineers triumphed over that of the health commissioners. That is, cities continued to dispose of raw sewage in adjacent waterways ignoring the externalities created for downstream cities, and protecting their water supplies if necessary by filtration and, after 1908, by chlorination.[1] Sewage treatment facilities were installed in cases of obvious nuisance but not to protect downstream water intakes. This approach seemingly represented a victory of those emphasizing cost-expedient and shorter-term approaches to pollution control and health over those who were concerned with a more holistic view of the environment and long-term health benefits in spite of considerably higher capital costs. The key to the victory of the sanitary engineer over the state public health officers was the reality of municipal budgetary limitations and the reluctance of municipal officials to take actions that would confer costs on their own municipalities but benefits on other cities. Given municipal cost constraints, only limited action was possible without state financial subsidies.

Underneath the positions of the two groups of "experts," however, lay important professional and value differences. Involved in the case was a clash between two professions, each trying to impose its standards and values on the field of public health. The dominant group in the field was the physicians. It was they who had been primarily responsible for founding the public health profession, they controlled its professional arm — the American Public Health Association — and they possessed legal control of all the state boards of health and most of the local boards in the country. The challenging group was the sanitary engineers, who had grown out of the nineteenth-century Sanitary Movement and who had come to play an increasingly important role in the area of public health.

Spokesmen for the sanitary engineering profession believed in the uniqueness of their profession and its special role within public health. For them it represented a combination of engineering knowledge and perspectives from sciences such as biology and bacteriology that could help solve public health problems. Sanitary engineering, they believed, was more useful for public health work than medicine because the "problem of curing disease is quite different from the problem of preventing disease." Sanitary engineers, insisted spokesmen such as George C. Whipple, had a superior conception of the "relative needs and values" of cities in regard to public health. In this context, sewage treatment was important in terms of "standards of comfort and decency" but of limited importance in regard to saving lives (Whipple 1912a: 805—806; 1912b: 899). In an editorial entitled, "A Plea for Common Sense in the State Control of Sewage Disposal," the *Engineering News* observed that public health physicians such as Dixon of Pennsylvania were "radicals," unwilling to "temper the ideal theories of the scientific laboratory with the practical experience of the engineer and the financier" (*Engineering News* 1912b: 412—413). In short, while ideally it would have been desirable to provide greater protection of drinking water supplies through sewage treatment, the realities of municipal finance limited this option and required that the citizens of downstream cities bear some risk in regard to drinking water quality.

Also at issue in this dispute was the question of what type of water quality standards should be utilized to best maximize social welfare in regard to waterway use. The position of the public health groups most imbued with the New Public Health was that it was a "very erroneous idea" that sewage disposal was a proper use of streams. They rejected the concept espoused by sanitary engineers of utilizing the assimilative power of streams for waste disposal as involving too high a risk in regard to infectious disease (Snow 1907: 266—283; *Engineering Record* 1909: 157—159). Minimizing exposure to health hazards and saving life was the highest value. In contrast to this, sanitary engineers maintained that water quality policy should maximize the assimilative capacity of waterways. "Waterways," argued the Committee on Standards of the National Association for Preventing the Pollution of Rivers and Waterways, a group dominated by sanitary engineers, could not be restored "to their original and natural conditions of purity." Waterway use for waste disposal, they argued, was not a health matter but rather an "economic question," and the "discharge of raw sewage into our streams and waterways should not be universally prohibited by law" (*Engineering News* 1912c: 835—836; *Engineering Record* 1912: 485—486).

In actual fact, both groups of experts had taken an "economic" approach but they differed in regard to the value they had placed on human life. For the public health physicians, the risk involved in using rivers for both sewage disposal and as sources of drinking water supplies bore too high a social cost. Hence, their orders to municipalities to remedy the situation. The consulting engineers, in contrast, represented the narrower and more limited interests of cities forced to operate under state imposed financial restrictions. Protecting human lives within these limitations was their objective. Thus each group of experts represented the particular interests of their clients or constituencies but the critical basis for the separation between their positions lay in the value differences imparted by their professional training.

II. Industrial Wastes and Public Health

Within the last two generations, questions about the health risks involved in the disposal of various industrial wastes in the environment have provoked heated debate within the public health community. In historical terms, this concern with industrial pollution as opposed to human wastes is a relatively recent phenomenon. Its beginnings actually date back to the years just after World War II. In contrast, most of the period from the late-nineteenth century through the 1940s, after initial interest, was characterized by a relative lack of attention toward industrial pollution except for a small group of interdisciplinary investigators. This section of our essay will explore the factors responsible for the low priority accorded to the health effects of industrial wastes in the pre-World War II decades and the shift of attitudes that occurred after World War II. The major themes that will be explored include the rationale for priority setting in regard to human and industrial pollution; the relationship between industrial wastes and water supplies; the development of indicators for the presence and measurement of organic industrial wastes; and, the changing state of the art in regard to analytical instrumentation.

Priority Setting: Sanitary and Industrial Wastes

During the late-nineteenth century urbanization and industrialization combined to produce water pollution problems in the United States of a scope not previously experienced. One type of water pollution, as has been discussed, was caused by human wastes, while a second type of pollution was a product of industrial processes, with the effluent discharged directly into streams or into sewers. Before the late 1880s and the 1890s, because of competing medical theories, restricted understanding of the etiology of waterborne disease, and methodological limits in determining hazard, there were no sharp priorities in dealing with either type of waste. Various investigations of river and water supply pollution undertaken by municipal and state authorities in the decades following the Civil War, for instance, often found industrial wastes to be as important, if not more important, than human wastes in producing foul and health-threatening conditions (Kirkwood 1876). By the turn of the century, however, both the relative components of river pollution and the methodology of water analysis had sharply altered. Numerous towns and cities had constructed sewerage systems and were discharging millions of gallons per day of untreated sewage into adjacent rivers and lakes. As noted in the previous case, the resulting water pollution caused large increases in morbidity and mortality from infectious waterborne diseases (such as typhoid) in downstream cities that drew their drinking water supplies from sewage-contaminated sources.

The epidemics resulting from sewage pollution of water supplies resulted in an intensive research effort directed toward understanding and coping with the problem. Building on critical progress in bacteriology, advances were made in providing uniform procedures for physical, chemical, microscopic, and bacteriologic methods of water examination (Prescott and Winslow 1931). In 1914, responding to these advances, the Public Health Service issued the first set of standards applying to interstate commerce (Borchardt and Walton 1971). These standards initially applied only to bacterial contamination. Thus, by the beginning of World War I, the public health and sanitary engineering community had developed a series of indicators, methods and standards, as well as filtration and treatment technologies, to help reduce risk from sewage contamination of drinking water supplies.

The priority placed by health officers and sanitary engineers on the dangers posed by sewage pollution reduced the amount of attention devoted to industrial wastes. The rationale for this prioritization was that, "from a purely pathogenic standpoint," industrial wastes were viewed as having a "remote" relation to sanitation (Leighton 1905: 29—41). Bacterial research had shown that sewage pollution could lead to epidemics of diseases with acute health effects, but the health effects of industrial wastes were generally not explored and poorly understood. The ordinary waterworks laboratory concerned itself mostly with the biological phases of the supply and certain routine chemical tests useful in controlling operation (Donaldson 1922: 420—421). While chemists and sanitary engineers in water supply laboratories often conducted analyses of a variety of chemicals dissolved in water, they were primarily attempting to determine the water's qualities for industrial use. Considerably more attention, for instance, was paid to the clearly observable effects of metals in water supplies on steam boilers than on human health (Babbitt and Doland 1931: 521). It was not until 1925 that maximum permissible concentrations were established in the Public Health Service Drinking Water Standards for lead, copper or zinc because of a failure of the advisory commission to agree on specific limits (Borchardt and Walton 1971).

The relative lack of attention paid by water quality experts to industrial wastes was reflected in the regulatory domain. The stringency of regulations varied from state to state, but because most states recognized stream control as a public health function, action was limited in regard to, as one authority noted, "serious pollution capable of working great economic injury, although of little or no public health significance" (Phelps 1921).

While industrial wastes often severely damaged stream life, causing fish kills and destroying vegetation, state departments of health seldom acted against the polluters because they did not consider aquatic life to be within their realm of responsibility. Some states, such as New York or Wisconsin, formed state conservation commissions in the 1920s to try to control the effects of industrial wastes on wild life but with limited success. Connecticut, Ohio, and Pennsylvania created boards with special responsibilities for industrial wastes whose purpose was to secure cooperation from industry in control of their wastes. Investigation and the supplying of technical advice rather than strict regulation was seen as their function, and their activities were only indirectly health related (Stevenson 1924: 201—216; Besselievre 1924: 217—230).

There were a number of times in the 1920s and 1930s when industrial pollution interfered with the potability of public water supplies, but these occurrences were not viewed as involving direct health effects.[2] Rather, the wastes were objected to because they had increased the difficulty and expense of water treatment processes such as coagulation and filtration and caused problems of color, turbidity and taste and odor. The addition of specific toxic chemicals, such as cyanide, was also of concern, but they received considerably less attention than did the former factors. Among wastes causing problems were those from sugar refining, coal mining and washing, gas and byproduct coke works, wood distillation, corn products, dye and munitions manufacture, oil-producing wells and refining, metallurgical processes and mining, textiles, tanneries, and paper and pulp mills (American Water Works Association 1923: 415—430). In the case of several of these industrial pollutants, such as mine acid drainage, tannery effluents, and metallurgical wastes, some health officers and sanitary engineers argued that they should not be excluded from streams because they had a germicidal effect on bacterial wastes. Only when sewage treatment facilities were constructed for the major sources of sanitary wastes could they be controlled (Pennsylvania State Board of Health 1909: 9—30: Frost 1924: 159).

Indicators of Industrial Pollution

An important obstacle to controlling and regulating industrial pollution as compared to human wastes was the existence of a variety of different types of effluents, variations in the types of treatment required, and uncertainty in regard to effects. Without generalizeable indicators comparable to those that existed for bacterial wastes, these characteristics made the effective control and regulation of industrial wastes extremely difficult. Indicators of the oxygen-consuming characteristics of organic industrial wastes were especially critical in the context of the total assimilative capacity of the stream and its ability to assimilate human wastes. This concern dictated that industrial waste control would take place within the framework of the bacterial paradigm.

The first major advance in regard to the development of indicators for organic industrial wastes was the famous Streeter-Phelps "oxygen-sag" curve, by which the self-purification characteristics of streams were understood in terms of the measurable phenomenon of dissolved oxygen (DO) and the biochemical oxygen demand (BOD) characteristics of various wastes under certain conditions of time and temperature (Streeter and Phelps 1925). Earle B. Phelps, the noted chemist and bacteriologist who shared in the development of this method, believed that industrial wastes constituted "an indirect menace to the public health in so far as they may draw upon the stream's natural purifying power, thereby delaying or preventing the ultimate disposal of directly infectious matter" (Phelps 1918). They thus constituted a risk enhancing factor in regard to the hazard of infectious waterborne disease.

The use of the Streeter-Phelps curve provided a quantitative model that circumvented the heterogeneous character of industrial wastes by agreement on a common characteristic that furnished a basis for pollution control. As the authors of a key 1924 Public Health Service study of the Ohio River noted, this was the "only procedure which permits any quantitative comparison, however imperfect, between industrial wastes and other sources of pollution in such a broad area as the Ohio watershed" (Frost 1924a/b: 80—84). Following the model's development, a standard approach utilized by many state health departments was established for estimating the comparative effects of the pollution of streams by industrial and human wastes. These procedures involved the determination of dissolved oxygen content through stream surveys, the calculation of the stream's assimilative capacity, and the requirement that municipalities and polluting industries reduce the BOD load of their wastes (Phelps 1944). Because the basic concern of the investigators was to measure the total assimilative capacity of the stream, they described industrial wastes in terms of population equivalents — e.g., some unit of the industrial waste was equivalent in its BOD characteristics to so many persons (Veatch 1927: 58—63). In some cases, DO and BOD studies would be combined with bacterial and biological surveys to form a measure of stream sanitation and the adequacy of dilution. In addition, chemical analysis in regard to factors such as suspended solids, hardness, acidity and alkalinity, and the presence of chlorine and various nitrogen compounds were also often performed, depending on the purpose of the survey (Phelps 1944).

The development of these pollution indicators marked an important advance in regard to stream sanitation, especially in regard to biological conditions and aquatic life. DO and BOD were also related to public health as indicators of the assimilative power of streams in regard to organic wastes. But, while representing an important step beyond the bacterial paradigm, they supplied no direct information concerning the presence or danger of toxic metals, and toxic organic and inorganic complexes (Donaldson 1930). (In some cases they served as surrogate indicators.) For many of these substances neither qualitative nor quantitative methods of analysis were known, and information was lacking on their effects on both aquatic and human life. Complicating the situation through this period was a lack of information about treatment requirements and procedures for a number of industrial wastes (Rudolfs 1931: 41—47; Fair *et al.* 1954: 867—868).

New Products, New Methodologies, and New Concerns

Developments during World War II and the years immediately after produced important changes in the field of industrial waste disposal. One important contributing factor was the large expansion by traditional industries (metals, coal, food, petroleum, etc.) without concommitant attention to pollution control because of wartime restrictions and a lack of legislation and enforcement. The second factor was the development and production of a range of new substances with large polluting potential. In retrospect, chemical products such as the chlorinated hydrocarbons (DDT, chlordane, benzene hexachloride, endrin, dieldrin, aldrin, etc.) and synthetic detergents (alkyl benzene sulfonate) were critically important in regard to stream quality, although other new products such as rayon and artificial rubber had more immediate effects. As one leading sanitary engineer wrote in 1945, the increase in both old and new industrial pollutants meant that

> after many years of indifferent or complete lack of attention, followed by slowly awakening interest in this important problem, it appears that in the postwar period the disposal of industrial waste is to be the leading topic (Symons 1945: 558—572).

A focus on industrial wastes implied an increasing perception of the risks to the public health and the environment involved in their disposal in waterways. In the 1940s and 1950s, knowledge of these possible risks was held almost entirely by a few experts from the various scientific disciplines concerned with waste disposal, water quality, and public health. But even among these professionals, there were large differences of opinion about the extent of risk involved with various industrial pollutants and what precautions were necessary to protect the public health (Faber and Bryson 1960; Dunlap 1981). As far as the public was concerned, it was largely unaware of the potential risks involved in industrial pollution.

A critical factor arousing the various expert communities to the potential hazards of industrial pollution was the development and application of new and improved methods of water pollution sampling and analysis.

These changes were related to advances in the fields of analytical chemistry and zoology. From approximately the turn of the century to 1940, few improvements occurred in the instrumental techniques available to organic chemists aside from advances in polarography and column chromatography. In the 1930s, however, an "Instrumental Revolution" began that accelerated after the war (Skolnik and Reese 1976). Most important for organic pollutant study were advances in organic sampling and analysis. Sampling methodology was vastly improved by the development in 1950 of the carbon absorption apparatus. Mass spectrometry, although developed earlier, was first used in reference to water pollution in 1953. Infrared spectrometry, ultraviolet spectrometry, chromatography, and gas chromatography were all applied in the 1950s and 1960s, facilitating the identification of various pollutants (Rosen 1976: 4—14; Clarke 1968: 26—34). In addition, important advances (one authority called them a "revolution") occurred in the development of bioassay techniques used to assess toxic effects (Cairns 1966: 559—567). The result of these developments was to stimulate interest in nonbiological substances in water supplies and to raise a variety of questions about the relationship between both new and newly identified pollutants in water supplies, the public health and the environment.

During the decades after 1960, concern over the environmental and human effects of industrial pollutants accelerated, as did research and legislation in the area. Investigations of water pollution continued to utilize the traditional parameters of DO, BOD, suspended solids, and acidity, but very gradually water pollution investigators began to add new parameters and to monitor for various trace elements and toxics that were either new or could not have been identified in the past in very small quantities. In the 1940s and 1950s, for instance, colorimetric techniques made it possible to measure 10 parts per million (ppm) of DDT. By the late 1950s and 1960s, paper chromatrography permitted identification of 1 ppm, while in the 1970s gas chromatography permitted identification of a "few" parts per billion. Similar detection improvements occurred for a number of other organic and nonorganic compounds (Council on Environmental Quality 1979: 238).

The development of powerful new analytical instrumentation and the growth of the chemical industry coincided with a decline in the significance of infectious disease as a major factor in the health of Americans and a rise in the incidence of chronic diseases such as cancer. Increasingly, environmentalists and some scientists began to see causal links between the growth of the chronic and degenerative diseases and the development of the "chemical society" (Eckholm 1977). The ability of the new instrumentation to identify trace elements of toxic substances in water supply played an important role in the development of this hypothesis. Thus, an environmental health paradigm began to replace the older bacterial paradigm, and in a classic reversal, concern over the risks from industrial pollution supplanted those from human wastes for both the experts and the public in the 1970s and 1980s.

III. Alternative Perceptions of Risk at Love Canal

In the mid-1970s, the problem of the land disposal of hazardous chemical wastes attracted national attention because of the contamination of a residential neighborhood at the Love Canal site in Niagrara Falls, New York. This case is of particular importance in the context of this essay because it is a contemporary example of an environmental-public health controversy in which expert disagreement has played an important role in defining the issues and determining the course of events. Unlike the cases discussed previously, expert disagreement at Love Canal took place in the context of a highly politicized atmosphere and in the presence of a public highly mobilized around the issue. Those who believed themselves immediately affected by the hazard at Love Canal actively attempted to influence policy. All sides in the controversy attempted to use scientific findings and the work of experts to define and justify their positions.

The Problems of Chemical Waste

As was noted in the previous case, industrial pollution increased in volume and altered in character during and after World War II as a result of greater production and new product development. In the chemical industry, the development of synthetic organic chemicals, the production of which increased more than tenfold from 1945 to 1985, presented special disposal problems. Although useful for a wide variety of purposes, many of these chemicals are highly toxic, carcinogenic and persistent in the environment. Improper disposal of such substances (and more traditional poisons as well) poses a threat to the environment and to public health, the exact dimensions of which have yet to be completely determined. The previous case has discussed the lack of attention paid by public health officers to the health risks of water disposal of potentially hazardous industrial wastes until well after 1945. This neglect has been even more striking in regard to the land disposal of hazardous wastes. Even during the 1960s, a time of growing environmental concern, land disposal of hazardous wastes attracted little attention. Reasons why the problem was ignored during this period include the absence of a clear hazard or crisis in regard to landfills, ignorance of possible groundwater pollution, and the lack of analytical instrumentation necessary to trace certain contaminants from landfills or to detect extremely low levels of potentially hazardous substances (Tarr 1984: 25—26).

A national mandate for the regulation of the land disposal of hazardous wastes only came in 1976 with the passage of the Resource Conservation and Recovery Act (RCRA). RCRA was designed to close the gap in environmental law left by the Clean Air and Water Pollution Control acts of the early 1970s. Unfortunately, the framework of regulation mandated by the act itself contained an important gap. As propounded in 1976, RCRA lacked provisions to deal with the so called "orphan" dump sites or abandoned hazardous waste sites (Quarles 1982: 34).

The nationally publicized discovery that a residential neighborhood at Love Canal in Niagara Falls, New York, had been contaminated by a long abandoned toxic waste dump dramatically demonstrated the importance of this defect. The need for reform revealed by Love Canal as well as other hazardous waste sites inspired the passage of a federal "Superfund" bill in 1980 that was designed to fill some of the gaps in hazardous waste law.

The Creation of Love Canal and its Emergence onto the Political Agenda

The Love Canal is a rectangular 16 acre landfill located about a mile from the industrial area of the city of Niagara Falls. The site is straddled by a public elementary school (closed in 1978) that separates its northern and southern portions, and is bordered on two sides by private houses. The area began to be used as a disposal site for waste chemicals as early as the 1920s. In 1942, the Hooker Chemical Company negotiated an agreement with the Niagara County Irrigation and Water Supply Development Company allowing the company to dispose of its wastes in the canal; in 1947, the company bought the land outright for use as a dump.

By 1953, Hooker had completely filled the canal with 43.6 million pounds of 82 different chemicals either buried in 55 gallon drums or drained directly into the excavation. Many of the materials dumped were toxic and some are known or suspected to be cancer causing (New York Department of Health 1978: 12). After the canal was filled, Hooker covered the excavation with a clay cap and, in April of 1953 donated the land to the Niagara Falls Board of Education (at a token price of $1.00) for use as the site of an elementary school. During the 1950s and 1960s, the area near the dump site grew rapidly as a residential suburb. Streets and utility lines were cut through the landfill breaking the clay seal and allowing the chemicals to escape. Builders used soil from the site as fill for homes. No public issue developed during this period, however, even though exposed chemicals from the waste dump caused occasional physical injuries to workers and nearby residents (Levine 1981: 14, 15).

Love Canal emerged onto the political agenda during the mid-1970s. In part, this was the result of the development of environmental issues as important local, state and national questions. While growing production increased the actual amount and intensity of many forms of pollution, the improvements in the standard of living that resulted from this production allowed people "the comparative luxury" of concern over environmental problems (Davies 1970: 22; cf. Hays 1981: 719—720). Growing public concern over environmental degradation led to increased governmental efforts to study and control pollution problems. In turn, these efforts attracted additional public attention to environmental problems, leading to further demands for change. (Davies 1970: 23). Public concern was increasingly reflected in the growth of networks of environmentally concerned private organizations, academic researchers, and government agencies that possessed the ability and expertise to ferret out and publicize environmental problems.

Scientific Investigation, Administrative Inadequacies, and the Definition of Risk

Chemical pollution at Love Canal became a political issue in the fall of 1976 when the *Niagara Gazette* published a front page story describing the history of the Love Canal site and the complaints by residents regarding chemical wastes seeping into their backyards and basements. Analysis of the wastes from the basement of one home revealed that the chemicals came from the waste dump and contained 15 organic chemicals, three of which are known to be toxic (Pollak and Russell 1976: 1). Scientific investigation was to play an essential role in defining the issues at Love Canal. If the chemicals from the canal could be conclusively shown to be causing substantial health damage or other major risks to neighborhood residents, then the government's responsibility to provide protection from the risks (as well as the potential liability of Hooker for creating them) would be consensually accepted by most of the public.

The responsibility of government to protect citizens against threats to public health posed by garbage, sewage, and contaminated water supplies has been generally accepted as a municipal or state responsibility since the beginning of the twentieth century. As historian Barbara Rosenkrantz defines it:

> There has been a consensus in this century that the state is responsible for controlling disease, grounded on the normative view that public and personal health are, and of right ought to be, the birthright of Americans, and that whatever or whoever violates this trust is external and repugnant (Rosenkrantz 1972: 2).

In the specific case of Love Canal, New York Commissioner of Health, Dr. David Axelrod, noted that "the community looks to government to protect it from hazards of this kind, just as it protects its citizens from crime, and from other disasters, such as floods and tornadoes (*Subcommittee on Oversight and Investigations* 1979: 289).

But despite this concensus as to governmental responsibility for public health, administrative structures at the local, state and federal levels proved unable to meet the demands imposed upon them by the emergence of Love Canal as a public issue. At the local level officials avoided the problems posed by Love Canal both because of the ownership of portions of the dump site by the city of Niagara Falls and the Niagara Falls board of education, and the importance of the chemical industry for the area's economy. Even if local officials had wanted to deal with the situation more seriously, they possessed neither the expertise nor the financial resources required to cope with a chemical dump of the magnitude of Love Canal.

Local officials thus faced a dilemma. They were not able nor perhaps willing to remedy the Love Canal situation, yet they were entrusted with a consensually accepted responsibility to protect the public health. Their initial response was to belittle the risks posed by Love Canal to nearby residents. The first official action taken by the Niagara County Health Department in response to the news that chemicals were seeping into the basements of Love Canal houses, for instance, was to threaten to fine residents for pumping the foreign chemicals into the city's sewage system. The Niagara Falls school board also refused to close the 99th Street Elementary School located on top of the waste dump until forced to do so by the state or to fence the fields covering the canal. Local, city, and county officials rationalized this course by arguing that there was no clear evidence of risk (Pollak 1977: b1). Clearly the necessary evidence could only be obtained through systematic investigation.

At the state and federal levels, the situation was more complex. As in the case of the local government, state and federal agencies had little experience in addressing the specific problems posed by chemical waste dumps. They too faced bureaucratic and financial constraints that would eventually affect (and distort) their response to Love Canal in important ways. Nevertheless, political and institutional willingness and ability to confront environmental issues at these levels of government was greater than that of local government in Niagara Falls. The New York Department of Health possessed far greater resources than the local agencies in Niagara Falls and, in addition, had an institutional history of awareness of public health implications of environmental problems. Both New York State and the federal government had departments specifically responsible for detecting and addressing environmental problems. These agencies initially responded to Love Canal by defining it in the context of the type of pollution problem they most commonly encountered — contamination of surface waters. Thus, during the late fall of 1976 and early winter of 1977, the first governmental investigation of chemical waste dumps in Niagara Falls was undertaken by the New York Department of Environmental Conservation as part of an effort to determine the source of pesticide residues in Lake Ontario.

State and federal agencies first began efforts to assess the risks and dangers posed by the chemicals in Love Canal to nearby residents during the spring and summer of 1978 (Subcommittee on Oversight and Investigations 1979: 288). To determine the extent of the contamination at the canal site, the state departments of Health and Environmental Conservation as well as the federal Environmental Protection Agency (EPA) launched an intensified program of soil, air, and groundwater sampling. In June, state investigators began a house to house health survey of residents living in the first ring of homes, closest to the canal. The health investigation concentrated on determining the numbers of miscarriages and birth defects among residents because investigators considered these prime indicators of human toxicity (New York Department of Health 1978: 12).

The results of these studies seemed to confirm that a danger existed, at least to those residents who lived closest to the dumpsite. The state report, *Love Canal: Public Health Time Bomb*, released in August, found extensive chemical contamination of homes near the waste dump and elevated rates of miscarriages and birth defects among residents (New York Department of Health 1978: 6). Well publicized by the New York Department of Health, these findings constituted the scientific basis for state and federal agencies to publicly define the situation at Love Canal as unacceptably hazardous to nearby residents.

On August 2, 1978, New York State Health Commissioner Robert Whalen declared that a state of emergency existed at Love Canal. The emergency declaration ordered the closure of the 99th Street Elementary School located on top of the dump site and called for the evacuation of pregnant women and children under two years of age living in close proximity to the canal. In order to prevent further leaching of chemicals from the dump, work was ordered to begin on a multi-million dollar remedial construction project. In addition, the Department of Health promised to conduct epidemiological studies in order to better "delineate chronic diseases afflicting all residents who lived adjacent to the Love Canal Landfills site" (New York Department of Health 1978: 6).

On August 7, President Carter declared that a federal state of emergency existed at the Love Canal, making the state eligible to apply for Federal Disaster Administration funds for relief work. Expecting federal reimbursement, Governor Carey (under political pressure in the Democratic primary) announced on August 9 that the state would pay for the permanent relocation of all 236 families who lived in the two rings closest to the canal and would buy their houses. By August 29, 98 Love Canal families had been evacuated from the neighborhood and another 46 had found temporary housing and were ready to move. Work on the remedial construction project began on October 10.

With these dramatic and widely publicized actions, state and federal agencies defined Love Canal as a major public health hazard in the spring and summer of 1978. Over the next two years, however, administrative and funding mechanisms proved inadequate to address the continuing demands imposed upon them by this definition of the situation. For example, Federal Disaster Assistance Administration requirements that disaster relief measures be limited to short-term needs proved inappropriate to the situation at Love Canal because of the long-term nature of the health threat. On January 29, 1979, the agency announced that New York state's request for reimbursement for the money spent removing inner ring families from the Love Canal neighborhood and buying their homes had been refused. This decision meant that federal money would probably not be forthcoming for further evacuation of Love Canal residents even if additional health problems caused by the toxic wastes were discovered.

Conflict Over Risk Definition and Expert as Advocate: Love Canal 1978—1980

Conflict over risk definition at Love Canal resembled in some respects that between sanitary engineers and public health physicians described in the first section of this paper. In both conflicts, experts played important roles in the formulation and defense of particular policy positions. They legitimated this role on the grounds of their expertise and knowledge of specific scientific findings. As in the conflict over water quality policy, some experts at Love Canal took the existing availability of government resources as given, and defined the nature of the risk and formulated policy on this basis, while other experts defined risk in broader and more absolute terms and argued that the allocation of government resources should itself be determined by this definition. Like the sanitary engineers, professionals working for the New York health department took the first position. They adjusted their definition of the hazards posed by Love Canal according to the availability of government resources. Thus, during the summer of 1978 when substantial federal aid was believed to be available, New York officials emphasized the importance of miscarriage and birth defect data as indicators of danger to residents. Such data constituted an important part of the rationale for permanent relocation of residents who lived adjacent to the canal. As the possibility of substantial federal assistance receded, officials increasingly downplayed the role of miscarriage and birth defect data as indicators of potential danger. Instead they now argued that because these effects constituted the only conclusively demonstrated health damage from the waste dump, permanent relocation of additional families from the neighborhood was not justified. The risk itself and the criterion used to discover it had become conflated (Levine 1981: 61).

Sanitary engineers succeeded in their effort to define the problem of water pollution solely in terms of protection of drinking water supplies over the opposition of public health physicians who defined the danger in broader terms. The New York health department, on the other hand, tried but failed to narrow the definition of risk at Love Canal. In large part, the reason for this failure lies in changes in the political context. In the 1970s, unlike the previous cases discussed in this paper that occurred in periods of less environmental political activism, opposition to the imposition of pollution risks took place at the grassroots level. Love Canal residents who lived outside the evacuated area rejected the state's efforts to redefine their situation as less hazardous than that of the residents who had been evacuated. They feared the spread of the chemical contamination to their homes and doubted the efficacy of the remedial construction designed to contain the wastes. Organized over the summer of 1978 as the Love Canal Homeowners' Association, the remaining residents demanded that the state pay for their relocation.

The Love Canal Homeowners' Association sought primarily to attract sympathetic attention from the media and the public. Its members realized that the appearance of governmental hard-heartedness and irresponsibility in the face of widely and credibly reported health hazards could prove politically costly to elected officials and thus stimulate them to positive action. Demonstrations, picketing, appearances on television shows and well-publicized meetings with public officials were all used as part of the effort to maintain public attention. Lois Gibbs, the head of the association, observed that:

> We had to keep the media's interest. That was the only way we got anything done. They forced New York State to answer questions. They kept Love Canal in the public consciousness. They educated the public about toxic chemical wastes (Gibbs 1982: 96).

For this strategy to be convincing, residents had to make a plausible case that the chemicals actually endangered their health. The primary focus of debate between the home owners' association and state officials, therefore, was on the extent of hazard facing the residents rather than the government's responsibility to act. Since the amount of risk could only be determined through research, arguments concentrated on research methodology and interpretation of findings. In actual fact, however, the disagreement really concerned a policy issue: should additional residents be evacuated from the neighborhood at government expense?

Findings of health damage, even if of uncertain validity, or additional chemical contamination, served the residents' cause in two ways: they supported the logic of the residents' case for relocation and provided news items that attracted media attention to their plight, thereby helping to keep the issue before the public. The Love Canal Homeowners' Association used the research findings of Dr. Beverly Paigen, a cancer research biologist working for the Roswell Park Cancer Institute in Buffalo, to support its demand that the state pay for the relocation of additional Love Canal residents. Dr. Paigen had conducted a telephone survey in September 1978 at the urging of the association designed to investigate unusual clusters of illness that appeared to exist in certain areas of the Love Canal neighborhood outside of the evacuated area. Her study found large differences in the incidence of birth defects, nervous system disorders, urinary diseases and respiratory problems between residents of homes built atop swales (filled streams and drainage ditches) and residents of homes built on dryer ground (Subcommittee on Oversight and Investigations 1979: 15).

The leadership of the homeowners' association attempted to publicize widely these and other research findings in order to lend credibility to their demand for government supported relocation. State officials, on the other hand, attempted to discredit Paigen's work and disclaimed responsibility for such dangers that they did admit existed. The head of the state research effort at Love Canal, Dr. Nicholas Vienna, called Paigen's findings "totally, absolutely, and emphatically incorrect". On the basis of its own research effort, the state claimed that the only substantiated illnesses found in remaining residents were higher than normal miscarriage rates and numbers of children born with low birth rates and birth defects. No evidence was found in their data, said the state officials, to support Paigen's findings of increased rates of disease among neighborhood adults (Levine 1981: 100). On the basis of these findings, the New York Department of Health maintained that the relocation of additional residents was unnecessary excepting families with pregnant women and children under two years old.

Despite the massive nature of the state research effort and the lack of conclusive evidence for Paigen's findings, residents won the battle to define the hazards posed by Love Canal in the arena of public perception. The health department's refusal to reveal its research methodology and to make available for independent review major portions of the data and analysis used in its research was one reason for this outcome (Subcommittee on Oversight and Investigations 1979: 16—20). A second critical reason is that disputes over scientific findings took place in a context of extensive national media coverage. The *New York Times*, for example, published 36 articles on Love Canal over the course of 1979; articles also appeared in national periodicals such as *Time, Newsweek, Business Week*, and even *Redbook*. Portions of two 1979 television documentaries dealt with Love Canal.

Media coverage naturally emphasized the dangers posed by the chemicals, the hardships undergone by the residents, and the drama and conflict of the case. The state health department's attempts to narrow the definition of risk posed by the canal (and to set limits on its responsibility for addressing those risks that it did admit existed) appeared callous and irresponsible in the face of these portrayals and the doubts cast on the credibility of the department's research findings. During the third week of August, Commissioner Axelrod announced that the state would not provide for even the temporary relocation of families contemplating pregnancies because they were now aware of the risks of miscarriage and birth defects and could decide for themselves whether or not to accept them. Axelrod defended this position as follows:

> But under what authority do we evacuate everyone? Do we evacuate all of Harlem because it has an infant mortality rate 4 times the state average? (McNeil 1979)

This position proved politically untenable. On November 1, 1979, the State Legislature appropriated $5 million dollars for use at the Love Canal. The bill's passage was a symbolic victory for Love Canal residents because it provided, in theory, for the state to purchase the home of any Love Canal resident who wanted to leave. The bill, however, appropriated less than a quarter of the necessary funds. In addition, the funds were supposed to be used to "stabilize and revitalize" the neighborhood as well. By June of 1980, state and local Niagara Falls officials had still not determined a method to administer the relocation and revitalization effort mandated by the bill. No government agency wanted to face a law suit resulting from people who had suffered adverse health effects by settling into an evacuated area of the neighborhood. By this point, however, the state's dominant policy-making role at Love Canal had been superseded.

Science, Publicity and Policy: Federal Policy-Making at Love Canal

The final crisis which brought about massive federal intervention into policy-making at Love Canal encapsulates the mixture of "science", publicity and politics that, by this point, had come to dominate events. On May 17, 1980, the announcement of the results of a small and not very conclusive pilot study of chromosome damage among Love Canal residents sparked a crisis that led to a federal order to evacuate the neighborhood. According to Dr. Dante Picciano, the cytogeneticist who performed the study for the EPA, 12 of the 36 Love Canal residents examined showed signs of genetic damage. Picciano pointed out, however (and critics of his study scathingly confirmed) that the inferences that could be drawn from the work were very limited because of the absence of a contemporary control group. (The EPA had denied him funding for such a group.) The study was neither intended nor designed to provide a basis for public policy. Nevertheless, because of the charged atmosphere surrounding events at Love Canal and the absence of more conclusive scientific findings, it had this effect.

The EPA tried to limit the impact of Picciano's study by declaring that no decision would be made concerning further evacuation of the neighborhood until his findings were reviewed by a panel of independent cytogeneticists. Love Canal residents, however, demanded immediate action, and on May 19 a crowd of angry residents held two visiting EPA officials hostage for five hours. Both the results of the study and the actions by Love Canal residents received extensive national coverage. On Wednesday, May 21, two days before the expert review of the Picciano report was scheduled to be completed, Barbara Blum of the EPA announced that President Carter had declared a second state of emergency at Love Canal that would permit the temporary relocation of the approximately 700 remaining families. Officials held that a decision on permanent relocation was contingent upon the results of additional health and environmental studies by federal agencies.

The permanent relocation of Love Canal residents, however, did not wait for the completion of these studies. Over the summer, the U.S. Congress approved an emergency appropriation allowing the President to spend $20 million for the relocation. Spurred by Congressman LaFalce and New York Senators Javits and Moynihan, a complex financial arrangement was worked out between the State Interagency Task Force on Love Canal and the Federal Emergency Management Agency in late August of 1980 in which the federal government agreed to loan and grant $15 million to the state of New York for the purchase of Love Canal homes. The final agreement was signed by President Carter during a campaign swing through Niagara Falls on October 2, 1980. Most Love Canal residents accepted the state offers for their houses and left the area. Although it is still official state policy to "revitalize" the neighborhood by resettling people in sections more distant from the canal, the safety of such a move remains in doubt and most of the houses are empty and derelict.

Scientists, engineers, and physicians played an important role in actually making policy at Love Canal as well as in advocating particular (and sometimes conflicting) policy alternatives. They legitimated this role on the grounds that their specialized knowledge and skills in scientific investigation were needed both for determining the nature and severity of the environmental hazard posed by the waste dump and for discovering the alternatives available to address it. Despite the seemingly technical nature of these tasks, the role of experts at Love Canal was clearly circumscribed by the political context. The waste dump itself surfaced as a policy issue, not only because of physical signs that a risk existed, but because of changes in the political context that resulted in these signs being taken seriously by the public and by policy makers. Expert disagreements on technical questions over correct interpretation of research findings at Love Canal were closely linked with disagreements over values and policy concerning the level of risk that residents of the community should be made to bear.

Love Canal appeared on the political agenda at a time of increased public concern over environmental issues and of growing sophistication by a wide variety of actors and organizations in the tactics needed to attract media attention to grassroots grievances (Dunlap 1981: 109). Members of the public directly affected by the hazard at Love Canal actively tried to influence policy both by obtaining access to experts who could legitimate a different definition of the problem than that formulated by government agencies and by using the media to publicize their plight. Their efforts succeeded despite a lack of conclusive evidence for findings of severe health damage caused by the waste dump. In the face of widely reported revelations of possible danger to residents trapped in the neighborhood (whatever their validity), efforts by officials to limit expenditures to levels they regarded as reasonable became politically untenable. In effect, administrators and "in-house" experts at Love Canal lost control of the process of risk definition. They therefore lost much of their ability to control the course of events and determine policy as well.

Conclusion

While these three case studies of environmental risk have dealt with separate problems at different points in history, they share certain characteristics. In all the cases, but most strikingly in the first case, disciplinary perspectives and conflicts affected public policy choices and outcomes. Clearly different types of professional training have involved alternative value systems and these values have resulted in different estimates of risk. Thus, "experts" with training as physicians have often taken one position on environmental hazard while another set of "experts," from the profession of engineering, have taken a different position. While it is possible to argue that client relationships and organizational affiliations played an important role, professional values imparted through training seem to be the critical variable.

A second theme illustrated in the cases involves the importance of indicators as a basis for the estimation of risk and the formation of public policy. Without indicators there was a limited basis for generating public policy, but the existence of the indicators themselves reflected a problem definition that defined the scope of the attack on the problem. Thus the efforts of engineers and scientists to deal with the problems of epidemic disease caused by sewage pollution of water supplies resulted in the development of indicators for coliform bacteria as a way to identify hazard in regard to drinking water supplies. The focus on the bacterial paradigm, however, limited the amount of attention devoted to industrial wastes and delayed investigation into their possible health effects. Essentially, it was not until the risk of infectious waterborne disease had been controlled that attention began to focus on industrial pollution. Here the production of new chemical products with highly toxic characteristics as well as the development of means of analysis for detecting trace elements of toxic industrial wastes were critical factors driving the change.

A third important theme illustrated in these cases concerns the roles of government, politicians, and the public. Government was a participant in all three cases as an actor or as a sponsor, but only in the last case, that of Love Canal, was the public an important actor. In the first two cases, the decision-making process was dominated primarily by the experts and by policy makers but in the third case the public, believing itself exposed to unacceptable risk, became an important participant. Clearly the different values held by the public in the 1970s over exposure to environmental risk compared with the earlier part of the century played an important role in their response. And here the changing nature of American society in recent decades — its heightened sense of environmental dangers, the tendency of groups who see themselves as wronged to use direct action rather than the established political process as a means to secure redress, and the distrust of "experts" — all came into play.

Finally, these cases have raised questions about how society sets its priorities and agenda in regard to dealing with environmental risk. Clearly, in the pre-World War II period, the environmental agenda was set by a narrower group of professionals or experts, economic interest groups that were directly affected, and a small group of politicians. On the whole, the public was only involved on a limited and local basis. The basic value shift that has generated greater public environmental concern in the last two decades has also produced a heightened sense of risk in regard to environmental health hazards. Thus the public and the media have helped set the recent agenda in a way that was not true in the past. Simultaneously, however, disputes among "experts" as to the extent of actual risk posed by various potentially hazardous situations have also amplified, causing confusion in regard to the correct policy "paths" to be taken. Thus, in the environmental domain, although knowledge has increased exponentially, certainty in terms of risk appraisal remains elusive.

Acknowledgements

The first two cases presented in this chapter, I "Professional Cultures in Conflict" and, II "Industrial Wastes and Public Health", were originally published by the senior author in somewhat longer form in the *American Journal of Public Health.* See Joel A. Tarr, Terry Yosie, and James McCurley III, "Disputes Over Water Quality Policy: Professional Cultures in Conflict, 1900—1917", *American Journal of Public Health*, April, 1980, 70: 427—435 and Joel A. Tarr, "Industrial Wastes and Public Health: Some Historical Notes, Part 1, 1876—1932", *American Journal of Public Health,* September 1985, 75: 1059—1067.

Notes

1. Some cities acquired new and purer upcountry sources.

2. In 1922 the Committee on Industrial Wastes in Relation to Water Supply of the American Water Works Association reported that industrial pollutants had damaged at least 248 water supplies in the United States and Canada.

References

American Water Works Association 'Progress Report of Committee on Industrial Wastes in Relation to Water Supply,' *Journal of the American Water Works Association*, 1923, 10, 415—430.

Babbitt, H. and J. J. Doland *Water Supply Engineering*. New York: McGraw-Hill, 1931.

Besselievre, E. B. 'Statutory Regulation of Stream Pollution and the Common Law,' *Transactions. American Institute of Chemical Engineers*, 1924, 16, 788—794.

Borchardt, J. and G. Walton. 'Water Quality,' in *American Water Works Association: Water Quality and Treatment — A Handbook of Public Water Supplies* (3rd edn.). New York: McGraw-Hill, 1971.

Cairns, John Jr. 'Don't Be Half-Safe — The Current Revolution in Bioassay Techniques,' in *Proceedings of the Purdue Industrial Waste Conference*, 1966, 559—567.

Chapin, Charles V. 'History of State and Municipal Control of Disease,' in Mazyck P. Ravenel (ed.), *A Half-century of Public Health*. New York: American Public Health Association, 1921.

Clarke, H. A. 'Characterization of Industrial Wastes by Instrumental Analysis,' in *Proceedings of the Purdue Industrial Waste Conference*, 1968, 26—34.

Council on Environmental Quality. 'Trends in Analytical Chemistry Detection Techniques,' Table 3—9, in *Environmental Quality. Tenth Annual Report of the Council on Environmental Quality*. Washington, D.C.: GPO, 1979.

Davies, J. Clarence. *The Politics of Pollution*. New York: Pegasus, 1970.

Donaldson, W. 'Industrial Wastes in Relation to Water Supplies,' *American Journal of Public Health*, 1922, 12, 420—421.

Donaldson, W. 'Industries and Water Supplies,' *Journal of the American Water Works Association*, 1930, 22, 202—207.

Duffy, John. *A History of Public Health in New York City: 1866—1966*. New York: Sage, 1974.

Dunlap, Thomas R. *DDT: Scientists, Citizens, and Public Policy*. Princeton: Princeton University Press, 1981.

Eckholm, Erik P. *The Picture of Health: Environmental Sources of Disease*. New York: W. W. Norton, 1977.

Engineering News. 'The Unusual Prevalence of Typhoid Fever in 1903 and 1904,' 1904, 51, 129—130.

Engineering News. 'The Rochester Sewage Disposal Case: Sewage Disposal by Dilution Strongly Endorsed, 1910a, 64, 154—155.

Engineering News. 'The Greater Pittsburgh Sewerage and Sewage Purification Orders,' 1910b, 63, 179—180.

Engineering News. 'The Discharge of Sewage into the Hudson River,' 1910c, 62, 502—503.

Engineering News. 'A Plea for Common Sense in the State Control of Sewage Disposal,' 1912a, 67, 412—413.

Engineering News. 'The Pittsburg [sic] Sewage Purification Order: Letters,' 1912b, 67, 548—552.
Engineering News. 'Standards of Purity for Rivers and Waterways,' 1912c, 68, 835—836.
Engineering Record. 'The Water Supply of Large Cities,' 1900, 41, 73.
Engineering Record. 'Sewage Pollution of Water Supplies,' 1903, 48, 117.
Engineering Record. 'The Pollution of Streams,' 1909, 60, 157—159.
Engineering Record. 'Conference on Pollution of Lakes and Waterways,' 1912, 66, 485—486.
Fair, Gordon M., J. C. Geyer and J. C. Morris, 1954. *Water Supply and Waste-Water Disposal*. New York: John Wiley.
Frost, Wade H. *A Study of the Pollution and Natural Purification of the Ohio River. Public Health Bulletin No. 143*. Washington, D.C.: GPO, 1924a.
Frost, Wade H. *A Study of the Pollution and Natural Purification of the Ohio River. II. Report on Surveys and Laboratory Studies. Public Health Bulletin No. 143*. Washington, D.C.: GPO, 1924b.
Fuller, George W. 'Sewage Disposal in America,' *Transactions of the American Society of Civil Engineers*, 1905, 54, 148.
Gibbs, Lois Marie. *Love Canal: My Story*. Albany: State University of New York, 1982.
Goodell, E. B. *Review of Laws Forbidding Pollution of Iland Waters in the United States*. Water Supply and Irrigation Paper # 152. U.S. Geological Survey. Washington, D.C.: GPO, 1906.
Gregory, George G. 'A Study in Local Decision Making: Pittsburgh and Sewage Treatment,' *Western Pennsylvania Historical Magazine*, 1974, 57, 25—42.
Hays, Samuel P. *Conservation and the Gospel of Efficiency*. Cambridge: Harvard University Press, 1959.
Hays, Samuel P. 1981. 'The Structure of Environmental Politics Since World War II', *Journal of Social History*: 717—720.
Hazen, Allen. *Clean Water and How to Get It*. New York: 1907.
Kerr, J. W. and A. A. Moll. *Organization, Powers, and Duties of Health Authorities*. Public Health Bulletin # 54. U.S. Public Health Service. Washington D.C.: GPO, 1912.
Kirkwood, James P. *A Special Report on the Pollution of River Waters. Annual Report. Massachusetts State Board of Health*. 1876; New York: Arno Press Reprint, 1970.
Leighton, Marshall O. 'Industrial Wastes and Their Sanitary Significance,' *Public Health Reports and Papers of the American Public Health Association*. 1905, 31, 29—41.
Levine, Adeline. *Love Canal: Science, Politics and People*. Lexington, Massachusetts: D.C. Health, 1981.
McNeil, Donald. 'Home Purchases by State Sought Near Love Canal,' *New York Times*, September 10, 1979.
Melosi, Martin. *Garbage in the Cities: Refuse, Reforms, and the Environment, 1880—1980* College Station: Texas A & M, 1981.
New York Department of Health. *Love Canal: Public Health Time Bomb*. Albany: New York Department of Health, 1978.
Office of Technology Assessment. *Technologies and Management Strategies for Hazardous Waste Control*. Washington: GPO, 1983.
Pennsylvania State Board of Health. 1909. 'The Germicidal Effect of Water from Coal Mines and Tannery Wheels Upon *Bacillus Typhosus, Bacillus Coli* and *Bacillus Anthracis*,' *Pennsylvania Health Bulletin*, 1909, 5, 1—10.
Phelps, Earle B. *Studies of the Treatment and Disposal of Industrial Wastes. Public Health Bulletin No. 97*. Washington, D.C.: GPO, 1918.
Phelps, Earle B. 'Stream Pollution by Industrial Wastes and its Control,' in M. Ravenel (ed.), *A Half Century of Public Health*. New York: American Public Health Association, 1921.

Phelps, Earle B. *Stream Sanitation.* New York: John Wiley and Sons, 1944.

Pollak, David. 'Seepage Threat Slight,' *Niagara Gazette* May 3, 1977.

Pollak, David and David Russell. 'Dangerous Chemicals Found Leaking from Hooker Dump,' *Niagara Gazette.* November 2, 1976.

Prescott, Samuel C. and C. E. A. Winslow. *Elements of Water Bacteriology.* (5th edn.) New York: John Wiley & Sons, 1931.

Quarles, John. *Federal Regulation of Hazardous Wastes* Washington, D.C.: Environmental Law Institute, 1982.

Rosen, Aaron A. 'The Foundations of Organic Pollutant Analysis,' in Keith, Lawrence H. (ed.), *Identification and Analysis of Organic Pollutants in Water.* Ann Arbor: Ann Arbor Science, 1976, 4—14.

Rosenkrantz, Barbara G. *Public Health and the State: Changing Views in Massachusetts.* Cambridge: Harvard University Press, 1972.

Rudolfs, Wilhelm. 'Stream Pollution in New Jersey,' *Transactions of the American Institute of Chemical Engineers,* 1931, 27, 41—47.

Skolnik, H. and K. M. Reese. (eds.). *A Century of Chemistry: The Role of Chemists and the American Chemical Society.* Washington, D.C.: American Chemical Society, 1976.

Snow, F. H. 'Administration of Pennsylvania Laws Respecting Stream Pollution,' *Proceedings of the Engineering Society of Western Pennsylvania,* 1907, 23, 266—283.

Stevenson, W. L. 'The State vs. Industry or the State with Industry?' *Transactions. American Institute of Chemical Engineers,* 1924, 16, 201—216.

Streeter, H. W. and E. B. Phelps. *Factors Concerned in the Phenomena of Oxidation and Rearation. Public Health Bulletin No. 146.* Washington, D.C.: GPO, 1925.

Subcommittee on Oversight and Investigations, Committee on Interstate and Foreign Commerce, United States House of Representatives. *Hazardous Waste Disposal.* Washington, D.C.: GPO, 1979.

Symons, George E. 'Industrial Wastes,' *Sewage Works Journal,* 1945, 17, 558—572.

Tarr, Joel. 'The Search for the Ultimate Sink: Urban Air, Land, and Water Pollution in Historical Perspective,' in *Records of the Columbia Historical Society of Washington D.C.* Charlottesville: University Press of Virginia, 1984.

Tarr, Joel A. *et al.* 'Water and Wastes: A Retrospective Assessment of Wastewater Technology in the United States, 1800—1932,' *Technology & Culture.* 1984, 25, 226—263.

The Outlook. 'Mr. Roosevelt and the People,' 1910, 96, 1.

Tobey, James A. *Public Health Law.* New York: The Commonwealth Fund, 1947.

Trauberman, Jeffrey. 'Compensation for Environmental Pollution: An Overview of Salient Legal and Practical Issues,' in Congressional Research Service, *Six Studies of Compensation for Environmental Pollution.* Washington, D.C.: GPO, 1980.

Veatch N. T. 'Stream Pollution and its Effects,' *Journal of the American Water Works Association,* 1927, 17, 58—63.

Whipple, George C. 'Fifty Years of Water Purification,' in Mazyck P. Ravenel (ed.), *A Half-century of Public Health.* New York: American Public Health Association, 1921.

Whipple, George C. 'The Training of Sanitary Engineers,' *Engineering News,* 1912a, 68, 805—806.

Whipple, George C. 'Sanitation More than Medicine,' *Literary Digest,* 1912b, 45, 899.

Wiebe, Robert H. *The Search for Order, 1877—1920.* New York: Hill and Wang, 1967.

Zuesse, Eric. 'Love Canal: The Truth Seeps Out,' *Reason,* February 16, 1981.

Chapter 14

OSHA's Carcinogens Standard: Round One on Risk Assessment Models and Assumptions

Frances M. Lynn

Introduction

This chapter uses the Occupational Safety and Health Administration's (OSHA) efforts in the late 1970s to adopt a generic way to regulate suspected workplace carcinogens as a vehicle for examining the role of values in cancer risk assessment. OSHA's attempt (a final standard was never implemented) to promulgate uniform guidelines for assessing cancer grew out of a frustration with arguing the same rules of evidence for each suspected carcinogen. This case-by-case approach resulted in the adoption of only 19 carcinogen standards, including 14 which had been regulated as a group shortly after the creation of OSHA in 1970.

OSHA's effort to adopt generic ways to identify and regulate carcinogens signaled the beginning of a public debate about risk assessment models and assumptions that continues to this day. OSHA's rule-making also stimulated a debate as to who was the most appropriate type of body to make those choices. Was it a regulatory agency like OSHA? Or were these the types of decisions more appropriately made by a disinterested scientific body? This in turn raised the question whether, given the uncertainties involved in cancer risk assessment, there could be disinterested scientists or even an objective science.

The chapter begins by focusing on the issue of the possibility of an objective science, examining how the issue was viewed within the emerging field of risk analysis but also looking at how sociologists over the last 30 years have viewed the issue of scientific objectivity. The chapter briefly examines some of the more controversial issues within the field of carcinogenesis and how they were viewed by the major actors in the OSHA cancer rule-making. The final section of the chapter reports the results of interviews with occupational health scientists working in different institutional settings. The purpose of the interviews was to explore the impact of nonscientific influences on scientists' selection among competing theories and models used to assess workplace and other suspected carcinogens.

B. B. Johnson and V. T. Covello (eds.), The Social and Cultural Construction of Risk, 345–358.

Objectivity in Risk Assessment

The possibility of an objective science permeated the early literature on risk. William Lowrance in his 1976 book, *Of Acceptable Risk*, set the tone when he divided risk analysis into two phases: risk assessment and risk management. Lowrance characterized the first phase, risk assessment, as an "objective pursuit" with decisions most appropriately made by scientists. He described the second phase, risk management, as "part of the policy process." Eventually Congress directed the National Academy of Sciences to conduct a study of the process of risk assessment. The National Academy of Sciences (1983: 10) had hoped to distinguish those components of a risk assessment

> that are scientific in nature from those that are value judgments or policy and to identify and describe those components that are neither strictly science or policy but a hybrid consisting of elements of both.

The study, directed by the Committee on Institutional Means for Assessment of Risks, was also to consider the feasibility of having a single organization to conduct risk assessments and to develop uniform risk assessment guidelines for all federal regulatory agencies.

Lawrence McCray (McCray 1983: 13) the Project Director for the National Academy Committee commented that in the case of risk assessment the underlying premise that "matters of science" might be segregated from "matters of value" and "left to an organization primarily responsive to scientific authority" was "naive." The National Academy Committee concluded that throughout the process of deciding whether or not a substance posed a threat to human health it was possible to make choices which would increase the likelihood that a substance would be judged to be a significant risk to human health. The National Academy Committee noted (NAS 1983: 6) "how difficult it is to disentangle the mixture of fact, experience (often called intuition) and personal values." The study of occupational health professionals described later in this chapter, while conceived and executed before the National Academy effort, empirically complements it by exploring what role social and political values as well as institutional affiliation play in scientists' choices. The theoretical background for the study of occupational health professionals drew from empirical work of sociologists of science. This literature is briefly reviewed in the next section of this chapter.

Sociologists Look at Science and Scientists

In the last 40 years sociologists of science have explored the impact of societal factors on scientific activities and scientific theories. The earliest

American work in the sociology of science supported the view that scientific objectivity was possible. Robert Merton argued that there was an "objectively toned comple of values and norms which is held binding on the man of science" (Merton 1973: 269). Merton's characterization of the scientist was that of a specialized autonomous expert following pre-established impersonal criteria and disinterested in power or financial gain. Subsequent empirical studies by sociologists of science challenged the Mertonian norm of the disinterested scientist (Kornhauser 1962; Cotgrove and Box 1970; Krohn 1971; Mitroff 1974).

William Kornhauser, in a 1962 study of scientists in industry, concluded that tensions exist between scientific norms of autonomy and industry's norms of integration and control. Kornhauser suggested that allegiance of scientists in industry to professional norms diminishes as they become oriented to the industrial incentive system of higher salaries and more challenging responsibility in managerial positions (Kornhauser 1962: 10—149). Kornhauser posited, however, that generally some form of accommodation is reached by scientists in industry between the norms of their profession and the organization within which they work. Moreover, industry will never totally try to control scientists since it depends on them for innovation (Kornhauser 1962: 74).

Kornhauser implied a greater commitment on the part of university and government scientists to professional norms because the basis of their promotions more often rely on professional recognition (for example publications). Kornhauser argued that scientists working within government share characteristics both with their university-based and industry employed colleagues because of government's diversity of functions. Promotions within government rely more heavily on publication than in industry. However, government provides greater opportunities for higher salaries and advancement than the university (Kornhauser 1962: 133). Dean Schooler (1971: 16, 149) suggested that government scientists and their industrial counterparts may face pressures "to provide scientific justification for actions politically based and motivated."

The university scientist has traditionally provided the norm against which government and industrial scientists are compared (Krohn 1971; Cotgrove and Box 1970; Schooler 1971). With the advent of industry sponsored research, the issue of conflict of interest has arisen when these scientists testify as disinterested witnesses before governmental bodies (Fritschler 1970; Carpenter 1976; Ladd 1965). Similar questions have been raised about scientists who work for independent consulting firms (Fritschler 1975; Blume 1974).

Carcinogenesis Controversies

The main issue facing OSHA was deciding what combination of evidence was sufficient to establish human carcinogenesis and to consider regulatory action. OSHA officials confronted issues such as what to do in the face of negative epidemiological data, but positive animal studies; whether to combine benign and malignant cancer and what mathematical models to use to extrapolate from animal data to humans. These are decisions that confront any effort to assess the risk of a substance to humans whether in the workplace or on the pantry shelf.

Scientists use several methods to identify carcinogens including epidemiological studies and animal bioassays. Each of these methods has limitations.

Epidemiological studies are often considered the most valid and direct evidence for identifying hazards (Maugh 1978; State of California 1982; NAS 1983). The drawbacks of epidemiological studies are both practical and ethical. The long latency periods associated with some chemicals (5—40 years) make it difficult and expensive to find those affected or to follow them long enough to notice effects. In addition, people are exposed to a number of hazardous substances both in the workplace and the general environment and it is therefore difficult to identify a single specific cause or to be able to account for the synergistic affects of exposure to more than one substance. A fundamental problems with epidemiology is the difficulty and expense involved in assembling a large enough number of cases to satisfy statistical requirements for excess risk. The State of California pointed out (1982: 24) that

> with a study population of less than 1,000 it may be impossible to identify an agent that increases the risk of a specific cancer by a factor of less than 5 to 10 . . . even large scale studies may fail to detect increases of less than 50 percent cancer incidence.

A major objection to the primary reliance on epidemiological studies is ethical. Few chemicals currently in use have been tested using epidemiological methods. Ethical issues arise when a hazard identification policy is premised on waiting 20 years for a reaction to occur.

For these reasons, animal bioassays are the most common means of hazard identification. The National Academy of Sciences (1983: 22) points out that

> the inference results from animal tests are applicable to humans is fundamental to toxicological research; this premise underlies much of experimental biology and is logically extended to the experimental observation of carcinogenic effects.

Nonetheless, the drawbacks of animal studies are well known. To detect a reaction to a substance with a low incidence and to avoid having to test tens of thousands of rats, it is necessary to administer high doses to animals. David Rall (1978: 9), Director of the National Institute of Environmental Studies, and a proponent of the use of animal studies for risk identification for humans, points out that even at low doses it is uncertain whether or not a reaction to a carcinogen's long-term development can be measured in the 2 to 2½ year life span of the mouse or rat.

Another major debate within the field of risk assessment for carcinogens is among the mathematical models which describe the relationship between the dose of a substance and the response in a human, the so-called dose-response curve. This is a multi-step process for it often involves extrapolating from high doses to low doses in animals and then from animals to humans. The two types of models that command the most attention are the linear and threshold.

A threshold model assumes that there is a dose below which cellular or tissue damage does not occur. Under such models one assumes that the body has mechanisms which withstand toxic events at low doses. A linear model suggests, especially in the case of carcinogens, that a single interaction with a single cell can trigger a toxic reaction. The National Academy of Sciences concluded (1983: 25) that there was no conclusive biological evidence to support either type of model. The State of California (1982: 19), under the Brown administration, chose to take the more cautious posture and concluded that

> it is not appropriate to apply the concept of "thresholds" to carcinogens unless convincing evidence is presented to demonstrate the existence of thresholds for a specific carcinogen.

This was also the position adopted by OSHA in its carcinogens standard. By contrast, the first draft of the Reagan Administration's Office of Science and Technological Policy (OSTP) carcinogens' policy, in addition to placing less reliance on animal tests, also assumed there were thresholds.

OSHA's Generic Cancer Standard

The debate that erupted around the OSHA cancer standard had as much to do with OSHA's authority to promulgate this type of rule as it did over the substance of the proposal itself. The question was whether the determinations that OSHA, a regulatory body, had made would be more appropriately decided by a group of disinterested scientists operating in a less highly charged forum. The American Industrial Health Council (AIHC), a coalition of over 100 corporate members and over 80 additional trade associations, formed in 1977 in reaction to the OSHA proposal, felt that the "determination of carcinogenicity . . . (is) . . . a scientific rather than a regulatory matter" (AIHC 1978: vii). The AIHC felt that there was a "sound," "unbiased" and "good" science for the identification of carcinogens (AIHC 1983). In its "Recommended Alternatives to OSHA's Generic Proposal," the AIHC (1978) proposed that Congress establish a central science panel outside of regulatory authorities like OSHA for the identification of carcinogens.[1]

OSHA, while perceiving itself to be operating on the "frontiers of scientific knowledge," also felt that "policy determinations in a legal framework could not be based solely upon scientific fact where such factual data were often incomplete" (USDL 1980). Where there was scientific uncertainty OSHA made a conscious choice to be "prudent" and to err on the side of protecting human health. In a draft proposal which was circulated in early 1977 OSHA hoped that "if new evidence or future scientific advances show it to be in error . . . (that) . . . the error was being too careful and overly protective of the human being for the alternative is unacceptable."

In the final standard issued on January 22, 1980, OSHA said that it "emphatically agreed" with the testimony of government scientist Richard Bates who said that

> if human disease and death is to be prevented it is often necessary to control exposure for which there is some evidence of hazard before that evidence has reached the point that scientists would universally regard as conclusive. The alternative, to continue exposure until there is conclusive evidence of human hazard, is a form of human experimentation that our society finds increasingly unacceptable (USDL 1980: 5022).

The AIHC (1978: 8) contended that the OSHA standard exaggerated the extent of occupationally caused cancers and questioned

> whether the public is being well-served by being led to believe that industrial chemicals are the overwhelming cause of cancer in this country. A public so convinced would expect, if not demand, the concentration of governmental "corrective" action be focused upon industrial activities in the expectation of preventing the great majority of malignancies.

The AIHC committee felt OSHA gave "too much emphasis to the uncertainty and difficulty of knowing the safe level of exposure to any known or suspect carcinogen with any degree of confidence." They objected to the assumption that "there is no safe level of exposure." They felt that implicit in such ideas was the notion that industry or any other useful activity can occur on a completely safe, risk-free basis. Instead they felt that it was important, even in an emotional context such as cancer, that sound public policy must take into account the inevitability of some risk and, the necessity of evaluating such risk not only against alternative risk but also in light of the benefits of the substance being regulated.

What became apparent in reading the testimony surrounding the OSHA standard was that, scientific issues notwithstanding, conflicts over regulatory goals and judging acceptable levels of risk in society and in the workplace were also being debated. While OSHA focused on protection, industry suggested the need for balancing the benefits of economic activity against the risk to human health. A comparison of the risk identification and risk assessment methods and assumptions that the two institutions supported revealed a pattern of selection of assumptions and techniques reflective of these values.

OSHA, when it released its 1977 draft of the carcinogens standard, said that it based its action upon three propositions. The first proposition stated that it was possible although difficult to define the term carcinogen. The second proposition stated that

> a toxic material that is confirmed as a carcinogen to a mammalian test animal species, as defined, is to be treated as a policy matter as posing a carcinogenic risk to man.

The third proposition stated that "there is presently no means to determine a safe level of exposure to a known carcinogen. OSHA went on to recommend that "the permissible exposure . . . be set as low as feasible or not permitted in certain cases" USDL 1977: 2—3).

The AIHC labeled OSHA's proposal as "indiscriminate in attributing significance to mammalian test data and, for all practical purposes automatically extrapolating therefore to human exposure." In particular, the AIHC and other industry scientists were annoyed that OSHA proposed to use animal test data "without regard to the dose used . . . (and . . .) the overwhelming of normal detoxification mechanisms." OSHA, however, responded by rejecting as "not scientifically sound and/or convincing" the concept of metabolic overloading from testing at high doses, despite experimental data from Dr. Perry Gehring of Dow Chemical Company (USDL 1980: 5093)

OSHA also chose to consider nonpositive epidemiological results only if they met very stringent conditions and to weigh positive results in animals more heavily. The AIHC disagreed with OSHA. They proposed instead that negative evidence from epidemiological studies in humans should be given more weight than positive results in animal tests.

The AIHC also produced evidence to suggest that a "threshold" or "no-effect" level of exposure was possible for carcinogens. However, OSHA concluded, after reviewing "the very extensive evidence in the Record" that the consensus of scientific opinion was there is "no presently acceptable way to reliably determine a threshold for a carcinogen" (USDL 1980: 5137). OSHA concluded that

> variability among humans makes it very difficult to have a great deal of confidence that an observed "no effect" levels of exposure in experimental test animals or even in a specific human population will be applicable to any given human population at risk. A large number of factors (e.g., age, sex, race, nutritional status, previous exposure to other substances) could affect individual susceptibility.

In reading the comments stimulated by the OSHA proposal one was struck by the fact that on a number of seemingly scientific issues industry and government scientists consistently lined up with opposing beliefs and conflicting evidence. Both sides at times claimed to be representing the dominant scientific consensus or well-established scientific principles. What seemed clear to a nonscientist reading the testimony as well as articles in scientific journals was a lack of consensus on the methods and models in the rapidly emerging field of carcinogenic risk assessment. This suggested that at least in the field of cancer, risk assessment paradigms were competing for acceptance and raised the question of what role non-science factors were playing in this seemingly scientific process. To look at this issue in more detail interviews were conducted with 136 health professionals, mainly occupational health professionals and industrial hygienists.

Attitudes of Occupational Health Professionals toward Risk Assessment Assumptions

Respondents participated in hour long interviews, answering questions which tapped social and political values, perceptions about risks and attitudes toward risk management strategies. In analyzing the data these attitudes were correlated with two of the main assumptions involved in quantitative risk assessments: the use of animal data and the existence of thresholds for carcinogens. The interviews were conducted with the occupational health scientists in the winter of 1981, a year after the release of the final OSHA standard.[2]

The study found strong links between the occupational health professionals' place of employment and their perceptions of risk and scientific beliefs. Even after controlling for the influence of standard demographic variables, including age, sex, region, religiosity and family background, scientists employed by industry tended to be politically and socially more conservative than government and university scientists and their political and social beliefs correlated highly with their scientific judgments. Consistently scientists employed by government chose those scientific assumptions normally seen to be more protective of human health but which would result in more costly control measures. Those employed by industry picked assumptions that decreased the likelihood that a substance would either be judged a risk to human health or ones which would find a substance regulated at a higher level of exposure. Academics fell in between, a finding which goes contrary to a more liberal image as postulated by Kornhauser. This might be best explained by the fact that a number of older academics (50 and above) had forged their careers when the workplace environment was much more dangerous than today. They had seen situations such as the construction of Gauley's Bridge where in a two-year period almost 500 men died and over 1,500 were disabled from exposure to silica. To some of these older academics today's risks paled. In fact, while age was not a significant discriminator *between* groups' attitudes, it did correlate significantly with attitude differences *within* the academic sample.

Figure 1 shows the response of those interviewed to the use of animal

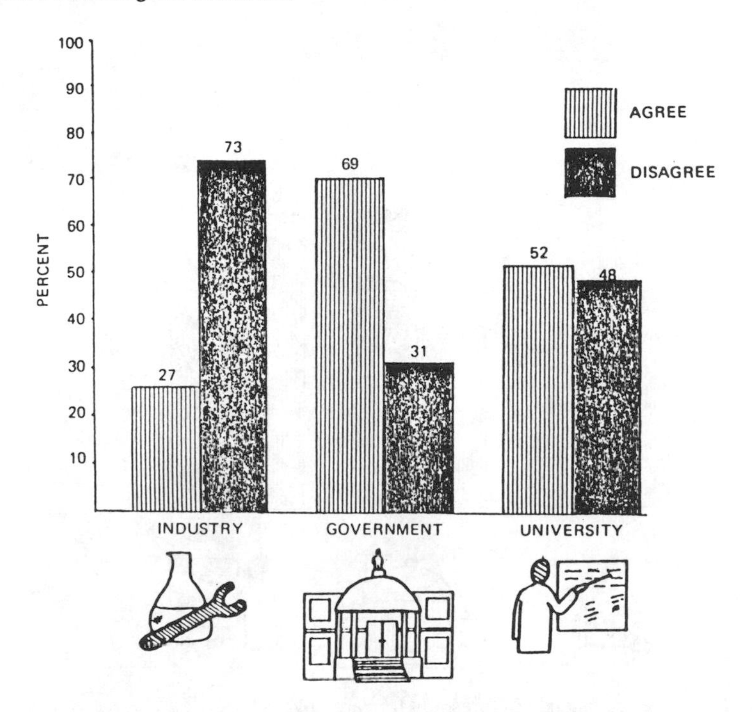

Fig. 1. Attitude: using animal data to identify risk to people.

data to predict responses for humans. Respondents were asked whether they agreed or disagreed that "a substance which is shown conclusively to cause tumors in animals should be considered a carcinogen, thereby posing a risk to humans." The majority of occupational health professionals who worked for government or the university (69 and 52 percent respectively) agreed with the statement. By contrast only 27 percent of the industry sample agreed.

Figure 2 shows the reaction of respondents to the question of whether or not "thresholds exist for carcinogens below which there are not negative effects." Eightly percent of the industry respondents agree that thresholds existed. Sixty-one percent of the academics agreed, while only 37 percent of government employees agreed.

Interviewees were asked whether or not they felt that "American society is becoming overly sensitive to risk and that we now expect to be sheltered from almost all dangers . . . (or) . . . simply becoming more aware of risks and starting to take realistic precautions." Table 1 shows that 85 percent of those who believe that Americans are overly sensitive to risk also support the idea that thresholds exist for carcinogens. Only 39 percent of those felt that Americans were "more aware of risks and taking realistic precautions" believe in thresholds.

Belief in the existence of thresholds correlated with vote in the 1980

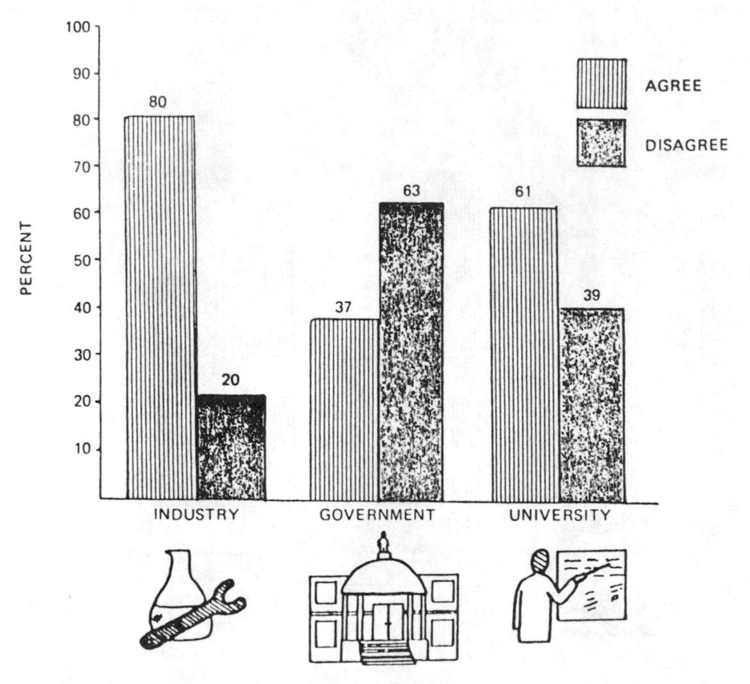

Fig. 2. Attitude: existence of threshold for carcinogens.

Table I. Relationship between attitudes toward the existence of thresholds and feelings about American society's attitudes toward risk

Attitudes about thresholds	Feelings about American society's attitudes toward risk		
	Overly sensitive	More aware	Both
Thresholds exist	85%	39%	67%
Thresholds don't exist	15%	61%	33%
Totals	100%	100%	100%
$N = 95$	(27)	(41)	(27)

$p > 0.00$.

presidential election (see Table 2). It also correlated with self-identification as a conservative. Moreover 74 percent of those who believe in thresholds also believe that the risks associated with advanced technology have been exaggerated by Three Mile Island (Figure 3). By contrast, only 39 percent of those who question the existence of thresholds believe that the risks of advanced technology have been exaggerated by Three Mile Island.

Table II. Relationship between attitudes toward the existence of thresholds and Carter-Reagan vote in the 1980 Presidential Election

Attitudes about thresholds	Vote in 1980 Presidential Election		
	Carter	Reagan	Total ($N = 75$)
Thresholds exist	36% (18)	63% (31)	99% (49)
Thresholds don't exist	69% (18)	30% (8)	99% (26)

$p > 0.02$.

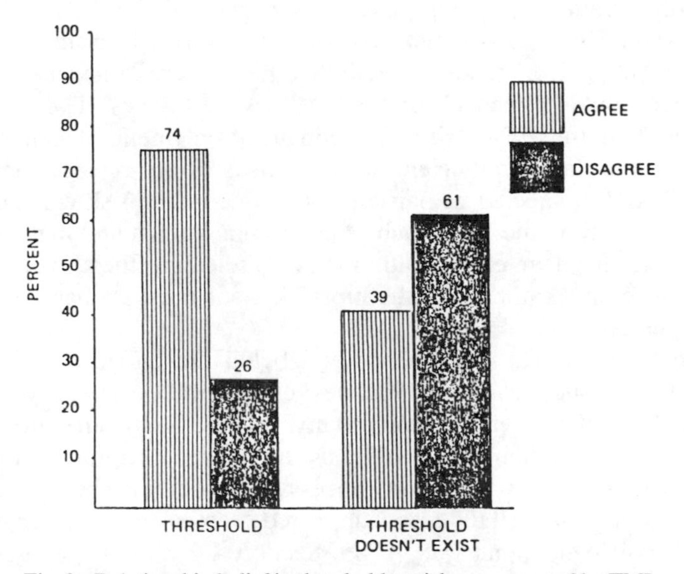

Fig. 3. Relationship belief in threshold to risks exaggerated by TMI.

Summary and Conclusions

The types of issues considered by OSHA in 1977 have been addressed in subsequent years by private and public bodies in their effort to influence how cancer risks will be assessed. Since OSHA issued its carcinogens standard other governmental bodies including the Environmental Protection Agency, the Consumer Product Safety Commission, the White House Office of Science and Technology Policy, the State of California, the National Academy of Sciences and the National Science Foundation have issued documents with the hope of effecting the way cancer risks are assessed and regulated. Assumptions such as use of a linear dose-response model or the weight to be given animal data and short-term tests have been joined by such issues as the incorporation of pharmokinetic and metabolic data, differentiation among genotoxic and nongenotoxic data, the role to be given to different routes of exposure and to tumor sites in animals not found in humans and whether or not to use best estimates or worse case characterizations of data.

The most forceful voice for American industry has continued to be the American Industrial Health Council who has responded to most of the above-mentioned guidelines and also sponsored a taskforce of eminent scientists to draft a state of the art for cancer risk assessment (AIHC 1985).

All parties to the drafting of these guidelines have become more sensitive to difficulty of separating policy and science. While the AIHC in its comments on EPA's 1984 risk assessment described scientific risk assessment as "objective," it also noted that there were "science policy choices" which could "enhance" or "reduce" risk estimates. The AIHC recommended that the EPA "should distinguish statement of generally accepted scientific fact and scientific consensus from science policy choices." The AIHC objected in particular to some of the risk enhancing assumptions selected by the EPA such as combining benign and malignant tumors, choosing a linear extrapolation model, selecting the 95 percent upper confidence limit rather than the most likely or most probable value to calculate unit risk.

Nonetheless, the AIHC soon thereafter labeled the Office of Science and Technology Policy cancer guidelines as generally reflecting "the positions held by the scientific community." These guidelines do not recognize any single mathematical model as the most appropriate for low dose extrapolation and would not necessarily combine benign and malignant tumors. The AIHC also supported Section 4 of Executive Order 12498 issued on January 8, 1985. Section 4 required risk assessments to be "scientifically objective and include all relevant information and adopt unbiased best estimates, not hypothetical "worse case" or "best case" and in particular said that risk assessment must be "scientifically objective."

The issues seemed to fall into the category of "transcience" a term which nuclear physicist Alvin Weinberg used in a 1972 article in the journal *Minerva*. Weinberg cited the determination of the biological effects of low-level radiation insults as an example of a transcientific issue. He suggested that the argument about low-level radiation insult would have been far more sensible had it been

> admitted at the outset that this was a question which went beyond science. The matter could have been dealt with initially on moral or aesthetic grounds (Weinberg 1972: 217).

Weinberg's observations in the early 1970s seemed to go unnoticed in much of the debate surrounding the OSHA standard and in the emerging field of cancer risk assessment.

Science grows through healthy debate and peer review. That is occurring in the field of risk assessment. However, what is also occurring is unhealthy posturing with words such as "objective" and "unbiased" being used to support theories and models when the basic disagreement is social and political and related to the very difficult choices that must be made in balancing the degree of protection we as a society find acceptable.

For the layperson — be he or she a regulator, an elected official, or a citizen who has been exposed to a potential carcinogen — what is important to keep in mind is that the science, which supports the conduct of a quantitative/scientific risk assessment is still unfolding. Paradigms are competing for acceptance. It is also important to remember that when the empirical base of science is sparse or uncertain (as it is in the case of cancer), institutional setting and social and political values can play a role in a scientist's or scientific organization's definition of what is true. Efforts to codify science prematurely or to place what are often social or ethical decisions in the hands of a scientific elite are not part of the tradition of science, and are not in the tradition of democracy.

Notes

1. In 1983, the AIHC convened a committee of eminent scientists including Nobel Laureate Joshua Lederberg who drafted a bill which called for establishing a Central Board of Scientific Risk Analysis under the National Academy of Sciences. This bill was introduced by Congressman James Martin (R-N.C.). It became Title II of House Bill 4192, the so-called Martin-Ritter. The bill was criticized by groups such as the American Chemical Society (1983: 2) who did not believe it was "appropriate for the Academy — a nonregulatory nongovernmental body to become involved." Martin subsequently withdrew the bill.

2. Interviewing took place in three geographic regions: metropolitan New York, Cincinnati, and the Research Triangle Park and Triad of North Carolina. The sample was drawn from the membership lists of professional organizations and from faculty and staff rosters supplied by government and university department heads. The interview consisted of original questions and questions from two other attitudinal studies of risk: "Risk in a Complex Society," conducted by the Louis Harris organization for Marsh and McLennan, an insurance brokerage firm and "Public and Worker Attitudes toward Carcinogens and Cancer Risk," conducted by Cambridge Reports for Shell Oil. Data were analyzed using the weighted least squares approach, an application of the general linear model to categorical data.

References

American Chemical Society, Department of Public Affairs (ACSDPA). Letter from Pres. Fred Basolo to Congressman Don Fuqua. Washington, D.C. December 8, 1983.

American Industrial Health Council (AIHC). Recommended Alternatives to OSHA's Generic Carcinogen Proposal. Scarsdale, New York: AIHC, 1978.

American Industrial Health Council (AIHC). Report to the Membership. Scarsdale, New York: AIHC, 1983.

American Industrial Health Council (AIHC). 'Criteria for Evidence of Chemical Carcinogenicity', *Science*, 1984, 225, 682—687.

Blume, Stuart. *Toward a Political Sociology of Science*. New York: The Free Press, 1974.

Cotgrove, Stephen and Steven Box. *Science, Industry and Society*. London: George Allen and Unwin Ltd., 1970.

Fritschler, A. Lee. *An Anatomy of Values*. Cambridge, Massachusetts: Harvard University Press, 1970.

Kornhauser, William. *Scientists in Industry*. Berkeley, California: University of California Press, 1962.

Krohn, Roger. *The Social Shaping of Science*. Westport, Connecticut: Greenwood Publishing Co., 1971.

Lowrance, William. *Of Acceptable Risk*. Los Altos, California: William Kaufman, Inc. 1976.

Maugh, Thomas H. III. 'Chemical Carcinogens: The Scientific Basis for Regulation,' *Science* 1978, 201, 1200—1205.

McCray, Lawrence. 'An Anatomy of Risk Assessment: Scientific and Extra-Scientific Components in the Assessment of Scientific Data on Cancer Risks.' Working Paper for the National Academy of Sciences Committee on the Institutional Means for Assessment of Risks to Public Health. Washington, D.C.: National Academy Press, 1983.

Merton, Robert. *The Sociology of Science*. Chicago: The University of Chicago Press, 1973.

Mitroff, Ian. *The Subjective Side of Science*. Amsterdam: Elsevier press, 1974.

National Academy of Sciences (NAS). *Risk Assessment in the Federal Government*. Washington, 1983.

Rall, David. 'Species Differences in Carcinogenesis Testing,' in *Origins of Human Cancer* (Holt, Hiatt *et al.*, eds.). Cold Spring Harbor, New York: Cold Spring Harbour Labs, 1977.

Schooler, Dean Jr. *Science, Scientists and Public Policy*. New York: The Free Press, 1971.

State of California. 'Carcinogens Policy: A Statement of Science as a Basis of Policy.' A report prepared by the Health and Welfare Agency Department of Health Services, 1982.

U.S. Department of Labor (USDL), Occupational Safety and Health Administration. 'Identification, Classification and Regulation of Potential Carcinogens,' *Federal Register*, 1980, 45, No. 15, Book 2.

U.S. Environmental Protection Agency (EPA). 'Water Quality Criteria Documents,' *Federal Register*, November 28, 1980, 45, No. 231.

Weinberg, Alvin 'Science and Trans-Science,' *Minerva*, April 1972, 10, 2, 209—222.

Chapter 15

Cultural Aspects of Risk Assessment in Britain and the United States

Sheila Jasanoff

Introduction

Concern about the effects of toxic chemicals is one of the most visible legacies of the environmental movement in modern industrial societies. Public pressure to regulate chemicals in the environment and the workplace, in consumer products and in food, has forced the communities of scientists and policy makers into new and uneasy patterns of collaboration. Regulatory programs place unprecedented burdens on government agencies to incorporate scientific and technical information into their decision-making. At the same time, there is growing awareness that science cannot answer all of our questions about risk[1] and that both scientific and value judgments are involved in the processes of risk assessment and risk management.[2]

In the light of science's limited ability to predict harmful events, it comes as no surprise that evaluations of risk by technical experts frequently take color from their social context. Cross-national comparisons of chemical and drug regulation strikingly illustrate this phenomenon. For example, numerous studies have documented the fact that British and American scientists and regulators diverge in their assessments of the same toxic hazards.[3] Differences between the two countries are especially pronounced in the area of carcinogen regulation.[4] Carcinogens have attracted altogether less attention in Britain than in the United States. Few substances have been overtly designated as carcinogens, and the British scientific advisory system has approached animal test data on carcinogenicity with much greater caution than have regulatory authorities in the United States.

Differences such as these are cited by comparative policy analysts as evidence that institutional structures and political interests influence the use of science for policy purposes. The issue for comparative research has often been framed in the following form: Why do scientists working in different regulatory systems interpret the same science in different ways? But comparative analysis raises a more provocative set of questions bearing on the sociology of knowledge, as well as on the substantive outcome of policy decisions. Is it clear, in fact, that regulators in different

B. B. Johnson and V. T. Covello (eds.), The Social and Cultural Construction of Risk, 359—397.

countries and in different environments use the "same science" in addressing the same risk controversies? I argue in this chapter that many decisions that are conventionally regarded as "scientific," such as the weighting of competing studies and methodologies, vary across national boundaries. Specifically, differences in the assessment of toxic chemicals in Britain and the United States point to systematic variations in the way scientific knowledge is certified or validated in the policy systems of the two countries. In each case, the certification process seems to be conditioned by larger cultural assumptions about risk. These findings pave the way for an exploration, at the end of the chapter, of a question that interests many analysts of science policy: How much does science really matter in public decision-making about risk?

To establish the principal thesis of this chapter, I present four case studies comparing chemical risk assessments in Britain and the United States. The cases chosen for this purpose are asbestos, formaldehyde, 2,4,5-T, and airborne lead. My primary focus in each case study is on the development of the scientific base for decision-making. Where do policy makers turn for empirical data bearing on their regulatory problems? How do they fill in perceived gaps in the evidence? What criteria are used to evaluate conflicting data and to distinguish between good and bad information? What risk assessment methodologies are employed? Each case study also compares the policy outcomes in the two countries and discusses how far these outcomes are consistent with scientific assessments of risk. To create a context for the case studies, however, I first briefly describe the legal and institutional frameworks within which risk decisions are made in Britain and the United States.

Approaches to Risk Management

In two centuries of separate historical development, Britain and the United States have diverged sharply with respect to their traditions of public policy-making. As noted in several recent studies,[5] the British administrative process tends to be informal and confidential, relying upon frequent unstructured consultation within government and between officials and private interest groups. Law plays only a marginal role in this system. Toxic substance statutes, for example, confer broad authority to regulate chemical hazards, but they seldom spell out the government's obligations in precise detail. Rule-making procedures are rarely specified by law. Since legislation is usually silent on issues of implementation, administrators enjoy almost complete discretion in establishing regulatory priorities and timetables, in soliciting and releasing information, and in selecting the means of controlling particular chemical hazards. Judicial review of standards is virtually unknown in Britain, largely because the legal system does not provide the grounds for challenging administrative decisions of general application. Enforcement actions can, and sometimes are, pursued in court. But even in the area of enforcement, British officials prefer to negotiate agreements with noncomplying firms, turning to the courts only as a last resort.

The American administrative process, by contrast, relies heavily on legal rules and institutions. Many aspects of implementation that are left to agency discretion in Britain fall under explicit statutory mandates in the United States. These include, most significantly, the procedural requirements to be followed in making regulatory decisions. American administrators must comply with detailed congressional directives in deciding how to collect and analyze information pertinent to their regulatory objectives. The mechanisms for seeking public input are legislatively prescribed, either in the agency's organic statute or in the Administrative Procedure Act. In recent environmental legislation, Congress has laid down numerous procedures for establishing priorities among the 50,000 or so toxic chemicals that the executive agencies are responsible for regulating. U.S. statutes make ample provision for judicial review to ensure that agencies are held accountable for fulfilling their myriad legal obligations. In marked contrast to the situation in Britain, American regulatory proceedings often end in litigation, and judges are prepared to look very closely at the reasons provided by administrators to support either positive decisions or decisions not to take action.

These general characteristics of the British and American regulatory processes determine to a large extent how science is incorporated into risk management decisions. Disparities in the overall approach to regulation in the two countries are reflected in their procedures for accumulating and interpreting scientific data. In each country, however, there is a "standard" approach to building the scientific basis for chemical regulation, though administrators may diverge from this basic strategy if the occasion demands.

Most of the scientific advice required by British regulatory agencies is drawn from a network of technical advisory committees. Major regulatory institutions, such as the Health and Safety Commission (HSC) or the Ministry of Agriculture, Fisheries and Food (MAFF), rely on standing committees for reviews of the scientific literature and evaluations of the risk presented by toxic chemicals. Two such bodies have acquired special prominence in recent regulatory controversies: HSC's Advisory Committee on Toxic Substances (ACTS) and MAFF's Advisory Committee on Pesticides (ACP). The organization of these two committees reflects their fundamentally different roles. Established under authority of the Health and Safety at Work Act, Britain's worker safety statute, ACTS is a "tripartite" committee composed of representatives from industry, labor and government. ACP's membership, by contrast, is limited to academic scientists. This means, in theory, that debates about occupational hazards within ACTS can be linked to the larger political agendas of labor and management, whereas ACP's deliberations on pesticide risks are shielded from political influence. In practice, however, both ACTS and ACP delegate the responsibility for assessing specific chemicals to ad hoc working parties of experts with specialized scientific knowledge. This shared practice diminishes the potential for systematic differences in the behavior of the two advisory committees. In particular, reliance on technical working groups reduces the possibility of political influence on the assessment of occupational health hazards by ACTS.

In reviewing toxic chemicals, British advisory committees are expected not only to evaluate the available scientific information, but also to recommend control measures to the relevant ministry or cabinet department. Though conceived as purely advisory, these recommendations carry great weight with political decision makers. Instances of government officials overruling their technical advisors are extremely rare. For all practical purposes, the expert committee is a powerful source of policy guidance as well as of scientific expertise in the British system of chemical regulation.

British advisory committees enjoy the benefits of informal decision-making in that they have no legal duty to conduct open meetings or to explain their scientific determinations to the public. In the absence of statutory requirements, committee practices vary widely with respect to publishing the reasons for particular policy recommendations. On the whole, the advisory system is weighted toward confidentiality. In a large majority of proceedings involving toxic chemicals the public never has an opportunity to review or criticize the basis for committee recommendations. But in a growing number of cases, public concern has outweighed the government's traditional preference for confidentiality and has prompted the disclosure of advisory committee reports. All four substances discussed in this chapter — asbestos, formaldehyde, 2,4,5-T, and lead — gave rise to significant political debate, and each has figured in risk assessments released to the public by government agencies.

Chemical regulation in the United States begins with steps that are quite similar to procedures followed in Britain. Agencies must generally compile and review scientific information produced by research scientists in academia or in independent public institutions such as the National Institutes of Health (NIH) and the National Institute for Occupational Safety and Health (NIOSH). Supervision by independent scientific panels is sometimes required by law, as in the pesticide regulation and air pollution control programs of the Environmental Protection Agency (EPA).[6] However, even without such statutory requirements, U.S. agencies are accustomed to seeking advice from a variety of standing and ad hoc advisory bodies. These are ordinarily composed of academic scientists, although pluralistic committees like ACTS are not unknown in the United

States. One example is the legislatively authorized National Advisory Committee on Occupational Safety and Health (NACOSH).

U.S. committees, however, exercise less influence on risk assessment than their British counterparts. Agencies in the United States are usually well equipped with in-house expertise. As a result, their advisory committees are relegated to the subordinate role of reviewing studies and analytical documents prepared by or for the administrative agencies. The preparation of risk assessments, for example, generally falls to the agency's own staff scientists or to consultants selected by the agency. The development of policy recommendations, such as emission standards, residue limits, or threshold limit values (TLVs), lies outside the mandate of most advisory committees. This arrangement is consistent with the prevailing view that the scientific input to risk decisions (risk assessment) should be separated from the consideration of policy (risk management).

In contrast to British advisory committee meetings, those in the United States are generally open to the public, and interested parties may present evidence, hear the committee's deliberations, and subsequently obtain transcripts of the meeting. Moreover, after being screened by the appropriate committees, the scientific record developed by U.S. agencies often undergoes another kind of scrutiny that is essentially unknown in Britain. This is the public hearing authorized by law for most regulatory proceedings concerning toxic chemicals. The form of the hearing varies according to the nature of the agency's policy action. It approaches the model of a formal trial when individual interests are at stake, as in the case of pesticide registration. But all administrative hearings involving chemical hazards provide at least informal opportunities for the public to examine the decision-making record developed by the agencies and to comment on the agency's technical determinations. The results of such hearings must be incorporated into the agency's final decision, and this, in turn, can be challenged in court by parties who are adversely affected. Judicial review becomes the ultimate check on the quality of an agency's use of science. It is a vehicle for ensuring that the agency has engaged in "reasoned decision-making," and in applying this standard, courts may consider the completeness of the record, the clarity of the agency's explanation, and the adequacy of its response to significant issues raised by dissatisfied parties.[7]

The effect of open advisory committee meetings, hearings, and judicial review in the United States is to pull technical controversies out from the relatively closed environment of the scientific community into the arena of general public debate. This change of venue influences the character and content of risk disputes in several crucial respects. First, the uncertainties and expert conflicts surrounding risk assessment are discussed more fully and openly in the United States than in other Western countries. Second, public awareness of the uncertainties involved in risk decisions creates pressure for U.S. decision makers to make their risk assessment procedures more reliable and uniform, for example, through quantification and through issuance of general guidelines and protocols.[8] Third, and perhaps most important, the public discussion of science in the policy context leads to a sharper appreciation of the fact that there are gray areas between science on the one hand and policy on the other. A number of policy analysts have pointed to the existence of this middle zone — termed trans-science or science policy — where decisions depend partially but not exclusively on data and techniques controlled by scientists.[9] This insight has triggered numerous U.S. controversies about the proper institutional locus for making decisions about risk. Should they be left to scientists or should they rather be delegated to administrative officials because of their presumed policy or value component? Such controversies have arisen less frequently in Britain, where the expert decision-making process does not draw clear distinctions between science and science policy. My objective in the remainder of this chapter is to trace the impact of these contrasting developments on the interpretation of scientific information for regulatory purposes in the two countries.

Four Determinations of Risk

The four case studies presented here cover a varied range of risks to human health. They involve different routes of exposure: the workplace, the general environment, ambient air, and food. The health effects of two of the four substances, asbestos and lead, are known with considerable certainty. For the remaining two, formaldehyde and 2,4,5-T, the acute effects are reasonably well understood, but the chronic impacts are still vigorously debated. Asbestos is a known human carcinogen and formaldehyde, an established animal carcinogen, may also present a risk of cancer to man. Lead, however, is a neurotoxin, and 2,4,5-T has been associated with neurological and reproductive effects, as well as with cancer. Taken together, the four substances provide a reasonably full data base from which to assess how Britain and the United States evaluate science in the course of managing chemical-related risks.

Asbestos

For policy makers seeking undisputed evidence of a hazard to health, asbestos comes close to being the proverbial "smoking gun." The substance has been commercially exploited on a large scale since the end of the nineteenth century, with demand increasing rapidly after the first world war. Because of its binding and fire retardant properties, asbestos has been used in a wide variety of consumer products, including insulation materials, brake linings and other friction products, textiles and cement. As its beneficial uses multiplied, asbestos revealed itself as a potent threat to human health. Some of its disastrous effects — irreversible lung disease and cancer — were recognized more than seventy years ago. By the 1970s, asbestos was identified not merely as a hazardous substance, but as one of the most serious public health problems of the decade.

Asbestos presents a unique case for risk management policy, not only because its effects on human health are beyond conjecture, but because there is a substantial body of epidemiological evidence from which conclusions can be drawn about the degree of risk at different exposure levels. The problem for regulators in both Britain and the United States has been to design reasonable public health measures based on this large mass of evidence. This exercise clearly includes both a scientific and a socioeconomic component. Considerations such as the impact on employment and the cost and availability of substitute products have played an unavoidable role in both British and American strategies for regulating asbestos. But although these considerations are permitted to influence the pace and stringency of regulation, they should not in theory have any impact on the assessment of health risks. It is on this presumably apolitical phase of scientific assessment that this case study focuses.

Positive results in well-designed and well-conducted epidemiological studies provide regulators the strongest possible ammunition for controlling substances of proven benefit to society. In the case of asbestos, government agencies have had at their disposal dozens of studies linking the substance to asbestosis, lung cancer, and mesothelioma (a rare, asbestos-induced form of cancer) in exposed groups across industrial countries in Europe, North America, and South Africa. This supply of information is "universal" in the sense that the same studies are available for analysis by scientists and government officials in Britain, the United States, or any other country concerned about the health effects of asbestos. A cross-national comparison of asbestos risk assessments, however, shows that marked divergences appeared in the characterizations of risk that British and U.S. scientists constructed from the same studies.

In Britain, the task of reviewing the health risks of asbestos and recommending control strategies was delegated in 1976 to a specially appointed Advisory Committee on Asbestos (ACA), composed of independent scientists as well as experts nominated by labor, industry, consumer groups, and local government.[10] This committee in turn established five working groups to provide detailed information on medical issues, environmental monitoring, legal and administrative controls, production, and substitutes. ACA also requested a Toxicity Review from an ad hoc two-member panel and commissioned additional papers from outside experts on the "ill effects of asbestos on health" and on "asbestos control limits." The paper on ill effects proved especially influential in guiding government policy. Finally, the committee solicited information from the public and held three days of open meetings at which twelve of the outside contributors presented additional comments. All these submissions were incorporated into the committee's overall assessment of asbestos risks and its policy recommendations. ACA's final report was published in 1979 by the Health and Safety Executive (HSE), the principal implementing agency for worker protection policies in Britain.

In the United States, two federal agencies have been responsible for assessing the health risks of asbestos: EPA and NIOSH, a research agency located in the Department of Health and Human Services. One of NIOSH's most important functions is to prepare risk assessments and recommend exposure standards for toxic substances in the workplace. These "criteria documents" play a major role in guiding OSHA's regulatory priorities and standard-setting. The 1972 and 1976 criteria documents for asbestos, for example, substantially influenced OSHA's policies on the substance. EPA is primarily concerned with nonoccupational exposures to toxic substances and has taken the lead in assessing risks to the general public from asbestos in public buildings. The following discussion compares ACA's analysis of the scientific data on asbestos with comparable analyses by NIOSH and EPA in the United States.

In reviewing the epidemiological data, ACA concluded that different fiber types differ in their capacity to induce human disease. The committee based these findings on the toxicity review carried out by its two independent consultants, E. D. Acheson and M. D. Gardner. These experts read the epidemiological studies as supporting quite different hazard assessments for the three major fiber classes: crocidolite, amosite, and chrysotile.[11] The strongest indications of hazard were found for crocidolite and the weakest for chrysotile, although no fiber type was completely exonerated. The argument that disease incidence varies with fiber type seemed most convincing in the case of mesothelioma. Several of the studies reviewed by Acheson and Gardner appeared to confirm these patterns of variation unequivocally, and the findings seemed all the more conclusive because they recurred in four different exposure contexts (miners, process workers, neighborhood and domestic exposure, and the geographical distribution of mesothelioma).

In the field of health hazard assessment, however, evidence is seldom one hundred percent consistent, and British cancer experts had to deal with numerous conflicts in the data on asbestos. Their handling of one disputed study indicates how the interpretation of troublesome individual items was reconciled with the preferred "reading" of the larger body of data. A study by Enterline and Henderson found only one case of mesothelioma in a cohort of asbestos production and maintenance workers who had lived to retirement age. Since this group was primarily exposed to chrysotile, the study appeared to support Acheson and Gardner's general thesis that chrysotile presents a relatively low risk of mesothelioma. The

Enterline and Henderson data, however, were challenged by Borow, who claimed to have found many cases of mesothelioma in younger members of the same study population. While not dismissing Borow's observations out of hand, Acheson and Gardner discounted their significance in various ways. They noted, for example, that Borow gave "no data on the histories of exposure of the individual cases."[12] They also provided an alternate explanation for Borow's findings which accorded better with their own theories about the linkage between mesothelioma and crocidolite. Specifically, they suggested that cases of mesothelioma appeared in the study population only after the introduction of crocidolite into the factory under investigation, and then only after the passage of the predictable latency period.[13] Data on women in the workforce, who were exposed only to chrysotile, strengthened this thesis, since the women showed only minimal susceptibility to mesothelioma.

ACA was persuaded by Acheson and Gardner's analysis and endorsed the view that the risks of asbestos-related disease vary with the fiber type. For mesothelioma, in particular, ACA summarized the case as follows: "evidence from miners; from process workers exposed to a single fiber type; from the distribution of neighbourhood and domestic cases; and from the geography of mesothelioma, when combined, presents a powerful case from four different sources that crocidolite has been more dangerous than chrysotile and anthophyllite. The position of amosite may be intermediate between crocidolite and chrysotile."[14] These conclusions provided the basis for regulatory measures directed at specific fiber types, as we see below.

In the United States, by contrast, NIOSH's 1976 assessment of asbestos did not identify fiber type as a factor of particular relevance to the characterization of risk. The agency's position was summed up as follows: "Although some differences in both the fibrotic and the carcinogenic responses to asbestos fibers may depend on the type of fiber administered, all types have definitely shown both these kinds of action."[15] In other words, NIOSH was more concerned about the fact that *any* fiber type can produce a pathological response than about risk differentials associated with different fiber types. The agency's interpretation of the Enterline and Henderson study and of Borow's subsequent findings was consistent with this position. Both studies were reported in NIOSH's brief summary of epidemiological evidence on individual types of fibers, but unlike Acheson and Gardner, NIOSH scientists found Borow's conclusions more congenial. Accordingly, the NIOSH appraisal attributed the difference between the two studies to "methodologic variations," specifically, to the fact that "Enterline and Henderson had limited their investigation to men age 65 or over, while many of the mesothelioma cases reported by Borow et al. had died before that age."[16] This explanation preserved the credibility of the

Borow study, which happened to agree with NIOSH's basic determination that chrysotile, along with crocidolite, is definitely implicated in the etiology of mesothelioma, and should be treated with equal caution.

The 1979 ACA report and the 1976 NIOSH report also differed in their consideration of a threshold for asbestos-related disease. Both reports acknowledged that there is no safe level of exposure to asbestos; asbestosis, respiratory cancer, and mesothelioma can all occur even at extremely low exposures. But the British and U.S. experts arrived at this conclusion along different intellectual paths. In its first criteria document for asbestos, NIOSH focused primarily on the risk of asbestosis and recommended a standard aimed at preventing this disease. The hope was that such a standard would also provide adequate protection against asbestos-induced cancer. By 1976, however, both NIOSH and OSHA were more concerned about the risk of cancer. The evidence of risk at very low exposure levels supported the position taken by both agencies that it was practically impossible to establish safe thresholds of exposure to carcinogens. Thus, NIOSH recommended in 1976 that the asbestos standard should be set at the lowest detectable level, an approach that the agency now favored for controlling all carcinogens.[17]

ACA, on the other hand, based its acceptance of the no-threshold model for asbestos on empirical evidence rather than on a general theory concerning carcinogens. The British committee attached great weight to the fact that existing epidemiological studies showed increasing risk with increasing exposure, and that there were well-attested cases of cancer following limited contact with asbestos in homes and around mines, factories or docks. This approach suggests that British scientists were not *in principle* opposed to the discovery of a threshold. For example, in evaluating the evidence relating to cancers of the gastrointestinal tract, Acheson and Gardner reported that "there is insufficient data to determine whether or not a threshold exists."[18] The statement implies a readiness on the authors' part to admit that there could be a threshold for some forms of asbestos-induced cancer, a position that NIOSH and OSHA appeared reluctant to accept after 1976.

Scientific opinion in the United States and Britain also diverged in assessing the risks of asbestos exposure to the public at large. ACA took up the question in a limited form in its 1979 report.[19] The committee reviewed the studies pertinent to exposure assessment and identified some problems that might be encountered in estimating the risk to the general public. For example, ACA noted that discrepancies in measurement techniques make it difficult to compare data from environmental monitoring to data gathered from studies of industrial exposure. A more recent report prepared for the Health and Safety Commission (HSC) by R. Doll and J. Peto again emphasized the uncertainties in measuring environmental exposure, although the authors provided some risk estimates based on data compiled by a Canadian agency.[20] U.S. scientists, however, were quicker to accept the idea that a conversion factor may be used in order to bring the data from environmental and industrial studies within the same comparative framework. For example, William Nicholson of Mount Sinai School of Medicine, a leading U.S. center for asbestos research, testified on the use of this technique at a congressional hearing in March 1982.[21]

ACA made no attempt to develop a quantitative estimate of risk to exposed populations even in contaminated buildings, where the risk to the general public is believed to be greatest. The committee made a number of proposals for reducing public exposure to asbestos, but these were based on a general philosophy of keeping dust emissions under control.[22] Although the government has instituted monitoring programs to measure asbestos fibers in the environment and in buildings, these actions have not led to the establishment of exposure standards or to asbestos stripping programs in schools and other buildings. In the United States, on the other hand, EPA, under pressure to do something about asbestos in schools, ventured to prepare precise quantitative estimates of the lifetime risk and the number of premature deaths for students and adult employees.[23] In so doing, the agency was clearly prepared to bridge some of the methodological uncertainties noted by ACA. EPA's initiatives in this area point to a more widespread use and acceptance of predictive models in the United States than in Britain.

While it is difficult to be more than impressionistic on this point, the asbestos case suggests that there may be a predisposition in each country to rely more on domestic rather than foreign studies for purposes of health risk assessment. In estimating the environmental risk from asbestos, for example, Nicholson presented data on dose-response relationships for cancer from several studies of American industrial workers.[24] Acheson and Gardner's report also addressed the question of dose-response relationships[25] and, at least in the case of asbestosis, apparently preferred data from British studies at Rochdale to comparable work done in the United States and Canada under somewhat different exposure conditions. In the area of exposure assessment, both the Mount Sinai group and EPA have relied on air sampling in U.S. homes and buildings, including public schools with damaged asbestos surfaces.[26] Acheson and Gardner's account of exposure in the nonoccupational environment mentioned studies by Nicholson and other foreign researchers, but attached more significance to a study done by Byrom and his colleagues on asbestos-contaminated buildings in Britain.[27]

One possible reason for cross-national differences like these is that the phenomena being studied in the area of health risk assessment — disease incidence, mortality rates, building contamination — are so dependent on local conditions that study results cannot easily be transferred across national boundaries. Another explanation, with more implications for the sociology of knowledge, is the so-called "not invented here" syndrome,[28] which makes scientists more receptive to protocols and methodologies developed in their own country than abroad. If this is a genuinely important factor in the certification of scientific knowledge, then a national bias is bound to creep into health risk determinations whenever the methods of data collection and risk assessment are not fully standardized across national lines. I shall return to this point at the conclusion of this chapter.

Differences between the asbestos control policies adopted in Britain and the United States reflect, to some extent, the divergences I have identified in the two countries' scientific assessments of risk. Following ACA's recommendation, for example, British policy has distinguished among different fiber types in prescribing exposure limits. Crocidolite is most stringently controlled. ACA recommended a statutory ban on the use or import of this form of asbestos and a control limit in the workplace of 0.2 fibers/ml.[29] For amosite, the committee proposed a less stringent limit of 0.5 fibers/ml, and for chrysotile a still higher control limit of 1 fiber/ml. Action on these proposals was delayed for a time while Britain participated in negotiations at the European Economic Community (EEC) on an asbestos directive that would apply to all member countries. But under public pressure the HSC eventually decided to implement the ACA proposal even before the EEC negotiations were completed. The asbestos control limits were tightened further in 1983, not in response to new evidence, but because industry agreed to accept stricter standards.[30] Although HSC has expressed concern about risks to the general public, there has been no attempt as yet in Britain to regulate asbestos levels in the environment, though there is a requirement to monitor after asbestos stripping where necessary.

These policy outcomes contrast markedly with regulatory approaches pursued in the United States. U.S. agencies have not attempted to link their asbestos control measures to specific fiber types. The pace of standard-setting, however, has been slow, even in the workplace, where asbestos has long been recognized as a top priority. The substance was included in OSHA's first promulgation of standards in May 1971, at which time the workplace exposure limit was set at 12 fibers/cc. In 1972, however, NIOSH recommended a 2 fibers/cc exposure limit, following the lead set by the British Occupational Hygiene Society in 1968 after its survey of asbestosis among textile workers. Pressured by the unions, OSHA first issued an emergency temporary standard (ETS) of 5 fibers/cc in late 1971. A permanent standard of 5 fibers/cc was promulgated in 1972 and was lowered to 2 fibers/cc four years later.[31] The current U.S. occupational standard for asbestos thus continues the 1972 NIOSH recommendation, and does not reflect more current scientific knowledge about the risks of cancer from low-dose exposures. Despite an interim setback in court,[32] OSHA was prepared by late 1985 to propose a new standard that would bring regulatory policy more in line with available knowledge.

In recent years, EPA has become a major actor in regulating asbestos, primarily because of mounting concern about environmental exposure in schools and other public buildings. The agency has authority under the Toxic Substances Control Act (TSCA) to regulate substances that cannot be adequately controlled under other federal programs. During the early 1980s EPA not only carried out a quantitative risk assessment for asbestos in schools, but worked on a variety of proposals to collect information about the uses of asbestos, to warn certain user groups, to inspect buildings and require removal of deteriorating asbestos, and to ban or phase out some uses of the substance. These initiatives were frustrated for a time by interference from the Office of Management and Budget. Indeed, EPA announced in early 1985 that it lacked authority to ban asbestos under TSCA, and moved instead to have OSHA and the Consumer Product Safety Commission (CPSC) take over further rule-making.[33] However, this decision was rescinded following a public outcry, and in 1986 EPA proposed a ban on five major uses of asbestos and a phase-out of all remaining uses over the next ten years.[34]

Formaldehyde

Formaldehyde, one of the largest-volume chemicals in use today, is known to be a highly toxic substance. An extremely versatile compound, formaldehyde is used in a wide variety of manufactured products, from plywood, particleboard and home insulation to textiles, cosmetics and embalming fluids. The substance is genuinely ubiquitous. Human exposure to formaldehyde can occur through numerous pathways including food, tobacco smoke, ambient air, and occupational contact. It is well known that the chemical is extremely irritating to the eyes, skin, and mucous membranes, and can cause death at exposures above 50-100 ppm (parts per million). The acute effects of formaldehyde are so severe that in 1976 NIOSH recommended a reduction of employee exposure levels to 1 ppm.[35] Formaldehyde has also been shown to cause cancer in rats, a finding that triggered renewed regulatory concern with this compound in the early 1980s.

The first indication that formaldehyde is a carcinogen came from a long-term bioassay conducted by the Chemical Industry Institute of Technology (CIIT), a research institute established by the U.S. chemical industry to test large-volume chemicals. Preliminary results from a chronic inhalation study of rats exposed to formaldehyde became available in 1979, showing that 20 percent of the animals contracted nasal tumors following exposure at a high dose level of about 15 ppm.[36] This finding, which was confirmed by final results from the same experiment, instantly aroused fears that formaldehyde could also present a carcinogenic risk to man. The substance accordingly underwent thorough review by scientific advisory committees in both Britain and the United States to determine whether the risk warranted new regulatory measures.

In Britain, formaldehyde was the second chemical to figure in a series of toxicity reviews commissioned by HSE for the use of ACTS. The review confirmed the mutagenic and acute toxic effects of formaldehyde and accepted the fact, based on the CIIT study, that the chemical is a carcinogen in rats at two dose levels. The conclusions with respect to human carcinogenicity, however, were brief and indicated no grounds for concern: "There is at present no evidence suggesting that exposure to formaldehyde has produced cancer in humans. A number of epidemiological studies are in progress to investigate this aspect."[37] The 1981 publication did not discuss any of the mortality studies already carried out in industries where workers might be presumed to have been exposed to formaldehyde. Following the publication of the HSE toxicity review, an epidemiological study designed by Acheson to look at the carcinogenic effects of formaldehyde has been completed and published. British authorities accepted the Acheson study as showing no increase in the risk of human cancer, thus ending further scientific debate on this issue in Britain.[38]

By comparison, the U.S. debate concerning formaldehyde has been much more voluminous, protracted and intense. Most of the discussion has been carried on within or at the fringes of the regulatory system; the expert committees and workshops reviewing the toxicity of formaldehyde have either included government officials or have operated under the auspices of government agencies. One of the earliest reviews was carried out by a panel of scientists from various federal research institutions in response to requests from CPSC and the National Toxicology Program. Published almost simultaneously with the British toxicity review, the federal panel's report apparently reached the opposite conclusion: that formaldehyde "should be presumed to pose a carcinogenic risk to humans."[39]

The U.S. panel reviewed the epidemiological data more comprehensively than did its British counterpart, taking into account mortality surveys in industries where formaldehyde exposures could be expected, even though these studies were not specifically designed to look at the effects of formaldehyde. In the panel's judgments, these data suggested a need for caution, though they were not conclusive: "Although the above investigations were not designed to study health effects of formaldehyde, the results suggest the possibility of formaldehyde being carcinogenic in humans."[40] The panel confirmed, however, that additional epidemiological studies would be required to make a more definite evaluation of the compound's chronic effects. The absence of relevant human data was also cited in a roughly contemporaneous National Academy of Sciences (NAS) report on formaldehyde and other aldehydes.[41] The NAS panel, as well, concluded that there were no studies showing that formaldehyde is carcinogenic in humans.

Upon what scientific basis, then, did the U.S. federal panel on formaldehyde conclude that the chemical should be presumed to pose a risk of cancer to man? The answer comes in two parts which must be deduced from the report as a whole, since the panel did not explicitly explain its conclusions with regard to human risk. First, the panel seems to have approached the preliminary CIIT findings with an unstated assumption that animal carcinogens should be presumed carcinogenic for man in the absence of evidence to the contrary. This presumption in favor of human carcinogenicity was consistent with positions adopted by U.S. government agencies throughout the 1970s.[42] It is plausible to assume that the members of the federal panel, who were all government scientists, shared this precautionary view about the significance of animal data. In the case of formaldehyde, this group of experts found no contrary indications that were strong enough to set aside the presumption of carcinogenicity. In particular, the panel rejected the theory that cancer might have occurred in the rats through a secondary mechanism that would not be found in humans, such as through irritation of the nasal lining.[43]

Equally relevant to the panel's final judgment, however, was its skepticism about the predictive value of the epidemiological studies then in progress. In a brief survey of these studies, the panel repeatedly expressed the view that they were very likely to prove inconclusive.[44] In some cases, the ongoing studies focused on too small a population, and hence lacked sufficient statistical power; in others, the study design provided insufficient information about exposure or ignored important confounding factors. The panel's overall assessment was that these investigations would neither exonerate formaldehyde nor definitely establish it as a human carcinogen. The probable indeterminacy of the epidemiological studies no doubt reinforced the panel's decision that human risk should be presumed on the basis of conclusive results in the CIIT rodent bioassay.

Between 1982 and 1985, the carcinogenicity of formaldehyde became one of the most extensively debated regulatory issues in the United States. A focus of controversy was a decision by John Todhunter, EPA's assistant administrator for toxic substances, that despite the CIIT findings formaldehyde should not be considered for priority action under TSCA.[45] Todhunter's decision was based on a review of the compound's toxicity which his critics regarded as scientifically unsound and heavily biased in favor of the formaldehyde industry's interests.[46] In an effort to undo the charges of bias, EPA in 1983 sponsored a consensus workshop on formaldehyde. This event brought together about sixty experts from the United States and abroad to discuss in an objective and nonadversarial manner the major toxicological issues relevant to policy-making. To ensure a balanced review, experts were invited from a variety of institutional settings (academia, government, industry, and public interest groups). The participants were divided into eight issue-oriented panels, each of which addressed questions put to them by the executive panel in charge of the workshop. The executive panel, like the larger group of participants, included representatives of multiple institutional and political interests.

The document that emerged from the formaldehyde consensus workshop was both more comprehensive and more directly relevant to questions of regulatory interest than the earlier British and American toxicity reviews.[47] While a point-by-point comparison of the three reviews cannot be undertaken here, it is instructive to compare them on at least one specific issue: the interpretation of the epidemiological data on cancer. The epidemiology panel at the 1983 consensus workshop examined both completed and ongoing studies on definable populations exposed to formaldehyde. It also looked at the available evidence concerning each category of neoplasm associated with formaldehyde in the scientific literature. The panel's general conclusion was that "it is not possible from the available epidemiological studies to exclude the possibility that formaldehyde is a human carcinogen."[48] More specifically, the panel found that the evidence indicated no substantial increase in the risk of nasal cancer, but that the excesses of brain tumors and leukemia in studies of certain professional groups exposed to formaldehyde were suggestive and merited further research.

The Acheson study, which had proved decisive in ending British policy debate on formaldehyde, was discussed by the consensus workshop panel primarily under the heading of lung cancer. The panel noted that this study provided some evidence of a dose-response relationship between formaldehyde exposure and lung cancer, although Acheson later qualified this conclusion on the basis of follow-up data.[49] The Acheson study was probably more thoroughly scrutinized in the U.S. workshop than it ever was in Britain, where ACTS reviewed it without fanfare. Ironically, however, Acheson's influence on the workshop's findings was clearly much slighter than his impact on the British assessment of formaldehyde, even though he chaired the epidemiology panel at the American proceedings.

The exposure assessment and risk estimation panels at the consensus workshop reflected the U.S. regulatory system's preoccupation with quantitative risk assessment. With the exception of one representative from the British Medical Research Council's radiobiology unit, the members of these two panels were all Americans. The risk estimation panel considered one of the most contentious issues involved in quantifying carcinogenic risk: the choice of an appropriate high to low dose extrapolation model for data derived from long-term cancer bioassays. The "consensus" that emerged on this point was more an agreement not to disagree than a substantive meeting of minds. The panel's final report simply left the way open for further experimentation — and controversy — in the selection of extrapolation models.[50] At the same time, the panel provided no convincing reason for federal agencies such as EPA to change their stance on certain critical issues. For example, the panel found no clear evidence for a threshold response to formaldehyde, and thus saw no reason to prefer a threshold extrapolation model. It also affirmed the appropriateness of linear low-dose extrapolation as a way of setting a plausible upper limit on the risk to humans. Assuming linearity at low doses is consistent with EPA's policies as expressed in the agency's carcinogenic risk assessment guidelines issued in 1986.[51]

Turning to regulatory action on formaldehyde in Britain and the United States, one is struck by the absence of any clear-cut link between the scientific debates and the policy outcomes in either country. Until 1984 workplace exposure to formaldehyde in Britain was regulated by means of a recommended limit of 2 ppm, a standard adopted from the list of threshold limit values developed by the American Conference of Governmental Industrial Hygienists (ACGIH). In spite of the controversy generated by the CIIT cancer findings, British officials deferred more stringent action on formaldehyde until the data from Acheson's study became available. These findings were accepted as persuasive evidence against the hypothesis that formaldehyde is a human carcinogen. Nonetheless, ACTS agreed that formaldehyde should be subject to a "control limit" of 2 ppm.[52] In the British regulatory scheme, the presumption is that workers *must not* normally be exposed to levels higher than prescribed by a control limit. Recommended limits, which are usually based on less conclusive hazard data, can be exceeded with a lower probability of official sanction than control limits. The decision to move from the laxer to the stricter type of standard for formaldehyde was apparently motivated by an agreement within ACTS that the 2 ppm limit was "reasonably practicable"[53] for industry. The policy change certainly was not founded either on any new epidemiological findings or on quantitative estimates of risk derived from the CIIT bioassay.

In the United States, the CIIT bioassay results had a decisive impact on the actions of the CPSC. A series of complaints concerning urea-formaldehyde foam insulation (UFFI) installed in homes led the agency to regulate UFFI in 1980. The initial plan was to require manufacturers to provide potential buyers with written information about the adverse effects of UFFI. CPSC soon decided, however, that only a ban on further production of UFFI would adequately guard consumers against the risks presented by faulty installation.[54] The change of regulatory strategy was prompted largely by the discovery that formaldehyde was an animal carcinogen. Indeed, CPSC's final proposal to ban UFFI was supported by a quantitative risk assessment which predicted that the lifetime cancer risk to exposed individuals was 51 in a million. The formaldehyde industry challenged the UFFI ban in the federal court of appeals for the Fifth Circuit. Upon review, the court overturned the ban, finding it unsupported by "substantial evidence."[55] An important reason for this reversal was the fact that the agency risk assessment was based on a single animal experiment, but ignored the numerous epidemiological studies which, in the court's view, tended to clear formaldehyde as a hazard to human health.

OSHA and EPA responded to the CIIT study very differently from CPSC. Both agencies initially decided not to consider formaldehyde for further rulemaking. In OSHA's case, this had the effect of leaving in place the occupational safety and health standard of 3 ppm adopted several years before the completion of the CIIT bioassay. This standard was less stringent than the exposure levels recommended by NIOSH and ACGIH just on the basis of formaldehyde's acute toxicity. At EPA, Todhunter's decision not to accord priority to formaldehyde seemed to preclude further action under TSCA, but environmental and labor organizations sued both EPA and OSHA in the District of Columbia federal court to change this state of affairs. These lawsuits resulted in a court order requiring OSHA to reconsider its negative decision on formaldehyde,[56] and in December 1985 the agency proposed a revised workplace standard for the substance.[57] EPA, in the meantime, reopened its proceedings under TSCA, acknowledging that serious procedural improprieties had marred its prior assessment of the compound.[58] In sum, the CIIT bioassay data pushed formaldehyde to the top of the policy agenda for several U.S. agencies, but as of 1986 the change in the chemical's scientific status had not induced definitive changes in regulatory policy.

2,4,5-T

Trichlorophenoxyacetic acid, known as 2,4,5-T, is the active ingredient in a family of herbicides commonly used in Britain and the United States during the 1960s and 1970s to control broad-leaved weedy plants in forests, agricultural lands, and rights of way. Health concerns arose in connection with 2,4,5-T mainly because the compound is invariably contaminated with dioxin, one of the most potent toxic chemicals known to man. Dioxin is associated with a variety of health effects, including teratogenicity, carcinogenicity, and a skin disease called chloracne. Dioxin levels in commercial 2,4,5-T preparations have been regulated for many years. Nevertheless, regulators in both countries were asked to consider whether 2,4,5-T can be safely used as long as it exposes the public to even trace levels of dioxin. From the standpoint of risk management policy, the key question was whether 2,4,5-T containing minute amounts of dioxin was hazardous enough to warrant its withdrawal from the market.

During the 1970s, public misgivings were aroused not by relatively esoteric experimental findings, as in the case of formaldehyde, but by two massive environmental disasters that highlighted the alleged toxic effects of dioxin.[59] Agent Orange, a herbicide containing 2,4,5-T, was widely used as a defoliant by American troops in Vietnam. After the war, stories began circulating about abnormally high numbers of birth defects and other medical problems not only among Vietnamese civilians, but also among American veterans exposed to Agent Orange. The pooling of these stories eventually led to an enormous, highly publicized lawsuit by the Vietnam veterans against the manufacturers of the herbicide.[60] In Europe, fears about 2,4,5-T were heightened by an explosion at a chemical factory in Seveso, Italy, which released a cloud of dioxin over the surrounding countryside. In the ensuing months villages near Seveso were evacuated and abandoned, animals died, people suffered varied medical disorders, and women underwent abortions to guard against defective births.[61] Dioxin became a dreaded word and political pressure grew in favor of banning the environmental application of trichlorophenol and its commercial derivatives.

By the mid-1970s, 2,4,5-T was hardly a scientific unknown. Already in 1970 most food uses of the herbicide were banned in the United States on the basis of laboratory tests linking the chemical to birth defects in animals.[62] But growing political attention led to more epidemiological research on both 2,4,5-T and dioxin, including follow-up studies of the victims of Seveso and Vietnam, in an effort to clarify the risks of environmental exposure. Many animal experiments were also undertaken in order to distinguish between the effects of dioxin and 2,4,5-T, and to create a data base for assessing the risks of low-level exposures to the chemical.

Among the epidemiological studies, one that proved particularly influential was a survey carried out by EPA in Oregon's Alsea basin, the site of regular annual 2,4,5-T spraying. In the so-called Alsea II study, EPA purported to find a correlation between the spraying activities and peaks in the regional miscarriage rates over a six-year period. Taken together with extensive, though often equivocal, data from other epidemiological and experimental sources, the study appeared to confirm the existence of a human health risk. This, at any rate, was the conclusion initially reached by EPA's own scientific staff. The Alsea II study prompted the EPA Administrator to suspend 2,4,5-T for use on forests and rights of way.[63] However, further assessments of the herbicide continued not only at EPA, where it was reviewed by the pesticide program's standing scientific advisory panel, but also by Britain's ACP. These two bodies agreed on the key question of 2,4,5-T's safety; however, as described below, their scientific evaluations had only a limited impact on the herbicide's marketability.

In Britain, ACP periodically reviewed the toxicity of 2,4,5-T and concluded as recently as 1979 that the compound was safe if used as directed.[64] In normal circumstances, the committee might not have agreed to reexamine the substance the following year, but its hand was forced by the National Union of Agricultural and Allied Workers (NUAAW), which launched a campaign to get the herbicide banned in Britain. NUAAW's drive against 2,4,5-T was unconventional in both its goal and its tactics. The decision to campaign against a specific pesticide was highly unusual in a policy environment where the identification and control of technological risks are traditionally regarded as tasks for experts. The union's objective, however, was to supplant ACP's deliberate scientific judgment with an alternative, more populist assessment of risk. To accomplish this formidable task, NUAAW adopted the extremely effective strategy of building its own "dossier" on 2,4,5-T.[65] This document marshalled evidence and arguments in ways that ultimately undermined ACP's credibility and took the decision to regulate 2,4,5-T out of the hands of the experts.

The union's first point of attack was ACP's determination that 2,4,5-T was safe when used "in the recommended way for the recommended purposes." The dossier gave an account of field conditions showing that the recommended conditions for the use of the herbicide were rarely met in practice. Here, the union was on particularly strong ground because ACP's expertise clearly did not extend to the circumstances under which farm workers actually applied 2,4,5-T. Second, the dossier dwelt on the evidence that had led authorities in other countries — most notably the United States and Sweden — to ban at least some uses of 2,4,5-T. The data cited by NUAAW included reports of animal experiments, as well as the controversial Alsea II study and a Swedish study finding an apparent increase in soft tissue sarcomas among workers exposed to phenoxyacetic herbicides. The centerpiece of the NUAAW dossier, however, was a collection of sixteen "case histories" investigated to varying degrees by the union. Most of the cases involved allegations of miscarriages or birth deformities following contact with 2,4,5-T.[66] These reports brought the discussion about risk down from an abstract, theoretical plane to a much more personal and concrete, as well as infinitely more newsworthy, level. The 2,4,5-T story became a media favorite, and the NUAAW took full advantage of the opportunity to publicize its disagreements with the government's scientific experts.

In response to this challenge, ACP issued another report on 2,4,5-T in December 1980. Though much longer and more detailed than the 1979 review, the new report was virtually identical in its conclusions about the safety of the herbicide:

> What we have had to consider in this Review is whether there is any sound medical or scientific evidence that humans or other living creatures, or our environment would come to any harm if cleared 2,4,5-T herbicides continue to be used in this country for the recommended purposes and in the recommended way. We have found none.[67]

Although the committee reaffirmed its earlier scientific opinion, it paid serious attention to NUAAW's allegations. A lengthy appendix analyzed each of the case histories compiled by the union, as well as a few others that had come to the committee's notice during the second review. In each case the committee concluded that "either there was no exposure to 2,4,5-T herbicides or any such exposure could not have given rise to the stated effects."[68] The caveat about exposure data was not limited to NUAAW's informal case studies. ACP expressed reservations about the quality of the exposure information in some of the more authoritative epidemiological studies that purported to find adverse effects on human health. For example, in reviewing the Swedish study mentioned above ACP found fault with the investigator's use of worker recall as a basis for reconstructing exposure patterns. Similarly, a major reason for dismissing the Alsea II study was that "evidence of relevant exposure to 2,4,5-T was lacking."[69] The committee identified a number of shortcomings in EPA's data on exposure:

> The essential preliminaries are to demonstrate that the mother was exposed to the environmental agent concerned at a meaningful level and . . . at a relevant time of foetal development. This time factor is especially important with a substance like 2,4,5-T which, unlike dioxin, is excreted fairly rapidly from the body, i.e., exposure may therefore not occur at the time when the affected tissue or organ of the foetus is developing. The fact that none of these preliminaries was demonstrated is one reason why, for example, we do not regard the Oregon studies . . . above as giving grounds for supposing that 2,4,5-T herbicides present a risk to women of childbearing age.[70]

In NUAAW's judgment, however, ACP's scientific qualms only masked a basic split in the ways that scientists and workers perceive risks. David Gee, safety officer to the mammoth General and Municipal Workers' Union, argued that the two groups hold fundamentally different views about what constitutes proof of a health hazard. Specifically, scientists are more inclined to apply the "beyond all reasonable doubt" standard of the criminal law to the evaluation of occupational hazards, whereas workers are willing to settle for a lesser level of proof, something more akin to the "preponderance of the evidence" standard of the civil law. Consequently, unions are prepared to give greater credence to nonhuman evidence in evaluating risks, while scientists "want to see the human evidence first, before they are convinced."[71] By the same token, unions might also be willing to attach some weight to positive results in studies that scientists consider methodologically flawed. Whether these characterizations hold true across the board or not, they did seem to capture the essential differences in the attitudes of ACP and the NUAAW activists in the 2,4,5-T case.

Official British policy on 2,4,5-T followed the recommendations of ACP. It remains legal to use these herbicides as far as the government is concerned. However, NUAAW persuaded several powerful political actors to accept its assessment of the herbicide's risks. In an unprecedented show of support, the Trades Union Congress (TUC), Britain's labor federation, called for a total ban on the compound.[72] British Rail, one of the major users of 2,4,5-T, suspended further applications of the herbicide, and by 1982 more than eighty local authorities had followed suit. The popular conviction that 2,4,5-T was too risky to remain on the market thus won out over ACP's informed judgment that there was no good evidence that 2,4,5-T was harmful to human health.

In the United States, EPA reacted quickly to preliminary findings from the Alsea II study, and, as noted above, immediately suspended some uses of 2,4,5-T. The FIFRA scientific advisory panel, however, concluded in time that the study was sufficiently flawed to be useless as a basis for evaluating the safety of the herbicide. This conclusion paralleled that of ACP in Britain, but came only after the agency had already declared the herbicide unsafe for some purposes. The main reason for this anomalous outcome is that under U.S. law the task of determining acceptable levels of risk belongs not to scientists, but to administrative officials, who may, if circumstances warrant, act on less than a watertight showing of harm. For herbicides like 2,4,5-T, FIFRA directs EPA to ensure that such substances will not produce "unreasonable adverse effects" or "imminent hazards".[73] FIFRA does not contemplate that scientific assessments should wholly determine the policy outcome with respect to chemical pesticides. Though EPA has to seek advice from scientists outside the agency, judgments about the unreasonableness of a chemical's effects must ultimately be made by the administrator. In this scheme, which requires EPA to balance costs and benefits in deciding whether a risk is "unreasonable," the level of proof required by the agency can easily fall below the "beyond reasonable doubt" standard preferred by science.

Peer review by the scientific advisory panel did, however, influence the regulatory status of 2,4,5-T in the United States and partly undid the divergence from Britain. Convinced that it could no longer rely on the Alsea II study, EPA entered into negotiations with the manufacturers of 2,4,5-T to reach a voluntary accommodation about the chemical's remaining uses. Yet events conspired to keep 2,4,5-T alive as a political issue. The Agent Orange lawsuit dragged on with intermittent eruptions into the headlines. Dioxin, too, continued to attract unfavorable public attention, culminating in a spectacular pollution incident at Times Beach, Missouri, which forced the federal government to buy up the homes of residents in the contaminated municipality. Finally, in 1983 Dow Chemical, the leading U.S. producer of 2,4,5-T, requested EPA to cancel all its remaining registrations on the ground that profits from selling the compound were no longer high enough to offset the legal and administrative costs of defending its safety.[74]

The 2,4,5-T controversy had substantially different impacts on the scientific advisory systems of Britain and the United States. In Britain, NUAAW's campaign shook ACP's credibility and at least temporarily strengthened demands for transferring regulatory responsibility for pesticides to institutions in which labor was formally represented. In the United States, by contrast, EPA was widely perceived as having overreacted to scientifically inadequate data on risk. Largely in response to the 2,4,5-T debacle, Congress amended FIFRA to require more extensive participation by the scientific advisory panel in EPA's decision-making on pesticides.

Lead in Gasoline

The assessment and regulation of lead in gasoline offer an interesting counterpoint to the 2,4,5-T case because they illustrate the impact of law and politics on the organization of scientific knowledge about a substance that is much more widely regarded as hazardous. For ease of comparison, I am here considering only the case of leaded fuels, even though lead has been regulated in other environments, such as the workplace and the paint and food-processing industries, in both Britain and the United States. The comparison begins with the United States, since leaded fuels were regulated in this country some years before comparable action was undertaken in Britain.

As in the case of asbestos, a major problem for regulators in both countries has been to decide how stringently to control different sources of exposure to lead. Once the substance enters the bloodstream, its effects are relatively easy to monitor, although numerous controversies about behavioral and other subclinical impacts still remain to be resolved. But determining the precise amount of risk contributed by different sources of pollution presented much greater uncertainties. Prior to 1970, for example, there was general agreement that automobile emissions accounted for as much as 90 percent of airborne lead. Yet, from the standpoint of human health, there were many unanswered questions about the significance of airborne lead in relation to other sources such as diet. Given the ubiquity of lead and the multiple pathways of human exposure, it was inevitable that controversies would arise over whether lead in fuels was worth regulating, and if so, to what degree.

EPA's consideration of this question was moulded from the start by its organic statute, the Clean Air Act of 1970, which framed the agency's authority to regulate hazardous fuel additives in the following terms:

> The Administrator may, from time to time on the basis of information obtained under subsection (b) of this section or other information available to him, by regulation, control or prohibit the manufacture, introduction into commerce, offering for sale, or sale of any

fuel or fuel additive for use in a motor vehicle or motor vehicle engine (A) if any emission products of such fuel or fuel additive will endanger the public health or welfare . . ."[75]

EPA took the position that the "will endanger" standard could be met even without showing that lead from fuels alone would harm public health. Rather, EPA argued that it was enough to show that such pollution could "present a significant risk of harm to the health of urban populations, particularly to the health of city children."[76] The gasoline industry challenged this interpretation as a misreading of the risk standard actually intended by Congress. In industry's view, the "will endanger" formulation of the Clean Air Act required "a high quantum of factual proof, proof of actual harm rather than of a 'significant risk of harm'."[77] The industry argued, in effect, for the application of a standard of proof as rigorous as that used by British scientific committees like ACP. But in the end it was the EPA Administrator's reading of the law that prevailed, since the D.C. Court of Appeals agreed with the agency that "endanger means something less than actual harm."[78]

EPA's compilation and analysis of scientific evidence were consistent with its legal position. The agency used data from animal, clinical, and epidemiological studies to establish the existence of a risk to health and to identify automobile emissions as a substantial causative factor. EPA argued, in particular, that blood lead levels of 40 μg/100 g should be taken as dangerous, even though clinical evidence of harm to health was found only at much higher concentrations. EPA also concluded that airborne lead presents a direct threat to public health through inhalation, and that urban children are subject to a further risk through the deposit of airborne lead and its mixing with dust and dirt. The last point was supported by a chain of inferences showing, first, that dust in urban areas, especially near highways, is unusually contaminated with lead, and, second, that a high percentage of children are prone to pica, or dust eating, which renders them particularly susceptible to harm from such surface deposits of lead.

Almost every point in the agency's reasoning came under attack from industry on the general ground that there was no *evidence* for proceeding from one step to the next. Industry claimed, for instance, that EPA had not shown any adverse health effects at the 40 μg blood lead level, that EPA's estimates of direct absorption of lead from the ambient air were flawed, and that there was no evidence that children with pica in fact swallow dust containing excessive concentrations of lead. The reviewing court, however, interpreted the agency's burden of proof in less rigid terms, requiring from EPA only enough evidence to provide a reasonable basis for each of its major conclusions. Applying this standard of proof, the court was satisfied that the agency had properly met its combined burden of fact-finding and policy-making in deciding to phase out the use of lead in gasoline.

In Britain, by contrast, a permissible limit of 0.4 g/liter was still in effect for lead in gasoline in early 1981.[79] At that time, the Department of Health and Social Security (DHSS) commissioned a new study of lead-related risks to health from a working party of scientists chaired by Patrick Lawther. The Lawther report found causes for concern and recommended additional reductions in public exposure to lead, but it stopped short of calling for the complete elimination of lead from gasoline. On the basis of this report, the government announced in mid-1981 a plan to reduce the allowable lead level in fuel to 0.15 g/liter. This decision provoked a public campaign similar to the one organized against 2,4,5-T by NUAAW, only in this case the activist group was an ad hoc entity called CLEAR (Campaign for Lead-free Air).

Unlike most small, single-issue campaign groups, CLEAR possessed two advantages that made its position similar to that of a labor union. The first was the political sophistication of its organizer, Des Wilson, who brought to the task several years of experience forming and directing campaigns on housing and environmental issues. The second was the backing of a wealthy business executive who was sufficiently concerned about his children's health not only to move his family out of central London, but also to become CLEAR's principal financial sponsor. This support enabled Wilson to wage a more polished and visible campaign than would otherwise have been possible, but it did not define his basic strategy. Like the NUAAW activists, Wilson recognized that the challenge was to present his organization's risk assessment to the public and to relevant governmental officials so convincingly as to undercut the scientific judgment of the government's own group of experts, the Lawther committee.

CLEAR, like NUAAW, decided that it could attain this objective only through a campaign directly aimed at arousing public concern about the risks of exposure to lead. But the group's leaders also recognized that the campaign would have to achieve enough scientific credibility to change the minds of powerful actors within the political establishment. The campaign also had to defuse opposition from the gasoline industry, which characterized CLEAR's positions as emotional and insisted that there was no evidence linking low-level lead exposure to health. To build an effective rebuttal, CLEAR needed not only strong scientific support for the existence of a risk, but the means to bring this information dramatically before the public.

By good fortune, international science played into CLEAR's campaign strategy in two ways. First, influential studies published in Italy and the United States suggested that eliminating lead in gasoline did indeed cause a drop in blood lead levels. Establishing this linkage, however, was not enough to overcome Britain's traditional reluctance to acknowledge risk without actual evidence of harm to human health. The case still had to be made that lead in the ambient air can cause measurable adverse effects on health. Here, the campaign received vital support from epidemiological studies in the United States suggesting that children with relatively high levels of lead in their blood display lower IQ and other adverse behavioral symptoms. To publicize these studies, which established the desired causal connection between lead and health, CLEAR elected to hold an international scientific symposium.

In advertising the risks of 2,4,5-T, NUAAW had bypassed the methods and standards of orthodox science. The individual case histories compiled by the union were part of a self-conscious strategy to "tell it like it is," and the union counted on the emotional impact of its real-life stories to overcome the impersonal authority of established science. CLEAR's strategy was very different in that it borrowed the forms and procedures of official scientific investigations to pursue an anti-establishment policy. The model chosen by the campaign was the public inquiry, Britain's equivalent of the American administrative hearing. As in official inquiries, contributions were invited from experts on the issue, in this case the scientists who had produced major studies on the effects of low-level exposure to lead. The campaign also followed establishment practice in selecting as chair for the symposium an individual whose credentials were beyond reproach: Michael Rutter, a leading British child psychiatrist, a member of the Lawther committee, and a scientist who enjoyed high credibility within government. At the end of the three-day symposium, Rutter fully lived up to CLEAR's expectations. He concluded that "the removal of lead would have a quite substantial effect of reducing lead pollution, and the costs are quite modest by any reasonable standard."[80] Since Rutter had previously joined in the Lawther committee's recommendation against a ban, this statement marked an important change of position, and dealt a serious blow to the earlier report.

CLEAR's public rehearsal of the scientific data on lead and health was not the only factor that ultimately changed British policy on leaded gasoline. It was extremely significant, for example, that the Royal Commission on Environmental Pollution decided after its own review of the lead data that a ban on leaded fuels was advisable.[81] Nevertheless, CLEAR's scientific campaign must be given credit for at least part of the momentum that made the government announce in April 1983 that it would press for a Europe-wide ban on lead in gasoline. Des Wilson persuasively describes his organization's role in changing the public's — and indirectly the government's — perception of the risks of lead:

> The other side, particularly the industries, have their own version of events. Above all, they argued and would still argue that we exaggerated the health case and deliberately frightened people for no reason. It is, of course, true that we aroused concern all over the country about the health risk. But that was our task. It was precisely because the authorities refused to acknowledge it that we had to do this. I do not believe that people should

be denied the knowledge of research and studies that indicate health problems. The people of Britain are adult and mature enough to consider the facts objectively and reach their own conclusions. Their instinctive human response was that they did not care whether the evidence was conclusive or not, for they could see that there were sufficient organizations and people of substance who were prepared to state that it was a risk, and they wanted action on the basis of risk. Not for them conclusive proof, achievable only by what would in effect have been experiment with their children.[82]

Wilson's suggestion that British officialdom would have continued to wait for conclusive evidence of a risk to health unless CLEAR had politicized the issue is partly supported by events following the international symposium. Shortly before the government announced its policy change, the Department of Environment (DOE) released a study on the effects of low-level lead exposure on children. The study found a correlation between low IQs and high blood-lead levels, but suggested that other variables, such as social class, might adequately account for the findings.[83] DOE, in other words, seemed still to be looking for evidence that would establish the risk of lead beyond a reasonable doubt.

Analysis

The four comparative case studies reveal a number of reasons why science, despite its universality, is neither uniformly interpreted nor forces uniform outcomes in the risk assessment and risk management processes of different countries. Although more or less the same raw data were fed into the earliest stages of decision-making in Britain and the United States, divergences quickly appeared in the evaluation of this information.

The case studies underscore an overarching difference in British and American attitudes toward risk. In Britain, scientists and governmental decision makers are certain to recognize a risk only when there is persuasive evidence of actual harm, as in the case of asbestos, whereas in the United States a risk may also be acknowledged where there is no direct proof of injury to the public, as in the case of formaldehyde and human cancer. This basic difference underlies numerous discrepancies in the way the two countries analyzed data for purposes of risk assessment. It accounts for variations in the credibility accorded to particular studies, as well as for some notable divergences in risk assessment methodologies, and in the ultimate design of regulatory policy.

Consistent with its demand for tangible evidence of risk, the British policy process places much more importance than the American on studies of human populations. Without empirical proof from epidemiological studies, British risk evaluations often stop short of finding that there is a basis for concern. This emerged quite clearly in the formaldehyde case, where the British toxicity review panel concluded merely that "there is at present no evidence suggesting that exposure to formaldehyde has produced cancer in humans." Acting on much the same data, the U.S. federal panel on formaldehyde concluded instead that the substance "should be presumed to pose a carcinogenic risk to humans." While the epidemiology panel at the formaldehyde consensus workshop was more cautious, it too agreed that a cancer risk could not be excluded on the basis of the available human studies.

Britain's preference for basing risk regulation on positive epidemiological studies is linked to a high degree of skepticism about any study that purports to establish causal connections between disease and exposure to toxic substances. For example, the official British response to the U.S. findings on lead and children's IQ was quite cautious, with DOE commissioning its own study to see whether there was a real basis for concern. Even after this study was completed, DOE was reluctant to accept its superficially positive results as final proof of risk because they could also have been explained on nonpositive grounds, for example, as a function of social class. Similarly, in the 2,4,5-T case, ACP dismissed both the Alsea II study and the Swedish worker exposure study for methodological reasons. Criteria for validating positive epidemiological studies were articulated by committees such as ACP in Britain at a time when U.S. agencies were not yet prepared to develop systematic guidelines of this type.

By contrast, merely suggestive epidemiological evidence can arouse concern in the United States when it seems to confirm evidence from other sources, most notably animal studies. EPA's quick response to the Alsea II study was one example of this phenomenon. Another, perhaps more revealing, example was the concern expressed by the formaldehyde consensus workshop about studies showing a possible correlation between occupational exposure to the chemical and increases in brain tumors and leukemia. Knowing that formaldehyde was an animal carcinogen, the scientists at the U.S. workshop seemed prepared to regard any excesses in human cancer as potentially significant, even though they did not occur at the same site as in the tested animals. The workshop's epidemiology panel also accepted the preliminary findings from Acheson's formaldehyde study as suggestive. In Britain, however, the Acheson study was seen only as further confirmation of the fact that formaldehyde poses at most a remote threat of cancer to humans.

Summing up this discussion in legal terms, we can say that the presumptions with respect to epidemiological data are reversed in Britain and the United States. The starting assumption for British science advisors and policy makers is that such data are likely to err in a positive direction; hence they should be subjected to strict methodological scrutiny before being accepted as a basis for regulatory action. In the United States, on the other hand, the presumption is that epidemiological studies will most often tend to show false negatives. The federal panel's assessment of the ongoing epidemiological investigations on formaldehyde illustrates this official orientation. Thus, negative studies carry relatively little weight with policy makers, whereas merely suggestive positive studies are accepted as confirming the existence of a risk, even if they are not absolutely sound, particularly when they support findings of risk from experimental sources. One can explain this difference in part by noting that the U.S. expert panels often consist of government scientists; as a class, these experts can be expected to display a more "pro-health" ideology than scientists unaffiliated with a regulatory agency. Yet, as noted above, the formaldehyde consensus workshop, which included nongovernment scientists, also came to a more precautionary assessment of risk than did ACTS and its toxicity review panel.

A further important reason for these differences is that scientists as a class exercise more control over risk decisions in Britain than in the United States. In the four cases discussed here, all of which involved chemicals already in commerce, the British government first sought advice from an established or ad hoc expert committee: ACA for asbestos, ACTS (and its toxicity review panel) for formaldehyde, ACP for 2,4,5-T, and the Lawther committee for lead. The initial policy response in all four cases was simply to ratify the conclusions arrived at in these expert forums. In so doing, the government implicitly accepted the judgment of its scientific experts as to what constitutes an adequate evidentiary showing of risk.

In the United States, by contrast, the major regulatory statutes provide mechanisms, such as the imminent hazard provision of FIFRA, which grant administrators greater authority to act at their own discretion when confronted by new evidence of a threat to public health. Further, American statutes clearly assign the responsibility for making the final policy decision to a politically accountable commission or agency head, and the law envisions that this ultimate decision will be directed by factors other than the scientific assessment of risk. Accordingly, it is easier, both legally and politically, for regulators to judge the sufficiency of the scientific data by different standards from those of their expert advisors. The regulatory history of lead in gasoline most clearly demonstrates the influence of American health and safety legislation in promoting the more precautionary U.S. approach to risk control. The fact that EPA and the courts were ready to read the "will endanger" provision of the Clean Air Act in a precautionary way indicates a pervasive social awareness of risk and a willingness to accept regulation on something less than a showing of actual harm.

The British scientific establishment's interest in avoiding false positives helps explain the differences between the British and American assessments of health risks associated with asbestos. Of the four cases discussed here, asbestos is the one on which scientists from the two countries should have been in closest agreement, since the evidence that asbestos injures human health is no longer open to question. Yet British scientists read even this conclusive evidence as narrowly as possible in three significant respects. First, the assessment carried out for ACA distinguished asbestos risks by fiber type, so that all forms of the mineral were not viewed with equal alarm. Second, British scientists based the finding that there is no safe threshold of exposure to asbestos on *actual* data showing that disease can occur through even low-level and limited contact with the substance. Third, ACA was reluctant to take a position on the risk to the general public on the basis of available data concerning environmental exposure to asbestos.

All three conclusions can be contrasted with those reached in the United States. As we have seen, U.S. policy avoided classifying the risks of asbestos by fiber type. NIOSH's reading of the Enterline and Henderson data and of Borow's subsequent findings was consistent with this position and with an apparent reluctance to understate the scientific case for risk. Similarly, NIOSH adopted the no-threshold assumption for asbestos as a general policy rather than on the basis of empirical findings specifically concerning this substance. Finally, EPA's policy initiatives on asbestos were motivated by concerns about public exposure, and the agency showed that it would undertake fairly stringent action even without firm empirical evidence of risk. Such decisions point to an underlying cultural consensus that it is better to err on the side of false positives, or too much regulation, than on the opposite side.

It is clear from the four cases that scientists in the U.S. regulatory process confront a tension that does not affect their British colleagues to the same extent. Once harm is abandoned as the touchstone for recognizing risks to health, it becomes unclear what is an adequate quantum of proof for establishing that risk exists. Should one, for example, always accept positive experimental results, such as data from rodent cancer bioassays, as evidence of a risk to humans? Any such blanket principle can be challenged as "unscientific," given the variability among species and their differing biological responses to toxic chemicals. Yet efforts to explain away positive animal tests through arguments about species-specificity can also seem suspect, since such explanations often are advanced by scientists with a bias against regulation.

Another way of loosening the standard of proof might be to accept studies that do not meet the highest standards of methodological rigor. For example, scientific advisers might recommend action on the basis of preliminary studies that have not undergone peer review. Alternatively, they might even accept flawed studies as evidence of risk when the consequences of guessing wrong would be severe, as in the case of lead pollution and its possible impact on children. Any such compromises in the methodological standards applied to science tend to be seen by members of the scientific community as an abdication of professional norms. Yet scientists affiliated with U.S. government agencies have recognized that such a relaxation of standards may be necessary if one is to protect the public before harm actually occurs. For British scientists, the problem is less acute, since advisory committees are usually composed of independent experts who have no institutional commitment to minimize risk and are not directly accountable to the public. Such experts are relatively free to judge both the adequacy of particular studies and the strength of the overall case for risk by the same rigorous standards that they would apply to the assessment of science outside the policy context.

Judging the sufficiency of risk data causes additional problems for scientists in both policy systems because the methodologies for conducting relevant animal and human studies or exposure assessments are still evolving. This fluidity feeds a predisposition in each country to rely upon research done by its own scientific experts. The four examples discussed here reveal a preference in both Britain and the United States for epidemiological studies conducted within their own boundaries, as well as a relative skepticism about research done in other countries. For example, in assessing lead and formaldehyde, British policy makers waited for confirmatory results from domestic research instead of taking action on the basis of results obtained elsewhere. This version of the "not invented here" syndrome introduces a further variability into the science that different national governments are prepared to use for purposes of risk management.

Paradoxically, the pressures to relax the standard of proof have led in the United States to a greater preoccupation with rigorous, quantitative methods for estimating risks to human health. To offset the inherent shortcomings of using animal data, for example, U.S. scientists have attempted to develop firm guidelines for evaluating such studies and deciding whether they can credibly be used as a basis for predicting hazards to man. Other methodological concerns arise out of the practical limits on animal experimentation, particularly the need for testing at high doses. To estimate the risk at very low levels of exposure, U.S. scientists have had to develop mathematical models for extrapolating from the effects seen at experimental doses. Heavy reliance on such methods distinguishes current U.S. risk assessment practices from those in Britain, where the evaluation of risk remains more a qualitative than a quantitative exercise. However, as the experience of the formaldehyde consensus workshop demonstrates, risk assessment methodologies remain highly speculative even in the United States, and it is difficult to decide, except as a matter of regulatory philosophy, which of several competing models should be chosen in any given case. The preference expressed by EPA for the linearized multistage model in high to low dose extrapolation illustrates a continuing tendency on the part of U.S. agencies to pick conservative assumptions in risk assessment, that is, assumptions that are more likely to overstate than to understate the degree of risk.

Risk assessment in Britain is free from some of the methodological disputes that loom so large in the U.S. context. But the 2,4,5-T and lead cases illustrate a problem that is more likely to occur in Britain than in the United States: government agencies and their scientific advisers may lose control of the risk assessment process altogether, so that political considerations overrule science in the process of policy formulation. In both these cases, committed private interest groups succeeded in moving the discussion of risk out of the contained environment of the expert committee into the media and other overtly political forums such as the Trades Union Congress. One result of this process was that expert judgments about risk gave place to the public's less "scientific" appraisal of the threat to its health and safety.

From the standpoint of the scientific community, this is a troublesome scenario, since the actions of pressure groups like NUAAW and CLEAR can seriously endanger the scientists' traditional monopoly over defining "legitimate" science. CLEAR's intervention into the risk assessment process was, in this respect, less dangerous than NUAAW's. Although CLEAR created an alternative forum for the discussion of lead-related risks (the international symposium), the group sought legitimacy through an alliance with established science, in particular, through the enlisting of Michael Rutter. NUAAW, by contrast, flouted the norms of science by creating its own quasi-scientific record. The individual case histories, compiled without any pretense at methodological rigor, directly challenged the validity of traditional epidemiological and medical research. NUAAW's message to its audience was that those working with risk are in the best position to create the evidentiary "dossier" that should be used in risk management. In the context of Britain's heavily scientist-dominated approach to risk management, this message appears little less than revolutionary, and ACP was understandably critical of such an open attempt to politicize science.

Turning now to the influence of scientific assessment on policy outcomes, we find that the record in the two countries is equally mixed. Britain offers the clearest examples of policy decisions tracking the recommendations of scientific advisory committees. In the asbestos case, the regulatory standards adopted by the British government were precisely those proposed by ACA, although political pressure was eventually needed to obtain official implementation of the ACA recommendations. In setting control standards by fiber type, the government accepted one of the key recommendations of its advisory panel. On formaldehyde, as well, HSC ratified the action proposed by ACTS, but in this case the move from a recommended limit to a control limit — a step carrying important legal consequences — was evidently not motivated by new scientific findings about the risks of formaldehyde. Indeed, neither the toxicity review commissioned by ACTS nor the Acheson study provided compelling reasons for HSC to tighten its existing controls on formaldehyde. For both 2,4,5-T and lead, the government's initial decision was in accord with the recommendations of the relevant advisory committees, but both decisions were in effect overruled by political events. The government's refusal to ban 2,4,5-T was virtually nullified by the voluntary decisions of several municipalities and public corporations not to continue using the herbicide. In the case of lead, CLEAR's consciously political campaign caused an actual turnaround in official policy, and an implicit repudiation of the report prepared by the government's own appointed experts.

The four U.S. cases present an even less systematic picture. For asbestos, standard-setting did not keep pace with increasing knowledge about health hazards, particularly the risk of cancer at extremely low levels of exposure. In the case of formaldehyde, although numerous advisory bodies, including NIOSH, the federal panel, and the consensus workshop, affirmed the existence of a cancer risk to humans, official policy did not alter substantially over a five-year period. Litigation and protracted controversies surrounding rule-making initiatives were the main reasons for this lack of movement. For both 2,4,5-T and lead in fuel, relatively stringent policies were adopted by EPA and survived judicial review, but it is unclear how far these actions were consistent with scientific assessments of the two substances. The Alsea II study on which EPA based its initial cancellations of 2,4,5-T was rejected by scientists in both Britain and the United States. EPA's ban on lead in gasoline was implemented pursuant to a precautionary statutory mandate before the agency had persuasive evidence of health risks at low exposures. In this case, however, later scientific findings apparently bore out the agency's conservative assessment of risk.

The four comparisons made here suggest that the potential for controversy about risk is far greater in the United States than in Britain. The reasons have to do in large measure with this country's legal and political culture, which fosters competition and encourages litigation. But another reason surely is the willingness of both scientists and public officials to regard risk rather than harm as an adequate justification for governmental intervention. Agencies, usually with Congressional and judicial support, are prepared to impose large costs on industry without the "smoking gun" of people actually getting sick and dying. Yet the absence of convincing proof leaves the agencies vulnerable to challenge from dissenting scientists, courts and the general public, often forcing a retreat from regulatory decisions based on an overly precautionary assessment of risk. These dynamics account for the inconsistency, hesitation, and high level of scientific controversy that have marked U.S. toxic substance regulation in the past decade.

Conclusion

I have argued in this chapter that social and cultural factors play a crucial role in shaping even the more scientific aspects of risk assessment. The acceptance or rejection of particular studies, the development of evaluation criteria, the decision to wait for more evidence or to commission new studies are all colored by varying degrees of risk averseness in different countries. Scientists, no less than policy makers or the general public, share in the prevailing national attitudes toward risk, and these values are reflected in the way they filter and organize scientific knowledge. Cultural differences penetrate so deeply into risk assessment that it is difficult to isolate any phase of the assessment process as purely scientific or truly universal. Comparative analysis of risk decisions thus emphasizes the blurriness of the line between risk assessment and risk management, and illuminates the factors that dilute the impact of science on public policies about risk.

Yet the ideology of science continues to exert a profound influence on the risk management process in both Britain and the United States. Scientists, of course, are the primary advocates of decision-making grounded in "good" science. But private and governmental actors too feel the need to justify their assessments of risk in credible scientific terms, even when their judgments run counter to those of established science. As parties with different interests compete over definitions of what constitutes an adequate proof of risk, the authority of science can come under attack in two ways. If the standards for demonstrating risk are relaxed too far as a matter of policy, then risk management decisions begin to be made on the basis of theories and studies that the majority of scientists consider invalid. On the other hand, if scientists apply overly cautious standards to the identification of risks, then pressure grows to remove risk assessment from the hands of scientists, substituting a more political process for determining whether there is sufficient evidence of risk. The case studies discussed in this chapter illustrate how Britain and the United States have tried to cope with both types of assaults on science. Divergences between the two countries indicate basic differences not only in public perceptions of risk, but in the degree of autonomy each country is willing to grant to scientists in the policy process.

Notes and References

1. See, for example, Alvin Weinberg, 'Science and Trans-Science,' *Minerva*, 1972, 10, 209—222; 'Science and Its Limits: The Regulator's Dilemma,' *Issues in Science and Technology*, Fall 1985, 59—72; Giandomenico Majone, 'Science and Trans-Science in Standard Setting,' *Science, Technology &Human Values*, 1984, 9, 15—22.

2. National Academy of Sciences (hereafter NAS), *Risk Assessment in the Federal Government: Managing the Process* (Washington, D.C.: National Academy Press, 1983).

3. Brendan Gillespie, Dave Eva and Ron Johnston, 'Carcinogenic Risk Assessment in the United States and Great Britain: The Case of Aldrin/Dieldrin,' *Social Studies of Science*, 1979, 9, 265—301; Frances McCrea and Gerald Markle, 'The Estrogen Replacement Controversy in the USA and UK: Different Answers to the Same Question?' *Social Studies of Science*, 1984, 14, 1—26; Ronald Brickman, Sheila Jasanoff and Thomas Ilgen, *Controlling Chemicals: The Politics of Regulation in Europe and the United States* (Ithaca, N.Y.: Cornell University Press, 1985), 211—215 (regulation of 2,4,5-T).

4. Brickman *et al., op. cit.*, pp. 187—217. See also Sheila Jasanoff, *Risk Management and Political Culture* (New York: Russell Sage Foundation, 1986).

5. Brickman *et al., op. cit.* See also David Vogel, 'Cooperative Regulation: Environmental Protection in Great Britain,' *The Public Interest*, 1983, No. 72, 88—106; Joseph Badaracco, *Loading the Dice* (Cambridge, MA: Harvard Business School Press, 1985); Sheila Jasanoff, 'Negotiation or Cost-Benefit Analysis: a Middle Road for U.S. Policy?' *Environmental Forum*, 1983, 2, 37—43.

6. Nicholas Ashford, 'Advisory Committees in OSHA and EPA: Their Use in Regulatory Decisionmaking,' *Science, Technology &Human Values*, 1984, 9, 72—82.

7. Thomas McGarity, 'Judicial Review of Scientific Rulemaking,' *Science, Technology & Human Values*, 1984, 9, 99—100.

8. Brickman *et al., op. cit.*, pp. 180—182.

9. Weinberg, 'Science and its Limits,' *op. cit.* See also Thomas McGarity, 'Substantive and Procedural Discretion in Administrative Resolutions of Science Policy Questions: Regulating Carcinogens in EPA and OSHA,' *The Georgetown Law Journal*, 1979, 67, 729—810; Nicholas Ashford, C. William Ryan, and Charles Caldart, 'Law and Science Policy in Federal Regulation of Formaldehyde,' *Science*, 1983, 222, 894—900.

10. Health and Safety Commission, *Asbestos*, Final Report of the Advisory Committee (London: HMSO, 1979) (hereafter cited as HSC Asbestos Report).

11. *Id.*, Vol. 2, 47—49.

12. *Id.*, p. 29.

13. *Id.* A note appended to the main text provides the following information:

> The factory opened in 1917, but crocidolite was not used until 1929, since when it has been in continuous use by men making pipes. Almost all the cases have been in men and have occurred since 1960. In the large workforce of women employed making textiles from chrysotile only one case has been found.

14. *Id.*, Vol. 1, 53.

15. NIOSH, 'Revised Recommended Asbestos Standard,' Washington, D.C., December 1976, pp. 41—42.

16. *Id.*, p. 35.

17. *Id.*, p. 93. In a roughly contemporaneous document on asbestos, OSHA stated the emerging federal position on the threshold issue more explicitly:

> While some level, below which exposure to a carcinogen does not cause cancer, may conceivably exist for any one individual, other individuals in the working population may have cancer induced by doses so low as to be effectively zero. This is not to say that researchers will never find a threshold for a carcinogenic substance, but it does mean that the threshold concept for carcinogens is, at present, more a matter of responsible regulatory policy than a precise scientific determination.

OSHA, 'Occupational Exposure to Asbestos,' *Federal Register*, October 9, 1975, 40, 47656.

18. HSC Asbestos Report, Vol. 2, 41.

19. *Id.*, Vol. 1, 35—45.

20. Richard Doll and Julian Peto, *Asbestos: Effects on Health of Exposure to Asbestos* (London: HMSO, 1985).

21. U.S. House of Representatives, Committee on Energy and Commerce, Subcommittee on Commerce, Transportation, and Tourism, *EPA's Failure to Regulate Asbestos Exposure*, 97th Congress, 2nd Session, pp. 64—65 (hereafter cited as Asbestos Hearing).

22. HSC Asbestos Report, Vol. 1, 89—92.

23. Bureau of National Affairs, *Chemical Regulation Reporter*, Current Report, April 19, 1985, p. 69.

24. Asbestos Hearing, pp. 57—60.

25. HSC Asbestos Report, Vol. 2, 34—41.

26. Asbestos Hearing, pp. 51—52.

27. HSC Asbestos Report, Vol. 2, 42.

28. See, for example, McCrea and Markle, *op. cit.*, p. 15 and n. 96.

29. *HSC Asbestos Report*, Vol. 1, 74—78. ACA had already recommended a ban on the use of crocidolite in manufacturing industrial products. Import of this fiber type ended in 1970 through voluntary action by the asbestos industry.

30. John Locke, 'Fixing Exposure Limits for Toxic Chemicals in the U.K.,' unpublished paper presented at WHO Symposium on Risk Assessment and Risk Management, 1985, pp. 12—17. As a result of this action, Britain now has stricter standards for both chrysotile and amosite than required by the EEC. Those seeking an economic explanation for regulatory behavior should note, however, that formal bans on crocidolite and amosite were proposed only after use of these materials had largely ceased in Britain.

31. OSHA, 'Occupational Exposure to Asbestos,' *Federal Register*, October 9, 1975, 40, 47653.

32. *Asbestos Information Association v. OSHA* 727 F.2d 1137 (5th Cir. 1983).

33. Bureau of National Affairs, *Environment Reporter*, Current Developments, February

1, 1985, p. 1315. For an announcement of EPA's reversal of the referral decision, see *id.*, March 15, 1985, p. 1443.

34. 'EPA Proposed Ban on Asbestos Use,' *Chemical and Engineering News*, January 27, 1986, p. 6.

35. NIOSH, 'Criteria for a Recommended Standard . . . Occupational Exposure to Formaldehyde,' Washington, D.C., 1976.

36. J. A. Swenberg, W. D. Kerns, R. I. Mitchell, E. J. Gralla, and K. L. Pavkov, 'Induction of Squamous Cell Carcinomas of the Rat Nasal Cavity by Inhalation Exposure to Formaldehyde Vapor,' *Cancer Research*, 1980, 40, 3398—3401.

37. HSE, 'Toxicity Review,' *Formalldehyde*, 1981, 2, 11.

38. Cyril D. Burgess, Director of Hazardous Substances Division, HSE, personal communication, March 28, 1985.

39. 'Report of the Federal Panel on Formaldehyde,' *Environmental Health Perspectives*, 1982, 43, 165 (hereafter cited as Federal Panel Report).

40. *Id.*, p. 159.

41. NAS, National Research Council, *Formaldehyde and Other Aldehydes* (Washington, D.C.: National Academy Press), 1981, p. 17.

42. See, for example, Brickman *et al., op. cit.*, pp. 198—201.

43. Federal Panel Report, pp. 152—154.

44. *Id.*, pp. 159—162.

45. Frederica Perera and Catherine Petito, 'Formaldehyde: a Question of Cancer Policy?' *Science*, 1982, 216, 1285—1291.

46. See, for example, U.S. House of Representatives, Committee on Science and Technology, Subcommittee on Investigations and Oversight, *Formaldehyde: Review of Scientific Basis of EPA's Carcinogenic Risk Assessment*, 97th Congress, 2nd Session, 1982.

47. *Deliberations of the Consensus Workshop on Formaldehyde*, Little Rock, Ark., October 3—6, 1983.

48. *Id.*, p. 47.

49. *Id.*, p. 45.

50. *Id.*, pp. 138—139.

51. EPA, 'Guidelines for Carcinogen Risk Assessment,' *Federal Register*, September 24, 1986, 51, 33992—34003.

52. Cyril D. Burgess, personal communication, March 28, 1985.

53. The British Health and Safety at Work Act of 1974 imposes a general duty on

employers to ensure "so far as is reasonably practicable" that employees are not exposed to risks to their health and safety.

54. CPSC, 'Ban on Urea-Formaldehyde Foam Insulation,' *Federal Register*, April 2, 1982, 47, 14366—14419.

55. *Gulf South Insulation v. CPSC*, 701 F.2d 1137 (5th Cir. 1983).

56. *International Union, United Auto Workers v. Donovan*, 590 F. Supp. 747 (D.C.D.C. 1984).

57. OSHA, 'Occupational Exposure to Formaldehyde,' *Federal Register*, December 10, 1985, 50, 50412—50499.

58. EPA, 'Formaldehyde; Determination of Significant Risk; Advance Notice of Proposed Rulemaking and Notice,' *Federal Register*, May 23, 1984, 49, 21870—21897.

59. For an argument that the health risks of low-level dioxin exposure have been exaggerated, see Michael Gough, *Dioxin, Agent Orange* (New York: Plenum Press, 1986).

60. *In Re 'Agent Orange' Product Liability Litigation*, 597 F. Supp. 740 (E.D.N.Y. 1984).

61. Thomas Whiteside, *The Pendulum and the Toxic Cloud: The Course of Dioxin Contamination* (New Haven: Yale University Press, 1979).

62. Sarah L. Hartmann, 'Case History of 2,4,5-T Regulation,' *Congressional Record*, 96th Congress, 1st Session, November 29, 1979, H11294—H11295.

63. EPA, '2,4,5-T Decision and Emergency Order Suspending Registrations for Certain Uses', *Federal Register*, March 15, 1979, 44, 15874—15897.

64. ACP, *Review of the Safety for Use in the U.K. of the Herbicide 2,4,5-T* (London: HMSO, 1979).

65. NUAAW, 'The 2,4,5-T Dossier — Not One Minute Longer,' March 1980. See also Judith Cook and Chris Kaufman, *Portrait of a Poison: The 2,4,5-T Story* (London: Pluto Press, 1982), pp. 50—52.

66. Cook and Kaufman, *op. cit.*, pp. 34—41.

67. ACP, *Further Review of the Safety for Use in the U.K. of the Herbicide 2,4,5-T* (London: HMSO, 1980, p. 25.).

68. *Id.*, p. 12.

69. *Id.*, p. 7.

70. *Id.*, p. 13.

71. Cook and Kaufman, *op. cit.*, p. 69.

72. *Id.*, p. 63.

73. 7 U.S.C. 136a.

74. *Chemical and Engineering News*, October 24, 1983, p. 5.

75. 42 U.S.C. 1857f—6c(c) (1) (A).

76. See *Ethyl Corp. v. EPA*, 541 F.2d 1 (D.C. Cir., 1976), p. 12.

77. *Id.*

78. *Id.*, p. 13.

79. My account of the leaded fuels controversy in Britain is based on a case study of the CLEAR campaign in Des Wilson, *Pressure: The A to Z of Campaigning in Britain* (London: Heinemann, 1984), pp. 156—180.

80. *Id.*, p. 170.

81. Royal Commission on Environmental Pollution, *Ninth Report — Lead in the Environment* (London: HMSO, 1983). For an interesting account of the standards by which the Royal Commission evaluated the lead data, see T.R.E. Southwood, 'The Roles of Proof and Concern in the Work of the Royal Commission on Environmental Pollution,' *Marine Pollution Bulletin*, 1985, 16, 346—350. The reasons why the Royal Commission's assessment diverged from that of the Lawther Committee are worth speculating on, but lie beyond the scope of this chapter.

82. Wilson, *op. cit.*, p. 180.

83. *Id.*, p. 176.

Index of Subjects